OUR CELESTIAL CLOCKWORK

From Ancient Origins to Modern Astronomy of the Solar System

Other Related Titles from World Scientific

A Cabinet of Curiosities: The Myth, Magic and Measure of Meteorites
by Martin Beech
ISBN: 978-981-122-491-1

The Invisible Universe: Dark Matter, Dark Energy, and the Origin and End of the Universe
by Antonino Del Popolo
ISBN: 978-981-122-943-5

Chinese Astrology and Astronomy: An Outside History
by Xiaoyuan Jiang
translated by Wenan Chen
ISBN: 978-981-122-345-7

Cosmic Pinwheels: Spiral Galaxies and the Universe
by Ronald James Buta
ISBN: 978-981-121-668-8
ISBN: 978-981-121-747-0 (pbk)

OUR CELESTIAL CLOCKWORK

From Ancient Origins to Modern Astronomy of the Solar System

Richard Kerner

Sorbonne Université, Paris, France

World Scientific

NEW JERSEY • LONDON • SINGAPORE • BEIJING • SHANGHAI • HONG KONG • TAIPEI • CHENNAI • TOKYO

Published by

World Scientific Publishing Co. Pte. Ltd.

5 Toh Tuck Link, Singapore 596224

USA office: 27 Warren Street, Suite 401-402, Hackensack, NJ 07601

UK office: 57 Shelton Street, Covent Garden, London WC2H 9HE

Library of Congress Cataloging-in-Publication Data
Names: Kerner, R., author.
Title: Our celestial clockwork : from ancient origins to modern astronomy of the solar system /
   Richard Kerner, Sorbonne Université, Paris, France.
Description: Hackensack, NJ : World Scientific Publishing Co. Pte. Ltd., [2022] |
   Includes bibliographical references and index.
Identifiers: LCCN 2021017842 | ISBN 9789811214592 (hardcover) |
   ISBN 9789811215315 (paperback) | ISBN 9789811214608 (ebook) |
   ISBN 9789811214615 (ebook other)
Subjects: LCSH: Astronomy--History.
Classification: LCC QB15 .K42 2022 | DDC 520--dc23
LC record available at https://lccn.loc.gov/2021017842

British Library Cataloguing-in-Publication Data
A catalogue record for this book is available from the British Library.

For any available supplementary material, please visit
https://www.worldscientific.com/worldscibooks/10.1142/11674#t=suppl

Desk Editor: Ng Kah Fee

Typeset by Stallion Press
Email: enquiries@stallionpress.com

To the memory of my father Dr. Samuel Kerner,
who taught me how to learn,
and from whom I learned how to teach.

Acknowledgements

I would like to express my deep gratitude to Prof. Pascal Richet of the Institut de Physique du Globe de Paris for his constant ecouragements, for sharing his deep knowledge of the history of science, and for a throughout reading of the manuscript. His constructive criticisms and suggestions have seriously improved the text. Separate thanks are due to Evelyne Richet for suggesting a better title.

I am greatly indebted to Prof. Jan-Willem van Holten of the University of Leiden for his extraordinary act of friendship, consisting in the careful and thoughtful reading of the draft version of this book. His invaluable remarks and suggestions concerning the scientific content and the style alike enabled a substantial improvement of the book's final shape.

Thanks are due to Dr. Jerzy Karczmarczuk for helping me with the computer programming and Latex compilation.

Last but not least, this book could not be written without my wife Grażyna's loving care, patience and support.

Paris, December 2020

Preface

The present book is based on a series of lectures the author delivered at the Parisian *Université Pierre et Marie Curie*, which has been since then given the new name *Sorbonne-Université*. These lectures were intended to help to compensate the lack of elementary astronomical knowledge and history of its development among the young people freshly out of high schools, and who were about to start University studies in science — mostly physics and mathematics.

The lectures on the astronomy of the Solar System and its historical development since the earliest times were a part of a wider educational project for the sophomore students, proposed under the name "A bit of Scientific Culture". Five semester-long programs were proposed, including lectures and tutorials: besides the astronomy, elementary particles, cosmology, geophysics and nuclear physics. Simple exercises were proposed, which our students highly appreciated solving them during the tutorials.

The present book follows faithfully the structure of lectures delivered during those years, with substantial additions permitted by the book format. On the other hand, we cannot reproduce here several interesting animated pictures and schemes, like the accelerated images of solar and lunar eclipses, the motion of lunar shadow on the surface of the Earth, or the revolutions of planets and stars. Traditional illustrations are abundant though, and readers' imagination will fill the gap.

Great care is taken to explain how most of astronomical knowledge could be obtained first with the exclusively naked-eye observations and the most elementary geometry, based on two great theorems known from Antiquity: the theorem traditionally attributed to Thales of Miletus, used to evaluate sizes of objects at a distance by comparing the angles at which they are seen, and the theorem bearing the name of Pythagoras (known also earlier to the Babylonians), which is the basis of trigonometry. By developing the consequences of these first theorems, Euclid set the axiomatic basis of classical geometry, in use until now.

Although ancient Indian and Chinese civilizations have developed astronomy and mathematics, too, they are mentioned only sporadically, because the interaction with European and Middle Eastern civilizations was almost non-existent.

The geocentric Ptolemaic astronomical system, which was also canonically accepted by Islamic and European scientists alike, was shattered by Copernican revolution in 16th century. However, the Copernican heliocentric system was still explaining planetary motions using exclusively circular orbits, and along with Tycho Brahe's system, was unable to give better astronomical predictions than the Ptolemaic one. It was only after elliptic planetary orbits were discovered by Kepler, that the new era of astronomy really began.

In parallel, physical knowledge was encoded in Aristotle's monumental work, influencing Islamic and European science until the revolutionary changes introduced by Galilei.

These dramatic changes are the main subject of this book, with tribute paid to the genius of ancient wizards and modern scientists of revolutionary 16th and 17th centuries, including Newton and Huygens. Finally, we describe the important great leap forward in astronomy which occurred in 18th century, when the real dimensions of the Solar system were firmly established by Halley, Flamsteed and Cassini. Our narrative ends at the dawn of 19th century, when the Milky Way was considered as representing the Cosmos up to its ultimate limit. The development of astrophysics in 19th century and of extra-galactic astronomy and cosmology deserves a separate volume.

As far as possible, the history of astronomic discoveries and the evolution of our understanding of celestial bodies' motion is implemented with simple mathematical exercises enabling the reader to follow the reasoning of famous mathematicians and astronomers from Antiquity till the advent of modern science. For the most part, the examples make use of elementary mathematics known to ancient astronomers; however, there are cases, starting from Newton's laws of mechanics and gravitation, when a bit more sophisticated proofs are shown. These are marked with asterisks [* .... *]. Readers less acquainted with mathematical tools taught beyond the college level can easily skip the longer proofs.

Contents

Prologue

In one of his first science-fiction stories [Asimov (1941)] published in 1941, the author Isaac Asimov (1920–1992) described an extraterrestrial civilization living on a planet named Lagash revolving in a system of six stars, so that it is constantly illuminated. Night does not exist, and its inhabitants are unaware of the existence of other stars, galaxies, their own Milky Way and other planets as well.

The main message Asimov intended to convey is contained in the quotation from American poet Ralph Waldo Emerson:

"If the stars should appear one night in a thousand years, how would men believe and adore, and preserve for many generations the remembrance of the city of God!"

Asimov opted for a totally different scenario: when confronted for the first time in their lives with nightfall and the awesome starry sky above their heads, Lagash inhabitants fell prey to tremendous panic, starting to set fires everywhere in order to enlighten the unbearable darkness, and soon their entire civilization disappeared in ashes. The recovery took thousands of years (one could ask incidentally what a "year" would mean in a system with six Suns revolving around a common center of gravity, and the planet Lagash sneaking in between on a complicated trajectory).

The spicy side of the story lies in the description of group of Lagashian archeologists who discovered that their civilization is not the first one, and that the myth about the end of the world in fires and its subsequent re-birth are based on real events, just two thousand years old. When they begin to realize that the next astronomical event aligning all their suns in one line, which will leave half of their planet in total darkness, it is too late to explain the phenomenon to the population and avoid panic. The nightfall occurs, the inevitable chaos takes over the entire planet, and the historical cycle is about to complete itself again.

Reading Asimov's story makes one aware of the extraordinary luck we have got living on the third planet of our Solar System, with a unique star at its center, with stable planetary orbits, with the eternal ballet of seasons, days and nights, with the Moon illuminating our nights, yet not excluding the observation of distant stars and the Milky Way — in one word, the possibility to admire with a mix of awe and wonder the surrounding Universe.

Most of us are city dwellers, which means that our perception of stars and planets is seriously hampered by permanent and intense light illuminating our streets. Usually only the few brightest stars and occasionally planets can be perceived by our eyes, half-blinded by city lights and luminous publicity banners. We can consider ourselves lucky if we see Venus or Jupiter from time to time, more rarely Mars and Saturn. Among the stars, only the brightest few ones are visible. It has become almost impossible to observe the Milky Way, even in the suburbs. But in ancient times things were different, and our ancestors could admire the spectacular and magnificent celestial sphere, full of constellations of various stars, the phases of the Moon, the erratic motions of planets and the tremendous Milky Way extending its wings from one end of the horizon to another. No wonder that their curiosity was strongly stimulated, and that the beliefs of all kinds attributing divine qualities to Sun and Moon, to planets and the stars, were common to many civilizations, along with the conviction that regular motions of celestial bodies as well as some exceptional events in the sky above us have direct influence on the fate of humans, individual as well as collective.

What the inhabitants of Lagash could see once in 2000 years, our ancestors were able to observe almost every night, and gather more astronomical data with years and centuries passing by, especially after the invention of writing. Looking back, we can only admire their intelligence and observation skills. Astronomy played an extremely important role in ancient civilizations for two different reasons. The first one is quite obvious: since the advent of agriculture, it has become crucial to know with as much precision as possible, when to expect rivers' next flood, to foresee a rain season, to forecast the coming of spring for plowing and sowing, and summer for harvest. Without exaggeration one may say that for ancient farmers such knowledge was vital, because their very existence strongly depended on it. The second reason is of religious nature. For our ancestors the sky was the home of powerful gods whose influence on human fate was preponderant. The divine nature of Sun and Moon, of planets and stars was commonly taken for granted and acknowledged in ancient civilizations. Gods could be helpful or obnoxious, depending on their mood and how they judged human behavior, and to foresee their intentions became a matter of great importance.

The development of astronomy was also made possible because, contrary to the imaginary star system inside which Lagash orbited, our planet Earth belongs to a very peculiar Solar system, and by itself is exceptional. The more we investigate its peculiarities, the more we are convinced about many happy coincidences to which we owe not only astronomy, but also intelligent life on Earth. Let us enumerate the most important ones:

- **1.** Contrary to the imaginary Lagash our Solar system is particularly stable. This is due to the fact that our Sun concentrates 99.86% of the total mass; the planets, asteroids and comets evolving around account for the remaining 0.14%; among them, the four remote giant planets, Jupiter, Saturn, Uranus and Neptune, contain 98% of those 0.14%, so that the Earth and its nearest planets, Mercury, Venus and Mars, and the asteroid belt between Mars and Jupiter, amount all together for the remaining 0.002%. This circumstance makes our planetary system, and our Earth in particular, a very stable place, proper not only for the development of astronomy, but for the appearance of life in first place.

• **2.** The stability of Earth's orbit is enhanced by the very weak (although not totally negligible) influence of other planets, which in principle could have been much greater than it is if Jupiter was orbiting around the Sun closer to us, e.g. in the place of Mars. A massive planet like this (Jupiter's mass is 2970 times greater than the Martian one) would destabilize our own planet's orbit very strongly indeed; perhaps Earth itself would not become a big planet, remaining in the form of asteroid belt like the one that fills the gap between Mars and Jupiter.

• **3.** The Earth orbit is almost circular. It is an ellipse, of course, but its eccentricity is extremely low: $e_E = 0.0169$ which is less than two percent. In a drawing representing Earth's orbit on paper, with average radius 10 cm, the distance between the center and the focal point would be less than half a millimeter, barely visible to the naked eye.

That the orbit is practically circular can be concluded from careful measures of Sun's apparent size. It varies very slightly, from its greatest value $32'32''$ between January 3 and January 5, when Earth is closest to the Sun at its orbit *perihelion*, and the lowest value $31'28''$ when Earth reaches the largest distance from the Sun, between July 4-th and July 5-th every year, at its *aphelion*. The relative difference is of the order of 0.033, the double of the eccentricity, as it should be.

This adds to the stability necessary for the creation of life on Earth, but also to the early development of Astronomy. It took a lot of time, effort and genius to decipher the relative motions of Moon, Sun and planets, but it would be much harder to do this for Martian astronomers (had they ever existed) to meet the challenge presented by the extremely complicated planetary motions as they are seen from Mars, whose eccentricity of orbit is $e = 0.0934$, i.e. almost 10%.

• **4.** The presence of a unique giant satellite, our Moon, almost a "sister planet", represents another extraordinary asset for the development of astronomy. The role of the Moon in primitive perception of the surrounding Universe can hardly be overestimated. It is present in all religions and myths. Its light illuminating the nights

is sufficiently strong most of the time to make many of animal and human activities continue after sunset. Its phases imposed their rythm and incited our ancestors to introduce lunar calendars with well-defined weeks and months. And the happy coincidence of its angular size with angular size of the Sun makes total solar eclipses possible. The priests in Egypt reinforced their authority when they discovered the *Saros*, which is an approximate period between two consecutive solar eclipses. The timespan of Saros is 18.6 years, and it was one of the first successful applications of mathematics to astronomy. It was also known to the Babylonians.

This circumstance — the almost perfect equality between angular sizes of both luminaries — is even more intriguing since the discovery that due to the tidal effects the Moon's orbit slowly changes so that the average distance to our satellite grows by almost 4 centimeters per year. In a few hundred million years from now the visible size of the Moon will become smaller, so that no total solar eclipses would be possible anymore, only the annular ones. Conversely, a long time ago, Moon was closer to the Earth than it is now. The fact that the equality of apparent sizes of Sun and Moon occurred during the period when humanoids became Homo Sapiens capable to observe and understand celestial phenomena makes this coincidence even more amazing.

- **5.** The axial tilt (or obliquity) of $23°27'$. This is the angle between the Earth's axis and the plane of its orbit around the Sun (the *ecliptic plane*, or simply the *ecliptic*). It produces regular changes of weather called *the seasons*, which made astronomical predictions necessary to ensure a right timing for agricultural activities. The best way to appreciate at its right value the role of the moderate inclination is to imagine life on Earth would one of the extreme situations prevail: the axis of rotation perpendicular to the plane of the ecliptic, or the axis of rotation laying in the ecliptic plane. In the first case there would be no seasons, just different climate zones between various parallels with steady, unchanging temperatures varying only between night and day. Everywhere Sun's diurnal path in the sky would be always the same; on the other hand, supposing that the

Moon's orbit would be still close to the ecliptic, solar and lunar eclipses would occur more frequently.

In the second case (observed in our Solar System in Neptune), with an axis of rotation contained in the ecliptic plane, it would be hard to imagine life on Earth, least alone the development of astronomy. Instead of seasons, there would be long periods of permanent night interchanging with equally longer periods of permanent daylight. The existence of solar and lunar eclipses would depend on the orientation of the Moon's orbit with respect to the Earth. Were it close to the equatorial plane, they would probably remain observable from time to time; if Moon's orbit were close to the ecliptic, with the same obliquity as now (about 5.5°), they would be probably as frequent a phenomenon as they are now.

• **6.** The presence of five planets whose brightness may reach or even surpass that of the brightest stars. This is true for each planet, although not all the time, but periodically. The two brightest planets, Venus and Jupiter, can be easily seen at dawn or at sunset before any star can be seen in the sky.

Their wandering among the stars, sometimes with change of direction, was as appealing as it was puzzling. The overwhelming influence of Sun on life of all creatures on Earth was more than obvious; the Moon's influence was also important. The idea that planets may influence the events and human lives down on the Earth did seem natural, too, and it was indeed considered so in various human cultures. Wherefrom the development of astrology as science able to determine the forthcoming events, be they of general character (floods, draughts, wars, epidemics and other disasters) or of strictly personal nature, concerning someone's future fate. Since the time immemorial, astrology along with interpretation of dreams, was a common practice not only at every royal court, but also by common people. In ancient and not so ancient times it had a preponderant role in the development of astronomy and improvement of observational skills. Although astrology is considered as marginal superstition in modern times, it can not be denied that it played some positive role in the development of astronomy as a genuine science.

• **7.** The atmosphere of our planet provides a reliable shelter against deadly cosmic rays and obnoxiously strong ultraviolet radiation, which made life outside the primeaeval ocean possible. On the one hand, it is transparent enough to let the light of distant planets and stars penetrate without too great distortion down to the Earth's surface, so that they are easily visible with the naked eye. On the other hand, our atmosphere is a constant source of bewilderment due to the plethora of optical phenomena and apparitions including rainbows, solar and lunar halos, meteorites (interpreted as "falling stars"), the ever changing colors of sky and clouds at dawn and at sunset, the aurorae borealis and australis in higher latitudes, and many other wonders, transforming the skies into a permanent theater. All this served as mighty incitation to observe the skies with interest mixed with awe.

Last but not least, the emergence of continents lying on top of the very viscous mantle made life possible outside the ocean. Without this specific geological phenomenon water would cover the entire surface of Earth. It is more than probable that even if intelligent life evolved on Earth deprived of its continents, astronomy could never develop due to extremely poor observational possibilities under water.

In what follows, we shall present the development of astronomical observations and their interpretation since the dawn of our civilization.

Chapter 1

How we see the world

Our vision's limits - visible light as a tiny part of electromagnetic spectrum - ultraviolet and infrared waves - - the structure of the human eye - resolution capacity - sky as a sphere - diffusion and refraction of light - aberration - stereoscopic vision and the parallax - stellar magnitudes - incompleteness of our image of the world - Plato's cave metaphor

1.1 Preamble

During millenia, ancient astronomy was developed from observations made exclusively with the naked eye. The spectacular breaktrough occurred only in the beginning of the 17th century, when Galileo first, and Kepler soon after, produced the first telescopes. Nevertheless one of the greatest astronomical achievements of all times, the three fundamental laws of planetary motions, were deduced by Kepler from the precise observations and measurements made previously by Kepler's mentor Tycho Brahe, still in the 16th century.

In order to appreciate astronomical discoveries of our ancestors at their just value we should keep in mind the limitations nature has put on human perception, especially in what concerns our vision. As it appears from our modern perspective, ancient astronomers disposed of partial and limited information conveyed by visible light coming from the faraway heavenly bodies. That in spite of such obvious shortcomings they were able to discover the shape and dimensions

1

of Earth, to evaluate quite correctly the dimension of the Moon and its distance from the Earth, and even find out (although with error of one order of magnitude) the dimension of Sun and its distance, is an amazing achievement of human mind.

This introductory section is intended to remind the reader of the observational possibilities with which great astronomical discoveries were made in Antiquity, the Middle Ages, and Renaissance, all of them using exclusively the naked eye and a few simple mechanical devices enabling a better evaluation of angles.

1.2 Visible light

Our eyes are one of the most sensitive instruments Nature has produced after hundreds of millions years of evolution. After long accomodation in total darkness they are able to detect just a couple of photons, if not a single one, in the visible part of spectrum. We can also distinguish the slightest subtleties of different colors, shapes and motions. Nevertheless what we are able to grasp is a tiny fraction of the vast spectrum of electromagnetic radiation, which covers an enormous range of wavelengths, from radio signals to gamma rays, as shown in Figure 1.1.

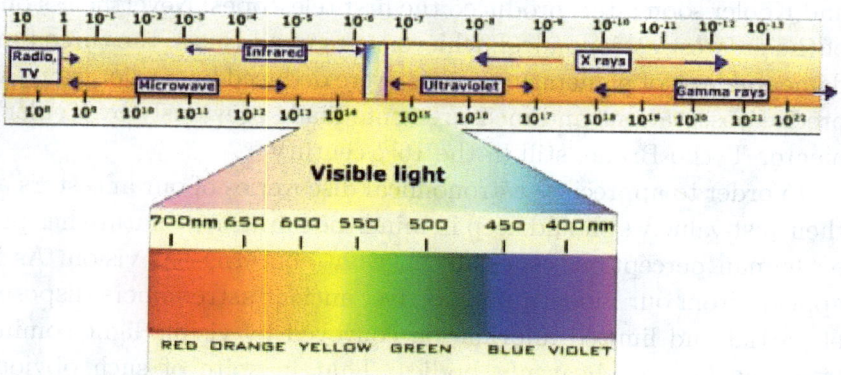

Fig. 1.1   The spectrum of electromagnetic radiation. Only a tiny part (one octave, from 380 (violet) to 750 (deep red) nanometers is accessible to our vision.

Not all those waves reach the Earth's surface, which ensures our survival. Our planet's atmosphere combined with the magnetic field of the Earth protects us quite efficiently from deadly cosmic radiation including strong ultraviolet radiation, X-rays and gamma rays. The infrared radiation is absorbed mostly by water vapor and also by carbon dioxide, but any heated object acts as infrared emitter, especially when the surrounding temperature falls down as compared to its own. Figure 1.2 shows the intensity of different wavelengths in the solar light arriving at the surface of the Earth.

Nevertheless we do not perceive ultraviolet rays, quite abundant on a sunny day — only our skin gets a tan when exposed to this part of solar spectrum. We cannot see the infrared radiation whose heating effects can be felt by us. Using laser emitted infrared radiation has become popular nowadays in devices capable of changing chains in TV sets at a distance, opening and closing car doors, etc. Also night vision optical devices transforming the infrared into visible part of the spectrum are widely used in the armed forces.

As it happens, the narrow limits on our perception of electromagnetic radiation have been imposed by biological evolution. The

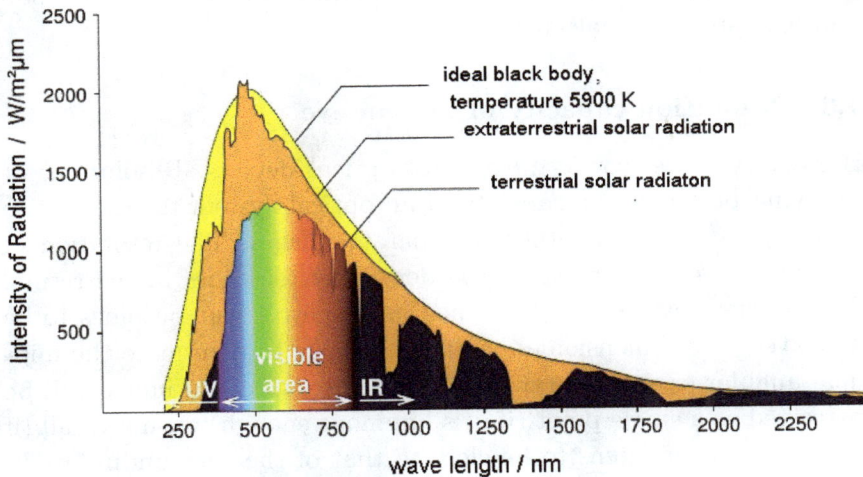

Fig. 1.2 The spectrum of solar radiation arriving at Earth's surface.

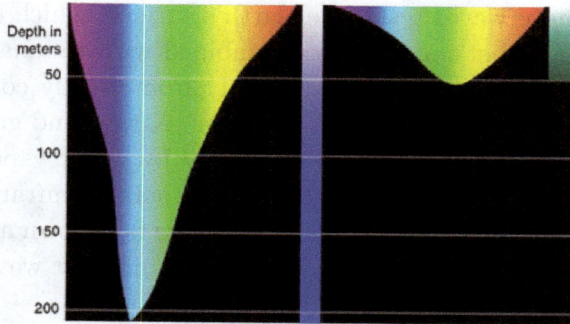

Fig. 1.3 Light penetration in sea water, in open ocean waters (left) and in coastal waters (right). Credit: Proceedings of the YOUMARRES Conference, DOI: 10.1007/978-3-319-93284-2_4.

earliest life on Earth developed in water, and stayed there for more than billion years before the first life forms started to colonize land. The *vertebrae* to which we belong stayed in the water even longer than many non-vertebrae which started to colonize the solid part of the surface. Our eyes are the direct descendants of the fish's eyes. Even one short look at Figure 1.3 can convince us that there was no use in developing optical sensibility to parts of the electromagnetic spectrum that could not penetrate into the water deeper than approximately one half-meter.

1.3 Resolution capacity of human eye

Human eyes are wonderful natural optical devices. Resolution, or resolving power, is the capacity of an optical system to resolve two points close to one another as separate images. The resolution of human eye is determined by the density of receptors in the retina. This means that the image formed on our eye's rear becomes to be perceived as a dimensionless point. It is easy to determine the minimal angular size of a faraway object under which its image will be perceived as a single point (if it is luminous enough), or not at all (if its luminosity happens to be close to that of the surrounding background). The eye is a globular organ filled with a viteous liquid called "humor", filling the main space between the lens and the retina; the

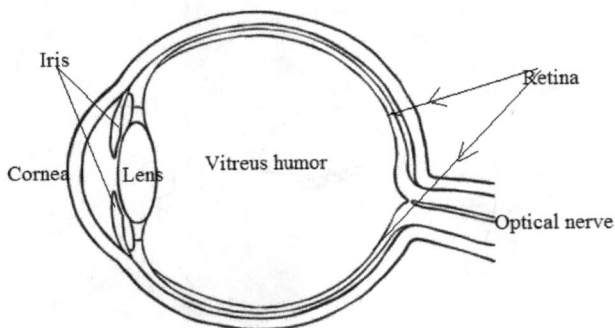

Fig. 1.4 Schematic structure of human eye. The average focal distance, from cornea to retina, is about 17 mm.

space between the transparent *cornea*, where most of the refraction occurs, is filled with an aqueous liquid, see Figure 1.4.

The retina is densely packed with two types of photosensitive cells, the *rods*, very sensitive to light but not distinguishing colors, and *cones*, less sensitive, but distinguishing three main colors, red, green and blue. The average size of receptor is about $2\,\mu m$, i.e. two microns, and at least three of them should be excited to give the impression of a colored spot. We can safely suppose that an elementary receptor containing three cones is about 5 to 6 microns of diameter. The average focal distance being about 17 mm, we can evaluate the minimal angular size starting from which our eye can perceive anything in form of a point-like object. This angle (in degrees) is given by the following formula:

$$\alpha_{min} = \frac{360\,x}{2\pi f}. \tag{1.1}$$

where $x = 5.5 \cdot 10^{-6}$ and $f = 1.7 \cdot 10^{-2}$ meters. Inserting these values, we get $\alpha_{min} = 0.0185° = 1.1'$. The resolution capacity of the human eye corresponds to one arcminute (one sixtieth part of one degree). This is how an object whose size is about 33 centimeters is perceived from distance of one kilometer.

The angular size, or angular diameter α of any object of maximal dimension d observed from distance D (Figure 1.5) is given (in

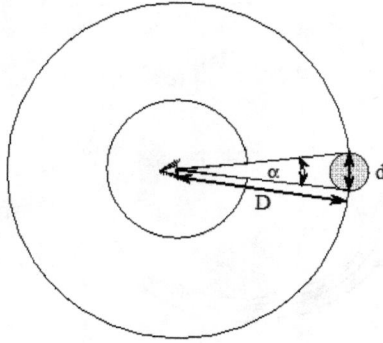

Fig. 1.5   Angular diameter α of an object whose real size is d, seen at a distance D.

degrees) by the following simple formula:

$$\alpha = \frac{d}{2\pi D} \cdot 360° \tag{1.2}$$

(The ratio d/D for $d \ll D$ is very close to angular diameter expressed in *radians*, 1 rad $= 360°/2\pi = 57°18'$). With our present knowledge of sizes and distances of planets and stars we can evaluate their angular diameters, which with no exception are under the natural resolution limit of human eye. This is why they all appear as points of different luminosity, but nevertheless, only points.

The situation is different when we see them through the telescope: the stars remain pointlike, while planets are magnified enough to be perceived as discs, even with some smaller details in the case of Mars, Jupiter and Saturn.

Let us evaluate the angular diameters of planets of the Solar System by means of the formula (1.2). In the table below we give angular diameters of Sun and Moon for comparison. For the planets we picked their minimal distances to Earth, which in the case of inner planets, Mercury and Venus, correspond to the lower conjunction, when these planets pass between the Earth and the Sun. Then they become invisible with the exception of their transit in front of the Sun, when they can be directly observed as dark spots slowly crossing the solar disc. For the outer planets, Mars, Jupiter, Saturn and beyond, the closest distance corresponds to the opposition, when given planet is

Table 1.1 The distances and angular sizes of the Sun, Moon and the planets.

Name	Diameter	Distance	Angular size
Sun	$1.4 \cdot 10^6$ km	$1.5 \cdot 10^8$ km	$31'27'' - 32'32''$
Moon	$3.57 \cdot 10^3$ km	$3.84 \cdot 10^5$ km	$29'20'' - 34'6''$
Mercury	$4.89 \cdot 10^3$ km	$7.7 \cdot 10^7$ km	$4.5'' - 13.0''$
Venus	$1.21 \cdot 10^4$ km	$4.5 \cdot 10^7$ km	$9.7'' - 30.1''$
Mars	$6.80 \cdot 10^3$ km	$5.5 \cdot 10^7$ km	$3.5'' - 25.1''$
Jupiter	$1.4 \cdot 10^5$ km	$5.88 \cdot 10^8$ km	$29.8'' - 50.1''$
Saturn	$1.16 \cdot 10^5$ km	$1.2 \cdot 10^9$ km	$14.5'' - 20.1''$
Uranus	$5.07 \cdot 10^4$ km	$2.57 \cdot 10^9$ km	$3.3'' - 3.84''$
Neptune	$4.92 \cdot 10^4$ km	$4.3 \cdot 10^0$ km	$2.2'' - 2.4''$

seen from the Earth on the opposite side from the Sun. Table 1.1 displays both minimal and maximal angular sizes of Sun, Moon and planets, corresponding to their maximal (respectively, minimal) distance from Earth. As it is easy to see, all planets' angular diameters are below 1 arcminute, i.e. below the average resolving power of the human eye.

1.4 The spherical illusion

Since time immemorial, in many cultures and civilizations, the starry sky was perceived as something material, and even stiff, due to the remarkable constancy of stars and constellations. The word "firmament" bears the trace of this conviction, as it evokes "firmness". The Earth in most ancient times was supposed to be flat. The space above was supposed to be home to deities such as Sun, Moon and planets. In the Bible, the canopy above was supposed to separate "upper waters", source of rains, from "terrestrial waters". In acient Greece philosopher Heraclitus of Ephesus supposed that celestial vault separates us from eternal fire, which we can see through tiny holes — the stars.

The spherical illusion results from the fact that the focal distance of human eye (about 2 cm) is infinitely small in comparison not only with astronomical distances, but even with much shorter distances on Earth. The two eyes provide us with two different images due

Fig. 1.6   An ancient perception of the Sky Vault. Wood engraving of an unknown author, first appeared in Flammarion's book on Atmosphere and meteorology.

to the phenomenon of *parallax*, which enables one to evaluate quite efficiently different distances to the surrounding objects. But as soon as those distances are greater than several hundreds of meters, the parallax becomes so small that it ceases to convey any valuable information, the images sent to the brain by each eye becoming identical. Everyone experienced this while looking at the faraway mountains: they seem like theatrical decorations painted on a flat canvas. The mountains that are farther are partly hidden by the ones that stand closer to us on the same line of sight, but they all seem flat and almost colorless.

The lack of any distinction between distances becomes even more spectacular when it comes to the astronomical objects that we see in the sky. They all seem to be at the same distance, conceived as infinite by our eyes. But who says infinite says also equal — as the result, we instinctively attribute *equal distances* to everything we see in the sky. Consequently, the geometry of this imaginary canopy stretched over our heads is the *spherical geometry*. The distances between any two visible celestial bodies is synonymous with the angle between them, or with the imaginary length of a *great circle* passing through these two points. And when in modern times people started to construct planetariums for pedagogical reasons, in order to show and explain celestial phenomena to the public, they were conceived as perfectly spherical white ceilings playing the role of a screen on which images of stars, planets and comets are being projected, creating an almost perfect illusion of starry skies.

The paradoxical effects of spherical geometry can be observed when the Moon and the Sun are visible in the sky at the same time, which occurs at sunset, when the Moon is in the last quarter and the Sun close to the horizon, or in the morning, when the Moon is in the first quarter, as shown in Figure 1.7.

When the phenomenon of simultaneous appearance of Moon and Sun occurs, it is worthwhile to proceed to a very simple experiment: place a tennis ball in the sunshine and observe it from a place from which you see it right under the Moon. Both sperical bodies, our satellite and the tennis ball, are illuminated in exactly the same manner. This suggests that the light rays falling on the Moon high in the sky and on the tenis ball here on Earth are parallel, or *almost* parallel. This means that the distance to the Sun must be quasi infinite — at least as compared with the distance between the tennis ball and the Moon.

Fig. 1.7   Left: the half Moon at the sunrise. The perpendicular to the half-crescent does not point towards the Sun. Right: Waxing gibbous Moon seen with morning Sun still close to the horizon, and a tennis ball on the ground illuminated by the same morning Sun. The Earth-Moon distance is negligible as compared to the distances of both to the Sun.

It seems nevertheless that a straight line perpendicular to the half-Moon pointing towards the source of light drawn mentally will be lost in the sky instead of ending up on the Sun. This illusion, as we already pointed out, is caused by the fact that we ignore the enormous, quasi infinite distance that separates us and the Moon alike from the Sun.

The most important result of spherical illusion is the necessity to describe apparent distances between celestial objects in terms of angles between the directions towards celestial bodies, which can be also translated into imaginary distances measured along the *great circles* joining the two objects. A great circle on a sphere is defined as the intersection of the sphere with a plane passing throught the sphere's center.

No wonder that ancient observers of skies above their heads arrived at the conclusion that all celestial bodies were firmly attached to rotating transparent spheres. They simply identified the apparent geometrical relations between visible luminaries with geometrical relations between material bodies as they could be reproduced by the hand-made models.

A first look at Figure 1.8 makes it obvious that the usual rules of Euclidean planar geometry do not apply. The sum of the angles of the

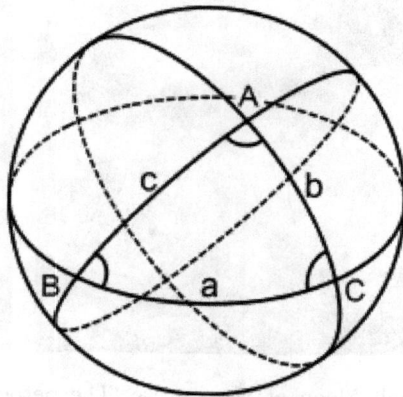

Fig. 1.8   Triangle on a sphere is made of segments of great circles. Its sides are (a, b, c) and the angles (A, B, C); are both measured in degrees.

triangle ABC drawn on the sphere is not 180° as it would be in the triangle drawn on a plane. For a spherical triangle big enough, e.g. an equilateral triangle made of three segments of 45° each would display the sum of three right angles equal to 270°! Obviously enough, the fundamental theorems of Thales and Pythagoras do not apply as well. On the other hand, alternative relations between angles and distances can be established, the most important one being the following:

• The sum of the angles of a spherical triangle is always higher than $\pi = 180°$. It is given by the simple formula

$$\sum_{i=1}^{3} \alpha_i = \pi(1 + 4p) = 180° \times (1 + 4p), \qquad (1.3)$$

where $p = \frac{S}{4\pi}$ is the fraction given triangle's surface cuts out of the total surface of the *unit* sphere ($S = 1$).

For example, if we cut the sphere by three mutually perpendicular planes passing through its center, we shall obtain *eight* identical equilateral spherical triangles. According to the formula (1.3), the sum of its three angles should be $\pi(1 + \frac{4}{8}) = \pi(\frac{3}{2}) = \frac{3\pi}{2} = 3 \times 90° = 270°$. Indeed, it is easy to see that such a triangle has three right angles.

Another example is provided by the maximal spherical triangle one can imagine, which is obtained when the great circle is cut into three equal parts. It cuts out exactly one half of the sphere, and according to the formula, the sum of its angles should be $\pi(1 + \frac{4}{2}) = \pi \times 3 = 540°$. Indeed, its three "angles" are all equal to $\pi = 180°$; For very small triangles the departure from rules of Euclidean geometry is negligible. Consider for example a huge rectangular triangle traced on the surface of the Earth, with sides 30, 40 and 50 kilometres long. Its surface area is, with good approximation, equal to 600 square kilometres. The total surface of our globe being 510 millions of square kilometres, the fraction p is extremely small, $p = 1.18 \cdot 10^{-6}$, and the sum of the three angles of this (quite big from human point of view!) triangle is $180° + \delta$, with $\delta \leq 1$ arcsecond, practically impossible to be measured with ordinary devices.

• The angles (A, B, C) and sides (a, b, c) of spherical triangles being measured in the same dimensionless units (degrees or radians), the following fundamental relation holds:

$$\frac{\sin A}{\sin a} = \frac{\sin B}{\sin b} = \frac{\sin C}{\sin c}. \tag{1.4}$$

• More general relationship between the angles and sides of spherical triangles, from which the "sine law" above can be derived, is as follows:

$$\cos a = \cos b \cos c + \sin b \sin c \cos A;$$

$$\cos b = \cos c \cos a + \sin c \sin a \cos B; \tag{1.5}$$

$$\cos c = \cos a \cos b + \sin a \sin a \cos C.$$

Before closing this section let us find out how big (or how small) are the areas in the sky occupied by typical constellations, the Sun and the Moon. There are two ways of measuring areas on celestial sphere: in *square radians*, or in *square degrees*. A *radian* is the angle subtended by an arc of a circle whose length is equal to that circle's radius. The circumference of a circle of radius r being equal to $2\pi r$, the full angle of 360° corresponds to 2π radians; therefore one radian is worth $360°/2\pi = 57.29°$ or $57°17'44''$.

The area A of a sphere of radius r is given by the formula $A = 4\pi r^2$. Now it is easy to evaluate the "angular area" of a sphere in radians squared:

$$A = 4\pi r^2 \rightarrow A = 4\pi (57.29°)^2 = 41253 \quad \text{square degrees.} \tag{1.6}$$

The sky is very big indeed! The Sun and the Moon seem quite big to us, but their diameter is only 0.5°. Their apparent diameters are so small that we can use planar geometry as good approximation, and apply the formula for area of a circle, $S = \pi r^2$, with r equal to 0.25°. This yields a surprisingly tiny result:

$$S = \pi \cdot (0.25°)^2 = 0.196 \text{ square degree.} \tag{1.7}$$

The stars' and planets' angular sizes are without exception lower than 1 arcminute in diameter, below the human eye's resolution power. Constellations are a different matter. They are usually many degrees wide and large, and their angular area can attain as much as 1000 square degrees.

1.5 Atmospheric refraction and aberration

When light penetrates into Earth's atmosphere under a certain angle, it is refracted according to the law discovered by Snell and Decsartes:

$$n_1 \sin \theta_1 = n_2 \sin \theta_2. \tag{1.8}$$

In this simple formula n_1 is the *refractive index* of the medium above the surface of separation, and n_2 is the refractive index of the medium below the surface. The angle θ_1 between the light ray and the normal to the surface is called *angle of incidence*, and the angle θ_2 is the *angle of refraction* (see Figure 1.9).

The refraction formula (1.8) permits to measure only the *relative refraction index* between two different media,

$$n_{12} = \frac{n_2}{n_1} = \frac{\sin \theta_1}{\sin \theta_2}. \tag{1.9}$$

At the time the law of refraction was discovered the relation between the refractive index and the speed of light in a given medium was not at all obvious. Descartes and Newton maintained that light is composed of tiny elastic particles. The corpuscular nature would explain the law of reflection, similar to the trajectory of perfectly elastic ball bouncing from a rigid plane, but refraction remained mysterious. Only after Huygens proved the undulatory nature of light, explaining not only refraction, but other phenomena as well, such as diffraction and interference. From Huygens' wave propagation model it became clear that the refractive indices are inversely proportional to the speed of light in given medium, so that the refraction formula

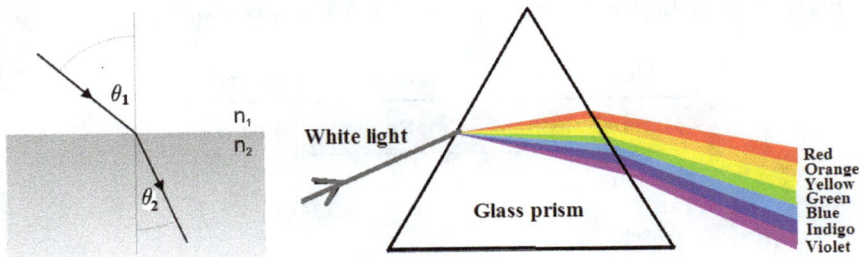

Fig. 1.9   Left: The Snell-Descartes refraction law; Right: Refraction dependence on light's wavelength results in color separation when a beam of white light refracts through a glass prism.

should be written as

$$\frac{1}{v_1}\sin\theta_1 = \frac{1}{v_2}\sin\theta_2. \tag{1.10}$$

Observing how a beam of white light refracts when it passes through a glass prism, Newton discovered that the index of refraction depends on color of given light beam, and that what we perceive as white color is in fact a mix of different colors, from red to violet — the entire spectrum.

Incidentally, it became possible to define an absolute refractive index by taking the speed of light in vacuum as reference, and attributing the value 1 to the symbolical refractive index of vacuum, $n_0 = 1$. In Table 1.2 the refractive indices of several common materials are given, including the air under normal conditions, i.e. at 20°C and under pressure of 1 atmosphere. The values in the table are given for the wavelength of 589 nanometers ($5.89 \cdot 10^{-7}$ meters) perceived by human eye as intense yellow color, close to its maximum sensitivity. In general, the refractive index varies slightly for different wavelengths, taking on higher values for blue and violet part of the spectrum, and sligthly lower values towards the red part.

Atmospheric refraction makes the celestial objects appear higher than they are in reality (see Figure 1.10). The dependence of the angle of refraction on celestial body's altitude above the horizon (given in degrees) is described by the curve shown in Figure 1.11.

Close to the horizon refraction attains 35', which is a bit above the angular size of the Sun or the Moon. This means that when we still see most of the red solar disc above the horizon, the real position of Sun is already below the horizon. At full Moon atmospheric refraction

Table 1.2   Refractive index of various materials.

Material medium	Refractive index n
Vacuum	1.0
Air	1.00067
Water	1.333
Crown glass	1.517
Flint glass	1.655
Diamond	2.417

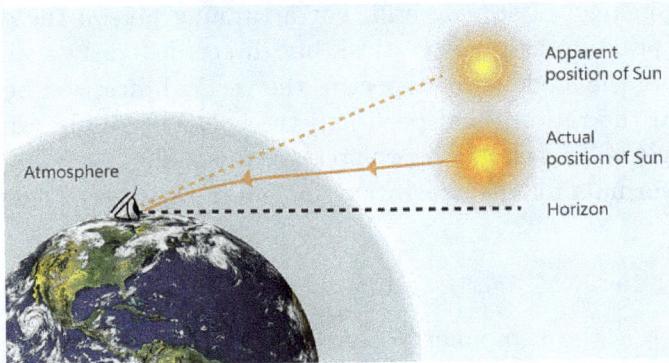

Fig. 1.10 Atmospheric refraction makes the celestial objects near the horizon appear higher than they are in reality.

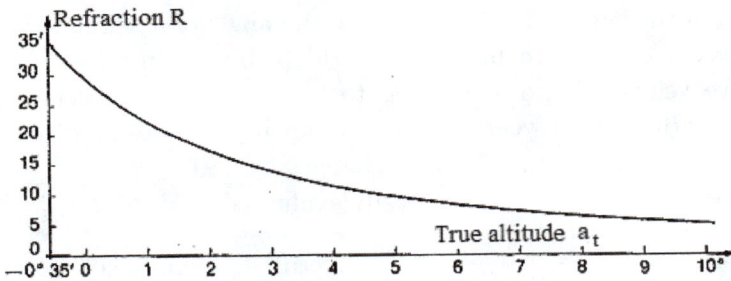

Fig. 1.11 Atmospheric refraction (in arcminutes) as a function of the altitude (in degrees).

can produce a rare phenomenon when both luminaries are visible just above the horizon, in opposite directions. But this is only an illusion: in reality, both are under the horizon. Tycho Brahe (1546–1601) was the first astronomer to take atmospheric refraction into account and correct systematically his observations (which had the precision up to 1 arcminute) of celestial objects with low altitudes.

Another distortion of the received image of stellar objects is caused by the finite speed of light and the motion of Earth around the Sun. This phenomenon is called *the aberration of light*, and can be easily understood be considering how the direction of rain drops falling vertically in absence of wind is perceived by someone who is running instead of staying still. For a running person the direction of falling rain drops is not vertical, but directed slightly against him, the sine of the angle with respect to the vertical direction being proportional the ratio of his speed and the speed of falling rain drops. This angle will reach 45° if two speeds are equal.

The formula for the aberration angle θ in the case of light rays is the same:

$$\sin\theta = \frac{V_\perp}{c} \qquad (1.11)$$

where $V_\perp$ is the component of velocity of the observer perpendicular to direction towards the object, and c is the velocity of light. The speed of light is tremendous by the everyday life standards, but becomes perceptible in astronomy due to Earth's velocity on her orbit, which is about 30 km/sec on the average. This is 10 000 times less than the speed of light; therefore the angle of aberration of stars observed close to the plane of Earth's orbit — in which case the relative velocity is perpendicular to the direction of observation — can vary during the year by twice as much, because the direction of Earth's motion after half a year changes by 180°.

The resulting shift can be easily evaluated:

$$\sin\theta \simeq \theta = \frac{2 \times 30\,\text{km/sec}}{300\,000\,\text{km/sec}} = 2 \times 10^{-4}\,\text{radian} \simeq 40'' \qquad (1.12)$$

i.e. about 20 arcseconds shift with respect to the average position in one direction or another during the year. Such small deviations could not be noticed by naked eye or by ancient astronomical devices before the invention of the telescope; but even after that it was confirmed and measured only in 1725 by English astronomer James Bradley (1693–1762). His precise measurements of aberration provided a fairly good estimate of Earth's velocity on the orbit, and confirmed once again without any doubt the validity of Copernican heliocentric system.

1.6   Why the sky is blue

An electric charge submitted to the action of electromagnetic field is accelerated and set into motion. When the electromagnetic waves encounter electric charges on their way, the latter are accelerated and start to oscillate under the influence of the electric and magnetic componenets of the field carried by the wave. But accelerated charges emit electromagnetoc radiation, as it follows from classical Maxwell's electrodynamics. An electric charge oscillating with a given frequency ω emits an electromagnetic wave whose intensity (in terms of power emitted) is proportional to ω^4, the fourth power of frequency, or to λ^{-4} in terms of wavelength of emitted radiation. The emission of radiation from electrons contained in atoms or molecules that have been put into motion by the incoming electromagnetic wave occurs in random directions; the result is the light scattering.

According to the inverse fourth-power law the ratio between the powers of scattered light belonging to the extremal parts of the spectrum, deep violet versus deep red, should be close to 16. However, as can be seen in Figure 1.2, the intensity of electromagnetic waves arriving from the Sun depends on their wavelength, with a clear maximum somewhere between yellow and green. The product of this curve with the curve λ^{-4} results in the maximum of intensity of scattered light at wavelengths close to intense blue, which is the color of very clear sky.

The λ^{-4} dependence is characteristic for the *Rayleigh scattering*, which occurs when the electromagnetic radiation interacts with atoms and molecules whose characteristic dimensions are much less than the wavelength of incident radiation. Rayleigh scattering is isotropic, equally inetnse in all directions.

When light scattering occurs on small spheres whose characteristic size is very close to the wavelength, which is the case of water vapor and very fine dust particles, the dependece on λ is different, close to the inverse square function λ^{-2}, and the maximum of intensity is shifted towards longer wavelengths. This kind of light scattering was studied by Mie (1908).

Finally, when light is scattered by particles hundreds of times greater than wavelenghts in the visible spectrum, all wavelenghts are scattered with equal intensity. This is why clouds composed of tiny water droplets floating in the air appear to be white.

Table 1.3   Dependence of scattering on the wavelength.

Scattering type	Sky color	Dependence on λ
Rayleigh scattering	Deep blue	λ^{-4}
Mie scattering	Blue	λ^{-2}
Selective	Light blue	λ^{-1}
Non-selective	Hazy	$\lambda^{-0.7}$
Non-selective	Very hazy	$\lambda^{-0.5}$

Fig. 1.12   Two effects at once: the Sun appears red at the sunset because violet and blue components are subject to stronger scattering than yellow and red ones; besides, atmospheric refraction being stronger closer to the horizon, the Sun appears squeezed horizontally. On the right: the Moon during a total lunar eclipse. Its red color is due to the refraction of solar light through the Earth's atmosphere. Only the orange and red components pass through the atmosphere to illuminate the Moon.

Chavez (1988) introduced five categories of atmospheric light scattering, with different power laws of the wavelength as shown in Table 1.3.

One can observe the preponderance of blue light scattering looking at distant mountain landscapes, where the visibility range can reach dozens of kilometers. The farther the mountain, the more its color becomes blurred by blue photons scattered along the distance between us and the remote objects, so that they appear as uniformly greyish-blue.

Consequently, the part of the spectrum that remains after the blue photons are scattered in all directions, contains mostly the wave lengths close to the red side. This is why the Sun at sunset becomes first orange, then intensely red (see Figure 1.12). This is also a reason for choosing red color for stop lights in cars and semaphors.

1.7 Stereoscopic vision and parallax

The image projected on our retina is a two-dimensional map of the surrounding three-dimensional world, in which we live and move. It is thus crucial for our survival to possess the ability to reconstruct the three-dimensional relations between surrounding objects, so as to evaluate distances as precisely as possible. Nature gave us wonderful means for this: the stereoscopic vision provided by doubling of image by the pair of eyes. As in most predators, our eyes occupy frontal position, both looking forward to pursue and chase the prey. Most of the herbivores, animals that may fall prey to predators, have their eyes on booth sides of the head enabling them to get as wide visibility range as possible, including easily accessible rear view.

It is not difficult to evaluate the maximal distance at which our stereoscopic vision still works. In order to be operative, it requires perceptible difference between the images created on the retinas of left and right eye (see Figure 1.13). As we already know, "perceptible" means the difference no less than one arcminute. When the difference between left and right image becomes smaller, our brain is unable to discern between the two. Average distance between two eyes d is about 7 centimeters. For an object to be far enough to become undistinguishable as seen by left or right eye, the angle between the directions pointing to that object from two eyes should

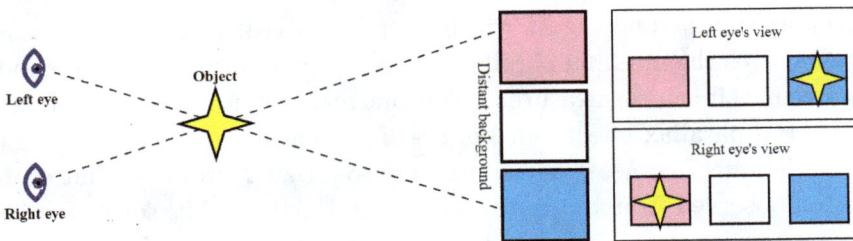

Fig. 1.13 The parallax scheme (left); The same object seen by left and right eye separately (right).

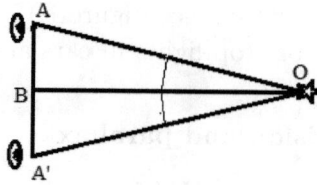

Fig. 1.14 The parallax angle here is $\alpha = 2\arcsin\frac{|AB|}{|BO|}$.

be less or equal to one arcminute. As shown in Figure 1.14 the relation between the parallax angle α, the distance betwen two eyes $d = |AA'|$ and the distance to the object $D = |BO|$ is given by the following formula:

$$\left(\frac{d}{2} : D = \frac{d}{2D}\right) = \sin\frac{\alpha}{2}, \tag{1.13}$$

where the angle α is expressed in radians (1 radian $= 360°/2\pi \simeq 57°18'$).

Applying this simple formula to the case of our two eyes separated by $d = 7\,\text{cm}$, and setting $\alpha = 1' = \frac{2\pi}{360\cdot 60}$, we get (with $d = 7\,\text{cm}$):

$$\frac{d}{2D} = \sin\left(\frac{\alpha}{2}\right) \;\rightarrow\; D = \frac{360\cdot 60\cdot 7}{2\pi}\,\text{cm} \simeq 240\,\text{meters}. \tag{1.14}$$

Resolution can be improved via increasing the distance between the observation points — instead of two human eyes two remote observers separate by distance d can note the angle at which they see the same object, and after comparing their data, find the parallax angle and determine the distance D.

The greater the separation d between two observation points, the greater the distance D at which parallax is still measurable — for naked eyes the minimal parallax value is one arcminute, but for good telescopes it can be measured below one arcsecond. Figure 1.15 shows how the parallax effect can be used for measuring distances to not too distant stars. Most of the stars are so far away from us, that even when observed from opposite sides of the Earth's orbit, separated by two astronomical units (A.U.), i.e. almost 300 millions of kilometers, they seem to stay at the same place, deserving the ancient name of "fixed stars"; but certain stars in the vicinity of the Solar System

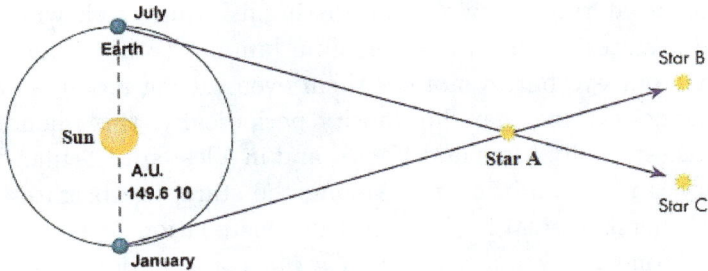

Fig. 1.15 Star parallax as observed from opposite points of Earth's orbit. The star *Star A* is seen either as *Star B* or as *Star C*. The distances even to the nearest stars are so great that maximal parallax observed is still far below one arcsecond.

are close enough to present yearly parallaxes of the order of fractions of arcsecond. The effects of parallax have been noticed by ancient Greek astronomers comparing observations of the Moon at different places on Earth. The conclusions concerning distance were rather shaky, taking into account that both observers on Earth and the observed object (the Moon) are in constant motion. The parallaxes of the nearest stars were measured for the first time by German astronomer Friedrich Bessel (1784–1846) only in 19th century, when telescopes became powerful enough.

1.8 Perception of brightness and stellar magnitudes

Our eyes can work in a wide range of light intensities thanks to the iris accomodation. The maximum sensitivity of the human eye during daylight is in the region of 550 nm wave length, corresponding to green color. During the night the maximum sensitivity shifts towards blue by about 45 nm, therefore close to 505 nm wave length. The sensitivity falls to zero on both sides, at 390 nm (far violet) and at 760 nm (deep red). The receptors working at weak luminosity do not distinguish colors anyway. The sensitivity threshold of human eye is remarkable: it corresponds to energy flux of about 10^{-9} erg/sec, an energy carried by thousand photons in optical region of the electromagnetic spectrum. This, of course, is possible only after long accomodation in complete darkness.

Looking at the starry sky, we distinguish quite well which stars are the brightest, which are of medium luminosity, and which are so dim that our eye hardly notices them even during moonless nights. The planets change thair luminosity periodically; this phenomenon concerns especially Mars and Venus, and in a less spectacular degree, Mercury, Jupiter and Saturn. Among the stars, the brightest of all (in both hemispheres) is *Sirius* in the constellation of *Canis Maior*; the next one is *Canopus* in the *Carina* constellation, then come, in decreasing order, α Centauri, *Arcturus* in the *Bootes* constellation, *Vega* in *Lyra*, *Capella* in *Auriga* and *Rigel* in the *Orion* constellation. Among these, *Canopus* can be observed in the Northern Hemisphere at latitudes not very far from the tropic of Cancer, and only in the wintertime, very close to the horizon. The third brightest star, α Centauri, although mentioned by Ptolemy who could see it from Alexandria, was forgotten by the Europeans, and rediscovered only in 16th century during the era of great geographical explorations.

In ancient times people often attributed greater apparent sizes to brighter stars. Later on it turned out that this is only an illusion, due to imperfection of human eyes. Light is diffused by crystalline body in the eye, resulting in apparent rays stemming from the star's center. This is why stars were always represented in this characteristic manner.

The first star catalogues were made by the Greek astronomers Hipparchus of Nicaea (190–120 B.C.E.) and Ptolemy of Alexandria (100–170 C.E.). They contained most of visible stars, classified in six categories, from 1 to 6, according to what could be interpreted as linear scale, the brightest ones being thus called "of first magnitude"; the dimmest ones being of sixth magnitude (see Figure 1.16).

The classification by Ptolemy was also used by Arab astronomers. When finally measured against a logarithmic scale, it turned out that the best appropriate base of logarithms to accomodate Ptolemy's classification was 2.54, meaning that the ratio between luminosities, defined as the energy flux carried by the light, between a given stellar magnitude and the next one, is 2.54. The luminosity ratio between stars whose magnitudes are separated by two units, e.g. between stars

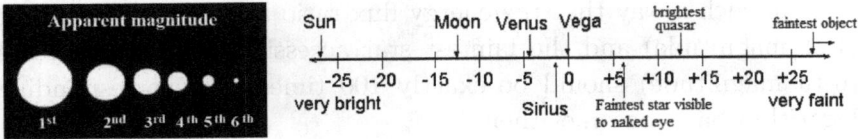

Fig. 1.16   Apparent magnitudes of visible stars, according to Ptolemy's classification.

Table 1.4   Stellar magnitudes of 11 most brilliant stars.

Star's name	Star's label	Magnitude
Sirius	α Canis Majoris	−1.44
Canopus	α Carinae	−0.62
Rigel Kent.	α Centauri	−0.28
Arcturus	α Bootes	−0.05
Vega	α Lyrae	+0.03
Capella	α Aurigae	+0.08
Rigel	β Orionis	+0.18
Procyon	α Canis Minoris	+0.04
Achernar	α Eridani	+0.45
Betelgeuse	α Orionis	+0.45

of second and fourth magnitude, will be then $(2.54)^2 \simeq 6.45$, and so forth.

Only about two thousand years later atronomers and physicists learned how to evaluate quantitatively the luminosity of an object by measuring the energy flux per unit of surface perpendicular to the ray, it turned out that the scale elaborated by Hipparchus and Ptolemy was proportional to the *logarithm* of the energy flux. The impression of linear scale is due to the Weber-Fechner law which stipulates that the intensity of sensorial impression is proportional to the *logarithm* of physical intensity received by our eyes (and also our ears — this is why the intensity of sounds is measured in *decibels* which are proportional to the logarithm of the energy flux carried by an acoustic wave).

In early 20th century astronomers slightly corrected the scale of magnitudes inherited from Ptolemy, choosing the basis of logarithmic scale in such a way that the energy flux ratio between the brightest (1-st magnitude) and the faintest star accessible to the naked eye (6-th magnitude) should be exactly 100 times. The corresponding logarithm basis becomes then

$$b = \sqrt[5]{100} \simeq 2.512. \tag{1.15}$$

Some examples are shown in Table 1.4.

1.9   Conclusion: Plato's cave metaphor

Our visual perception of the surrounding world turns out to be very limited indeed, as it appears from this chapter's previous sections. With our eyes, the unique detecting devices Nature gave us, we can register only one octave of wide electromagnetic waves' spectrum. The sensitivity, although quite high, stops at sixth-magnitude stars, leaving all fainter stars unnoticed. More than 2300 years ago ancient Greek philosopher Plato ($\Pi\lambda\alpha\tau\omega\nu$, ca. 427–348 B.C.E.) compared our situation with that of people dwelling in a cave and condemned to see only shadows of objects moving in full sunlight outside the cave, and projected on the rear wall.

Half jokingly, we can consider Plato as a precursor of modern multi-dimensional theories like the Kaluza-Klein model or string theories, which are based on the idea of existence of extra dimensions beyond the three space ones accessible to our senses.

The modern "Plato's cave" realized by people watching a television screen is even more radically distant from the reality than

Fig. 1.17   Left: Plato's metaphor of the cave (exposed in the "Republic"). Right: Modern version of Plato's cave. The metaphor has become reality.

shadows on the wall imagined by Plato to illustrate the imperfection and incompleteness of the information our senses give us about the surrounding world.

In fact, the two-dimensional images perceived by the spectators are a total illusion: they are an averaged result of the series of consecutive flashes of light emitted by pixels of the screen, one after another, in a one-dimensional temporal sequence. The information is conveyed in the form of a one-parameter (time) series of numerical data, with a rapidity that defies imagination. The screen is scanned 25 times per second (in modern computers even more frequently, up to 75 times per second). During each scan the pixels are activated one row after another, with diferent intensities of each of three colors, forming images that are composed of instantaneous flashes, superposing in our brain and perceived as a whole. The temporal sequences of slightly different images create the illusion of motion. Our brain is capable to distinguish between instantaneous images when they are separated by more than a tenth of a second; below that limit the illusion of moving objects replaces the sequence of different images, and at frequencies of image changes over 20 frames per second or more, the illusion of continuity becomes perfect.

As seen from a modern perspective, Plato clearly underestimated engineering and technical prowesses of human beings, who were able (although more than two millenia later) to enlarge the possibilities of their perception. In particular, we extended our sensibility to electromagnetic waves on both sides of visible spectrum, towards colors that had no names before, the ultraviolet and infrared, and beyond, towards X-rays and gamma-rays beyond ultraviolet radiation, and millimetric and radio waves on the inrared side. Images of stars and planets, galaxies and dust nebulae revealed new details and new phenomena filling the Universe.

And yet, in spite of all new findings, we are confronted with the situation Plato described in his metaphor of the cave. According to the latest analysis of star motions in galaxies and of accelerating expansion of the Universe at a cosmological scale, the Universe is filled with yet unknown type of substance called "dark matter", invisible to our telescopes and other observational devices, but creating extra gravitational pull needed to not letting the stars go away and to keep galaxies together. Apparently all visible matter in galaxies, including gas and dust clouds, is not enough to prevent the "evaporation" of its stars displaying average belocities too high to stay on closed orbit — unless there is extra gravitational mass accounting for the resulting gravitational field. Therefore what we see in form of stars and dust can account for only 20% of total mass of an average galaxy.

The discovery of acceleration of cosmic expansion poses another problem. According to Einstein's General Relativity, whose predictions agree with observation at least on the scale of our Galaxy, this acceleration must be caused by an unknown type of energy permeating the Universe, called *"dark energy"*, which should account for more than 70% of the total mass. The conclusion is that what is accessible to our observation is only 5% of the total content. Not much better than Plato's cave dwellers!

Chapter 2

The visible sky

Why observing the sky was important - Ancient versus contemporary observation possibilities - What can be seen by naked eye - Stars and constellations - Stars' diurnal motion - The Sun and its visible trajectory - The Zodiac - Sidereal and Synodic days - The Moon and its motions - The five planets - The Milky Way - Comets and meteorites

2.1 Preamble

From the dawn of human history two kinds of civilizations did appear: the sedentary ones, based predominantly on agriculture, and nomadic or seafaring ones, based on extensive grazing or fishing, and later on trade. Agricultural civilizations developed in fertile zones surrounding great river valleys, like Egypt or Mesopotamy, while seafaring civilizations arose along the coasts and on islands, like those of Minoans, Phoenicians and Greeks. For both types of civilization astronomy proved to be of vital importance, but for different reasons. To make growing crops successful farmers needed to know in advance when to expect the next flood of the Nile or Euphratus, or the next rain season in India or China, and this was made possible by careful observation of the Sun and stars; in one word, astronomy helped to master time. For nomads and sailors time was not as crucial as orientation in space: while on land merchant caravans or shepherds transferring their herds could leave characteristic marks on the ground, like huge stones or wooden poles, no trace can be left in open sea, far from the coast where at least lighthouses can indicate the direction. The sky then becomes a unique reference frame.

Ancient civilizations in which astronomy as we know it at present originated, developed in the Northern Hemisphere. The Babylonians and the Egyptians, ancient China and India, the Greeks, the Romans and the Arabs — all without exception observed the sky from places positioned North of the Tropic of Cancer. They could observe the Polar Star and the Great Bear[1] constellation, and could not see the Southern Cross and most of the stars visible in the Southern Hemisphere. This is why the contemporary astronomical system of positioning objects on the celestial sphere is adapted to the point of view of an observer looking at the sky as it is seen in the Northern Hemisphere. Even the dials of our watches and clocks imitate the right-screw orientation that is observed in the daily motion of the Sun during the day, and of the Moon during the night.

Until the invention of the telescope by Galilei and Kepler, all astronomical observations were made with the naked eye; still, some ancient astronomers were able to gather an amazing amount of data and grasp the fundamental periodicities in the motion of celestial objects. Some of their achievements may seem on the verge of miracle as seen from our own perspective. But this impression is mostly due to the modern life of city dwellers whose environment is not "observation-friendly", to say the least, and for multiple reasons. The atmosphere is often polluted by intense car traffic and nearby industries, the parts of the sky below 15° above the horizon are unaccessible due to the surrounding buildings; finally, the strong and abundant city lights forbid accomodation of the eye to the darkness, so that only a bunch of brightest stars remain visible, besides the Sun and the Moon. Nothing to do with what ancient people could admire almost every night, especially those living in such cradle of civilization as Egypt and Babylon, in valleys of mighty rivers surrounded by deserts, with scarce clouds and often clear skies. They saw almost every night what an average city dweller can see nowadays only in

---

[1] Also called Big Dipper in Britain and the USA.

a good planetarium. On the other hand, with few exceptions due to the precession of the Earth's axis of rotation, ancient astronomers saw the same motions and configurations of Sun, Moon, planets and stars as we observe them today.

Let us sum up the totality of astronomical phenomena in the skies that are accessible via observation made exclusivly with naked eye. This will give us an adequate picture of what was known to ancient astronomers who were able to find out most of the fundamental parameters characterizing our Solar System, including the motions of two luminous bodies, the Sun and the Moon, and the five planets with their periods and motions with respect to the Sun and to the fixed stars.

2.2 Stars and constellations

Independently of our whereabouts on Earth, all celestial bodies seem to be at the same distance, infinitely far away. This creates the spherical illusion, as mentioned in the previous chapter. The first written statement concerning this fact can be found in Greek geographer Strabo's comments on astronomy. During the day, when the sky is cloudless, the only visible heavenly body is the Sun. The sky is blue due to the diffusion of sunlight in the air; it is so bright, that the only other celestial objects that can be seen occasionally are the Moon and the planet Venus; all other atronomical objects can be seen only after sunset. When the night is definitely settled, the starry sky appears in all its majesty; we can also see the Moon, the planets, certain nebulae, and occasionally, comets.

The sky appears to us as a sphere surrounding the Earth, so we use angular measures to figure out distances on the sky. For example, a circle is divided into 360 degrees, with each degree further divided into 60 arcminutes, and each arcminute divided into 60 arcseconds (so there are 3600 arcseconds in a degree). These measurements are based on time, which is why we use "minutes" and "seconds", but to avoid confusion with real time measurement we put the word 'arc' in front to remind us we are dealing with arcs. So one can say, 'Those two stars are 4 degrees, or 240 arcminutes, or 14 400 arcseconds apart in the sky'.

Being able to measure apparent distances between two stars in degrees of the segment of the unique great circle joining them, we should also be able to measure surfaces of chosen portions of the sky. The circumference of a circle of radius r is equal to $2\pi r$, and the surface area of a sphere of radius r is equal to $4\pi r^2$.

The *radian* is the natural angular measure: it describes the angle of the circle cutting the length of one radius out of the circumference. As the total length of the circumference $2\pi r$ corresponds to 360°, the segment of the circle with the length of one radius will correspond to 360° : $(2\pi) = 57.3°$. Therefore we shall get the surface area of a sphere of any radius expressed in degrees if we replace r in the formula $4\pi r^2$ by one radian. This gives the surface area of full sphere expressed in degrees squared:

$$\text{Area} = 4\pi(57.3°)^2 = 42\,253 \text{ degrees}^2, \qquad (2.1)$$

which seems pretty big — more than forty thousand square degrees!

The first impression we get looking at the starry sky is the large number of stars, seemingly infinite, and their chaotic positioning in the sky. In fact the number of stars accessible to the naked eye does not exceed six thousand. This total refers to ALL stars, visible from both northern and southern hemispheres; if we restrict the count to the northern hemisphere only, there will be no more than three thousand visible stars. Their positioning with respect to each other does not change, at least in historical times. Only very precise angular measurements can show tiny motions of certain stars; the great majority does not show any relative motion at all. This enables an observer to take note and to memorize and recognize the brightest stars at first glance. As a matter of fact, stars are not motionless, and their velocities can attain hunreds of kilometres per second, but they are so distant that their relative positions as seen by an observer on Earth seem constant as if they were fixed to a solid dome, turning as a whole. The word "firmament" (containing the root "firm") applied to celestial canopy bears the trace of this

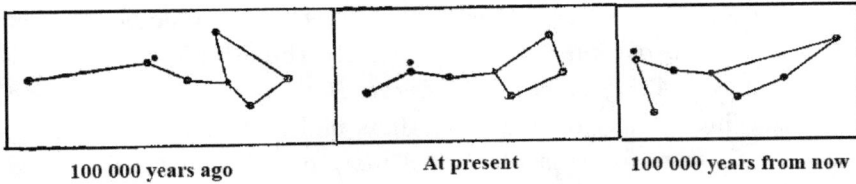

100 000 years ago At present 100 000 years from now

Fig. 2.1 Evolution of the Great Dipper over 200 000 years.

belief. Due to precise measurement of the Doppler effect,[2] velocities of many stars could be determined quite well, and we can reconstruct the shape of the Great Dipper in the constellation of Great Bear (*Ursa Major*) as it was in the past, and predict how it will look like in the future - the characteristic time span being 100 000 years.

In order to make the recognition easier, ancient people organized the brightest stars in *constellations*, or characteristic groups, although the choice of these groups varied from one civilization to another. At present astronomers all over the world agreed to adapt the constellations defined by Babylonians and Greeks to those observable in Northern Hemisphere; the constellations visible in Southern Hemisphere got their names much later, after being discovered in 16th century.

Traditional constellations bear the names of animals (Great Bear, Little Bear, Draco, Cygnus, Leo), the names of legendary characters of ancient myths (Perseaus, Andromeda, Hercules, Orion, Cassiopea, etc.), or the names of familiar objects (Triangle, Libra, Corona Borealis, etc.). Most of the brightest stars also have names, some of them of Greek and Latin origin, and many bear Arabic names given by Arab astronomers during the golden age of Islamic science. Since the end of the 18th century astronomers decided to label the stars by letters of Greek alphabet, in decreasing order accordingly to their visible brightness. The brightest star of all is *Sirius* belonging to the

---

[2]The Doppler effect consists in the apparent shift of the wavelength of incoming light depending on the relative velocity of the emitting object with respect to the observer: towards red for the receding objects, towards blue fot the approaching ones.

constellation of Great Dog (Canis Maior), therefore its official astronomical name is α *Canis Majoris*. Similarly, the bright star *Capella* is
identified as α *Aurigae*, *Vega* is α *Lyrae*, *Betelgeuse* is α *Orionis*, etc.

Observing the sky during a cloudless night, after certain time we
see that all stars follow a rotational motion which seems to be like
a rigid body rotation. Mutual positions of stars with respect to each
other remain the same, as well as the shape of constellations. But
not all of them can be seen after sunset and before dawn — the
remaining part disappears after sunrise. The diffusion of light by the
atmosphere makes the blue sky so bright that not only stars, but
even the Moon become invisible, their brightness being lower than
the brightness of the sky itself.

However, as the Earth changes position on her orbit, the direction
towards the Sun changes with respect to the fixed stars, and gradually different stars and constellations can be observed after sunset.
Thus, Orion, Taurus, Gemini, and Canis Major and Minor are high
in the sky to the South in the winter evenings as seen from North
America or Europe. On the summer evenings, Scorpio and Sagittarius
are most obvious in the southern sky. The "summer triangle" whose
defining vertices are at Altair, Deneb, and Vega, each of which is the
brightest star of its constellation (respectively Aquila, Cygnus, and
Lyra) is very high in the sky for locations in Northern Hemisphere
during June, July and August evenings.

Nevertheless, some part of the celestial sphere is unaccessible to
the eye if the observation point is not close enough to the equator.
Figure 2.2 shows three cases of an observer being on the equator, in a
place at 45° Northern latitude, and standing on the North Pole. From
any place on the equator all stars can be observed, the circumpolar
ones (those close to the North or South Pole alike) being found very
close to the horizon.

On the equator, during the year one can observe the stars situated
on both hemispheres, with positive and negative right ascensions.
Between the equator and the tropics, some stars and constellations
become invisible for certain period of time, the longer the farther
from the equator we stay, but still all can be observed at some time of
the year. But beyond the tropics, a growing part of celestial vault can

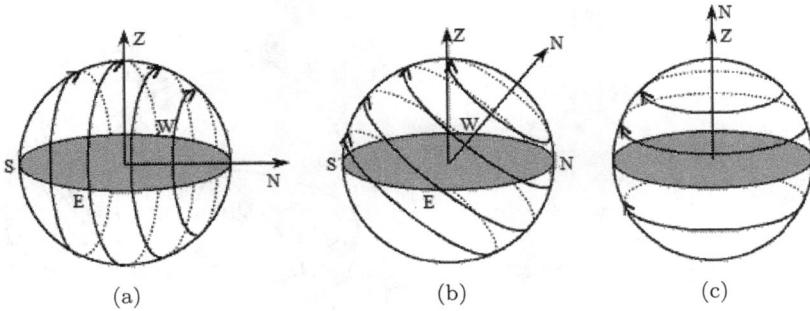

Fig. 2.2 Diurnal motion of stars as seen by observers in different latitudes: (a) on the equator, (b) at the 45° latitude, (c) on the North Pole. (Z = Zenith).

never be seen; the Southern Cross constellation remained unknown to the dwellers of the northern hemisphere until Portuguese navigator Vasco da Gama discovered the southern itinerary to India going around Africa way down south of Tropic of Capricorn. Similarly, the Great and Little Bear and Polaris star cannot be seen in Southern America, South Africa and Australia. It is worthwhile to remark that the notions such as *Equator, Meridian, North Pole, South Pole, Polar Circle* and *Tropics* were first applied to celestial sphere, before they became geographical terms denoting points and curves on terrestrial globe.

Figure 2.3 shows how a star's postion on the celestial sphere can be described by two angles, the altitude angle h and the azimuth angle α when the system chosen is that of *horizontal coordinates*.

At first glance horizontal coordinates are simple and natural, because they are defined with respect to the local horizon. They are called *altitude*, which is the angle at which given celestial object is observed, counted from horizontal plane along the meridian passing through the object, and the *azimuth*, which is the angle between the northern meridian and the meridian of the object, counted eastwards. Simple as it may seem, horizontal coordinates are not well suited to follow the motions of celestial objects, starting from the periodic diurnal motion caused by the Earth's rotation.

It turns out that the equatorial coordinate system is much better suited to describe stellar positions. As shown in Figure 2.3, any

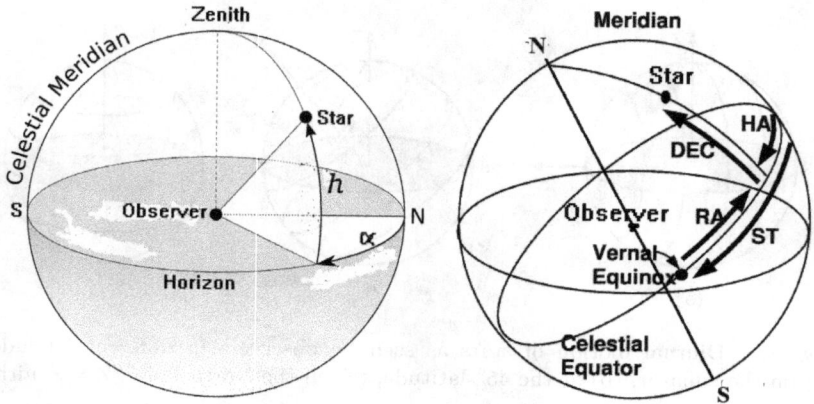

Fig. 2.3   Horizontal (left) and equatorial (right) coordinates. h = altitude, α = azimuth; DEC = declination, RA = right ascension, HA = hour angle, ST = stellar time.

object on celestial sphere is located by two angles, the *right ascension* α, often denoted as RA, and *declination* δ, often denoted as DEC. Declination is analogous to geographical latitude, and is the angular distance from celestial equator, measured in degrees, arcminutes and arcseconds. It is positive for objects observed north of equator, and negative for those south of equator. Table 2.1 displays equatorial coordinates of seven brightest stars of Northern sky and four stars in the Southern sky.

The right ascension is the analog of geographical longitude, and is measured in time units: hours, minutes and seconds. The conversion factor between the time and angular units is 24 hours = 360 degrees, so 1 hour = 15 degrees, 1 minute of time corresponds to 15 angular minutes. However, the circles of constant right ascension, which are analogs of parallels on terrestrial globe in geography, shrink when RA grows from low values close to celestial equator, towards higher values when approaching celestial poles. The shrinking factor is given by $\cos(DEC)$, e.g. the itinerary of a star whose declination is 45° is a circle with a circumference two times smaller than the length of the equator, because $\cos(45°) = 0.5$. Close to the pole the circles shrink almost to zero, like in the case of Polaris star, away only 1 degree

Table 2.1   Equatorial coordinates of some best known stars.

Star	Astroname	Right ascension	Declination
Polaris	α *Ursae Min.*	02 h 31 min 49 sec	+89° 15′ 50″
Dubhe	α *Ursae Maj.*	11 h 03 min 44 sec	+61° 45′ 04″
Mizar	ζ *Ursae Maj.*	13 h 23 min 56 sec	+54° 55′ 31″
Deneb	α *Cygni*	20 h 41 min 26 sec	+45° 16′ 49″
Vega	α *Lyrae*	18 h 36 min 56 sec	+38° 47′ 01″
Betelgeuse	α *Orionis*	05 h 55 min 10 sec	+07° 21′ 25″
Sirius	α *Canis Maj.*	06 h 45 min 09 sec	−6° 42′ 58″
Rigel	β *Orionis*	05 h 14 min 32 sec	−08° 12′ 06″
Canopus	α *Carinae*	06 h 23 min 57 sec	−52° 41′ 44″
Mimosa	β *Crucis.*	12 h 47 min 43 sec	−59° 41′ 20″
Acrux	α *Crucis*	12 h 26 min 06 sec	−63° 05′ 57″
Proxima	α *Centauri C*	14 h 29 min 42 sec	−62° 40′ 46″

from the North Pole, so that its rotation around the pole is practically imperceptible to the naked eye.

The shorter circumference of constant right ascension circles must be taken into account when we want to evaluate the distance between two stars in degrees of arc. For example, two stars on the celestial equator (DEC = 0) with RA = 0 (the vernal equinox point, in fact) and RA = 6 hours are separated by 90°, but two stars with similar right ascensions, 0 and 6 h, but with declination DEC = 45° will be only $90° \times \frac{1}{\sqrt{2}} \simeq 63°$ degrees apart.

Often the *hour angle* denoted also as HA is useful, because it relates local sidereal time S and right ascension RA via the simple formula:

$$HA = S - RA, \text{ or, equivalently, } S = HA + RA. \qquad (2.2)$$

This simple equation explains why it was deemed practical to measure the right ascension in units of time, hours and minutes, instead of degrees.

In Figure 2.4 the most important constellations surrounding the North Polar star are shown: *Great Bear* and *Little Bear* (Ursa Major and Ursa Minor in Latin), also called Big and Little Dipper. The Ursa Minor contains the Polaris, or the North Pole Star, which indicates

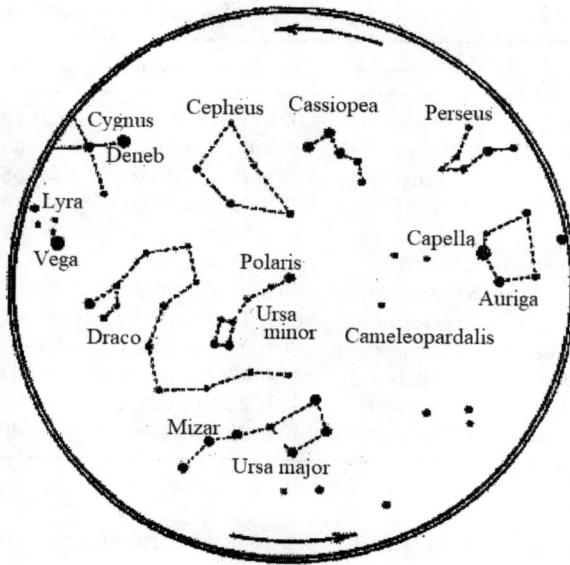

Fig. 2.4   Circumpolar constellations, with Polar Star in the center.

the North from wherever it can be seen. South of Tropic of Cancer the Polar Star can be seen only during a few summer months, and is unobservable in the southern hemisphere past the Tropic of Capricorn.

Between the two *Ursae* lies the long *Draco* constellation. Close by one can see the constellations of *Cepheus*, the characteristic *W* shape of *Cassiopea, Perseus, Auriga* and *Cygnus* (or the Swan). Besides the Polaris, a few brightest stars are indicated, too: *Vega* (α Lyrae), *Capella* (α Aurigae), *Deneb* (α Cygni). The vast but faint *Cameleopardalis* (Giraffe) constellation was unknown in Antiquity; it was not mentioned among 48 constellations defined by Ptolemy in his celestial atlas (second century C.E.). It was introduced by the Dutch astronomer Petrus Plancius and documented by the German astronomer Jakob Bartsch in 1624.

A person observing the stars from a point with a geographical latitude λ (in the Northern Hemisphere, therefore supposed positive), naturally divides the totality of the celestial vault into three zones:

a) The part of the celestial sphere from North Pole to the circle parallel to the equator, at angular distance $90° - \lambda$ from celestial North Pole. These stars are always visible, and they never rise or set; they cross the meridian twice in 24 hours, with an upper and lower culmination. During the upper culmination a star crosses the southern meridian from East to West, and during the lower culmination it crosses the northern meridian from West to East, as the rotation of celestial vault when looking in the direction of Polaris goes counterclockwise. In one word, the observer in the North Hemisphere sees all the stars whose declination (DEC) is higher than $90° - \lambda$. For observers from southern hemisphere a similar rule is valid: they can see all stars whose declination is contained between $-90° + |\lambda|$, their latitude λ being always negative.

b) The spherical strip containing the stars with declinations between $90° - \lambda$ and $-90° + \lambda$ (for Northern Hemisphere observers) These stars rise in the East and set in the West, and are visible only during some part of the night. For example *Canopus*, the second brightest (after *Sirius*) star in the sky, whose declination is almost $-53°$ can be seen in the northern hemisphere only from places with latitudes lower than $90° - 53° = 37°$. It never appears in the sky of northern and central Europe, but was known to all ancient Mediterrenian civilizations (the latitude of Babylon is 32.5°, the latitude of Alexandria is 31°, that of Cairo is 30°).

c) The stars with declination lower than $-90° + \lambda$ can never be seen by a Northern Hemisphere observer finding himself at (positive) latitude λ. They always remain under the horizon.

2.3 The Sun's diurnal and annual motions

Nothing is more important than the Sun in regulating our everyday life — and it was even more so in ancient times, where there were no street lights, and the only available sources of light during the periods of darkness were tally oil lamps, torches and candles. Due to Earth's obliquity, i.e. the tilt of its rotation axis with respect to the ecliptic plane on which the apparent solar motion takes place, the trajectory of Sun in the sky changes periodically: it rises and sets at different points of the horizon, and culminates at noon at different altitudes. Figure below shows how the Sun's trajectory varies during one full year. Every day at noon Sun crosses the southern meridian; it reaches its highest point in the sky at noon of June 21, the day of the summer solstice.

Our principal luminary rises exactly in the East only twice during the year, at autumnal (September 22) and vernal (March 21) equinoxes. During wintertime sunrises and sunsets occur in the South-East and South-West, respectively, while in summertime they occur in North-East and North-West, respectively, reaching extremal points on June 21 (summer solstice) and December 21 (winter solstice). These dates may vary slightly because of leap years. The trajectories of the Sun at different times of the year are represented in Figure 2.5.

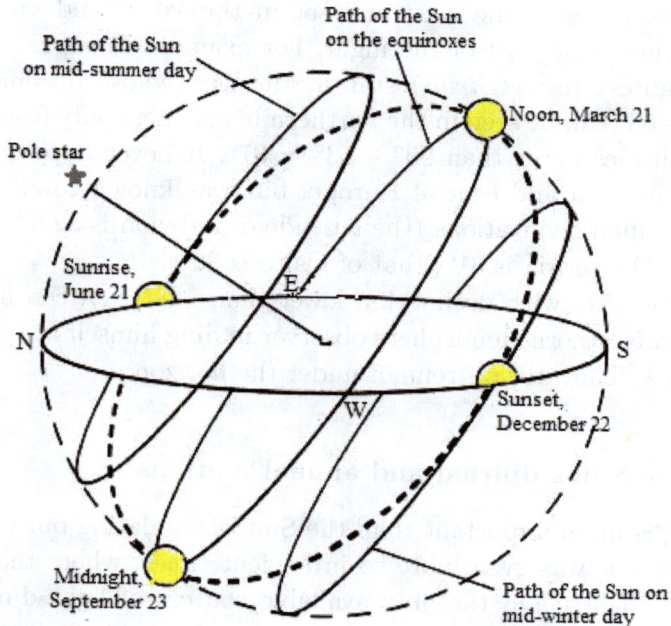

Fig. 2.5 Paths of Sun's trajectories as seen in the Northern Hemisphere (at a latitude about +45°) during different seasons. The extremal trajectories define northern and southern tropical circles. The trajectory during vernal or autumnal equinox traces the celestial equator.

The position of sunrise shifts along the horizon over the course of a year. The position of sunset also shifts by the same amount. How far this migration of the sunrise or sunset extends along the horizon depends on where the observer is based on the terrestrial globe. At the equator, the sunrise migrates plus or minus 23.5° from due East, while at a place on 45° latitude North, the position can vary about plus or minus 35° from due East. At the arctic circle the change is plus or minus 90° which is the maximum possible amount.

It is worth noticing that the concepts of equatorial and tropical circles were applied to the celestial sphere long before they became traced on the terrestrial globe. As we know at present, they are faithful projections of circles of the same name defined in geography as *parallels* of specific latitudes: polar circles, with latitudes ±66°33′, the Tropics of Cancer and Capricorn with latitudes +23°27′ and −23°27′ respectively, and the equator which is the parallel with 0° latitude.

It has been noticed since Antiquity that the Sun culminates (i.e. reaches its highest point in the sky) exactly on the southern meridian only four times a year, near solstices and equinoxes; the exact positions of those culminations depend on the observer's latitude. Between these special dates culmination occurs slightly eastwards (approximatively between winter solstice and vernal equinox, and between summer solstice and autumnal equinox), or westwards (between autumnal equinox and winter solstice, and between vernal equinox and summer solstice).

If Sun's positions in the sky were photographed at the same time of the day at regular intervals during the year, a closed curve looking like figure "8" would appear on the picture combining all photographs in one. The resulting closed curve is called an "analemma", and was known to solar dials makers in ancient times. We shall discuss its properties in one of the next chapters devoted to astronomical time measurements.

Usually, the analemma curves are produced by noting the positions of the Sun's culminations at noon, and display an almost perfect symmetric vertical form. If the Sun's positions are traced at any other hour of the day, e.g. in the morning, the analemma projected on the sky will be tilted, as in Figure 2.6 which shows the photographs

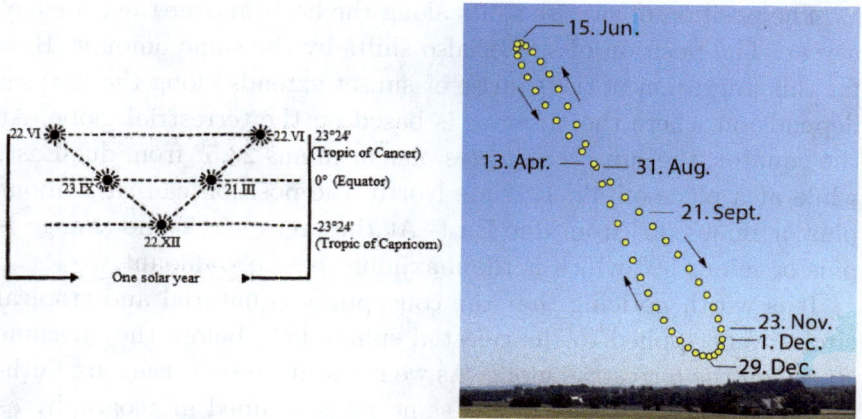

Fig. 2.6   Left: Schematic positioning of the Sun during the year, showing tropics and equator; Right: Position of the Sun at the same time of the day fixed at regular intervals during the year. The resulting closed curve is called "an analemma"

taken during an entire year at nine o'clock in the morning in central Germany.

Stars rise and set invariably at the same points of the horizon, but at different hours, but we can observe them only during the night, because as soon as the Sun rises, they become invisible due to its brilliance which illuminates the entire sky with diffused light brighter than the stars. It did not escape ancient astronomers' attention that the Sun *does not* rise and set at the same point of the horizon, changing slowly but constantly its position with respect to the fixed stars. To determine Sun's position on the map of the starry sky seems to be quite a challenge: how can one know this for sure when we see either the stars after sunset, without the Sun, or on the contrary, we can see the Sun very well, but not the stars! This point of view prevails among modern city dwellers whose horizon is hidden behind the surrounding buildings, who can rarely observe sunrise or sunset in full splendor.

As a matter of fact, noticing the Sun's whereabouts during its slow displacement among the stars was not that difficult for ancient

Egyptians and Babylonians, as well as for other civilizations after them, and this was for several reasons:

1) Their countries were situated mostly in flatlands, with a very good circular view of horizon;
2) Often enough the sky was cloudless, providing excellent observational conditions; besides, no street lights existed polluting nocturnal vision.
3) Both Egypt and Babylon are situated not far from 30°-th parallel. At these latitudes both dawn and dusk are much shorter than in Europe. This phenomenon is known to every traveler from the North, who is amazed discovering how quickly the Sun rises in the morning, and how rapidly the sky becomes dark after sunset. It was not difficult to notice which constellation was still visible near the horizon where the Sun was about to rise, or similarly, to notice shortly after the sunset which constellation appeared where Sun went down under the horizon. And even if it was not exactly the same constellation, it was the closest to it.

In the absence of an atmosphere, the sky would become totally dark a few seconds after the sunset until a few seconds before sunrise. But our atmosphere is there, and due to the diffusion of light it creates a twilight zone hundreds of kilometers wide. If we could see twilight from outer space, we would find that it is not marked by a sharp boundary on Earth's surface. Instead, the shadow line on Earth (sometimes called the *terminator line* is spread over a fairly wide area on the surface and shows the gradual transition to darkness we all experience as night falls, or in the reverse order, from darkness to daylight before the sunrise.

The terminator forms a great circle on Earth's surface, perpendicular to the ecliptic plane, so that a terrestrial observer crosses it at different angle depending on the latitude of his position. Two factors contribute to the duration of twilight: linear velocity and the angle of crossing. On the equatior, the linear velocity due to Earth's rotation around its axis is

$$V_{eq} = \frac{40\,000\,\text{km}}{24 \cdot 3600\,\text{sec}} = 463\,\text{m/sec}. \qquad (2.3)$$

At the latitude λ linear velocity is always directed along the local parallel eastwards, but its absolute value decreases with latitude according to the formula $V(\lambda) = V_{eq}\cos\lambda$, so that, for example,

$$V(50°) = V_{eq}\cos 50° = 463 \cdot 0.643 = 298\,\text{km/sec}, \qquad (2.4)$$

becoming even less for the observers living close to the polar circle.

At the equinoxes, while crossing the terminator, the linear velocity of any point on the Earth is perpendicular to it; but before and after, especially close to winter or summer solstices, the angle of crossing depends on the position of the observer. At higher latitudes this angle is so small that both dusk and dawn can last up to few hours, during which the sky is too bright to see the stars. But close to the tropics and on the equator the transition between day and night is very rapid indeed.

The precise determination of the duration of twilight for a given latitude and given day of the year is a very difficult problem of spherical geometry. It was resolved by Portuguese mathematician Pedro Nunes (ca. 1502–1578) in his work *De Crepusculis*, published in 1542 in Lisbon. [Nunes (1542)]

In any case, after the Zodiac constellations were identified and made easily recognizable, the alternative way to determine Sun's position with respect to the fixed stars is to look at the nocturnal sky and see which one of twelve zodiacal constellations culminates on local meridian at midnight, the place in the sky where the Sun was at noon, twelve hours earlier. This constellation is the *opposite* one to actual constellatiion hosting the Sun at the moment. For example, on March 21 (vernal equinox) the zodiacal constellation that culminates at midnight is Virgo, the opposite to Pisces (see Table 2.2).

Both the Latin and English names of the twelve Zodiac constellations, along with the celestial coordinates of their centers and their area in square angular degrees are given in Table 2.3.

Table 2.2 The Sun's zodiacal positions during the year.

Spring	Summer	Autumn	Winter
Pisces ♓	Gemini ♊	Virgo ♍	Sagittarius ♐
Aries ♈	Cancer ♋	Libra ♎	Capricorn ♑
Taurus ♉	Leo ♌	Scorpio ♏	Aquarius ♒

Table 2.3 Coordinates and angular areas of zodiacal constellations.

Sign	Latin	English	R.A.	DEC	Area (deg)^2
♓	Pisces	Fishes	0.85 h	+11.08°	889.42
♈	Aries	Ram	3 h	+20°	449.40
♉	Taurus	Bull	4 h	+15°	797.25
♊	Gemini	Twins	7 h	+20°	513.76
♋	Cancer	Crab	9 h	+20°	505.87
♌	Leo	Lion	11 h	+15°	946.96
♍	Virgo	Maiden	13 h	0.0°	1294.5
♎	Libra	Balance	15.21 h	− 15.59°	538.05
♏	Scorpio	Scorpio	17 h	− 40°	496.78
♐	Sagittarius	Archer	19.11 h	− 25.8°	867.43
♑	Capricorn	Sea Goat	21.02 h	− 21°	413.95
♒	Aquarius	Water Bear	22.71 h	− 10.19°	979.85

The pictorial interpretation of zodiacal constellations that prevailed in European astronomy is based on Greek and Roman tradition. Among the twelve imaginary drawings, only one is attributed to an inanimate object (Libra, the Balance), and only one is easy to be identified with what it is supposed to represent (Gemini, the Twins). Other constellations need more imagination to evoke real or mythical beings they are supposed to represent. Figure 2.7 represents one of the possible graphic interpretations of zodiacal constellations. The planets, whose motion in the sky follows paths close to the ecliptic circle, appear as rising or falling when the night comes with respect to the horizon, depending on which zodiacal constellation they happen to visit. The ascending constellations are: Sagittarius, Capricorn, Aquarius, Pisces, Aries and Taurus. The descending ones are: Gemini, Cancer, Leo, Virgo, Libra and Scorpio. The Moon follows a similar ascending and descending cycle, but it lasts one sidereal month, i.e. 27 days 7 hours and 43 minutes. The *ascending node* is the point

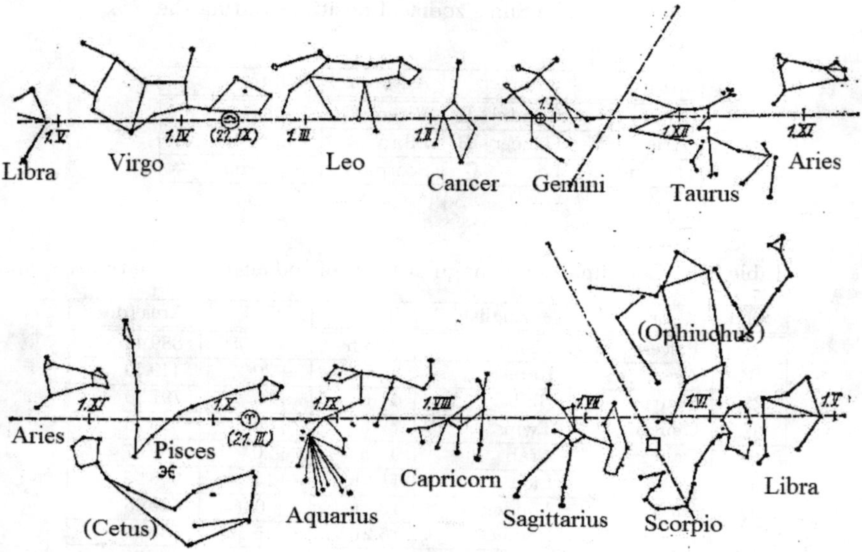

Fig. 2.7 The Zodiac constellations. A non-standard representation imagined by an anonymous amateur astronomer.

where the Moon moves into the North with respect to the celestial equator.

Let us follow Sun's yearly travel along the ecliptic:

• On March 21, the Sun is positioned at the vernal equinox. By definition, its right ascension (RA) and declination (DEC) are equal to 0. Everywhere across the Earth, the Sun rises exactly in the East, and sets down exactly in the West, and day and night last equally 12 hours each. Then the Sun starts to move towards the point of summer solstice; increasing both its right ascension and declination.[3]

• On June 22, about three months later, the Sun arrives at the summer solstice point. Its right ascension is then 3 h, and its declination is $23°27'$ (North). The astronomical summer begins in the

[3]This is of course an approximation; the Sun crosses the celestial equator at a very specific time during the day; when it rises/sets it is before/after the true equinocturnal point.

northern hemisphere (long days and short nights), and the astronomical winter, respectively, in the southern hemisphere (long nights and short days). The Sun's right ascension continues to increase, while its (North) declination starts to decrease.

• After another three months, on September 23, the Sun arrives at the autumnal equinox. Its right ascension is then 6 h, its declination is 0. The astronomical autumn begins in the northern hemisphere (long days and short nights), and the astronomical spring, respectively, in the southern hemisphere. Again, it rises exactly in the East and sets down exactly in the West, everywhere across the Earth. Day and night have again equal length, but the days start to shorten, and nights to lengthen. The Sun's right ascension continues to increase, while its declination starts to decrease towards the South.

• On December 22 the Sun arrives to the winter solstice point. Its right ascension is 9 h, and its declination reaches $-23°27'$ (South). This is the date of the longest night in the year in the northern hemisphere, and the longest day in the southern one; from this point on, days tend to become longer again, until the vernal equinox is reached, and the annual cycle completed.

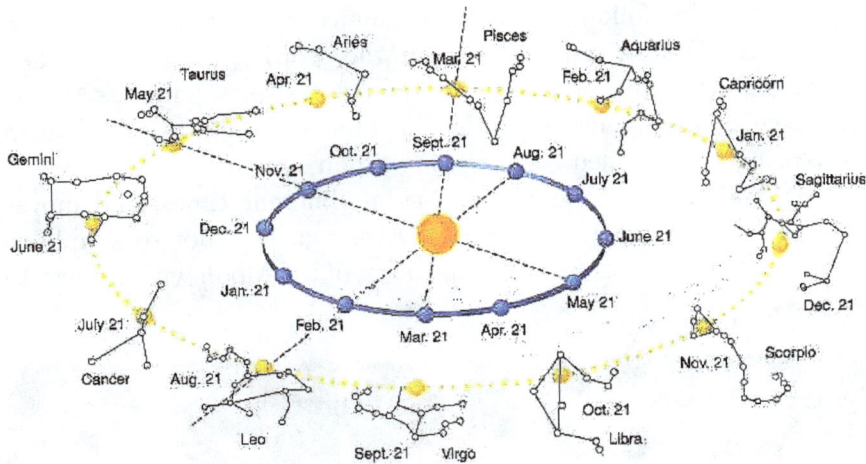

Fig. 2.8 The Sun's changing positions with respect to the fixed stars. To each of twelve months corresponds one of the Zodiac constellations.

The variations of the Sun's equatorial coordinates during the entire cycle are not strictly uniform. With a good approximation, already established in ancient Babylon, one notes that the Sun's right ascension increases by 1° per day. Its declination changes by approximately 0.4° during the month before or after the equinoxes, by 0.1° during the month before or after one of the solstices, and by approximatly 0.3° during all other months of the year. This variation approximates the sinusoidal shape of its declination varying with time.

2.4 The Moon, its motion and phases

Lunar phases are the most accessible time marker. Their cycle defines the *synodic month*, which lasts 29.56 days separating two identical phases, e.g. two full Moons, or two new Moons, or any other phase we prefer to choose as starting point.

As it appears in Figure 2.9, the lunar months are naturally divided into four periods easy to be seen by everybody: from new Moon to the first quarter, from first quarter to full Moon, from full Moon to third quarter, and from third quarter back to new Moon. Dividing 29.5 days by four gives 7.37 days, which is very close to 7. This is the reason why, following the Babylonians, we divide a month into four weeks. During the first quarter the growing crescent is called "waxing Moon", during the second quarter we see a "gibbous Moon" in growing phase, then after the full Moon, the gibbeous Moon in decreasing phase called also "waning Moon".

Night after night the Moon rises at different times, and moves eastwards by about 13° every day. Although it is not so simple to measure this angle precisely, the position of the Moon with respect to

Fig. 2.9 Lunar phases, the full cycle from one waxing Moon to the next one. The names of the phases: waxing Moon, full Moon, waning Moon, new Moon.

the stars can be determined quite well. This period of full rotation of the Moon with respect to the stars is called the *sidereal month* and lasts 27.3 days. From this we can evaluate more exactly the angular path the Moon covers between two consecutive nights: $360°/27.3 = 13.157° = 13°9'25''$. During all this time the Moon follows a path that is close to the ecliptic, but not exactly on it: it crosses the ecliptic plane only twice, and the points at which crossing takes place are called *nodal points*. The distance from the Moon to the ecliptic is never greater than $5.5°$, so it is always found within one of the constellations of the Zodiac.

Figure 2.10 shows two positions of the Moon separated by one sidereal month. Note that although the Moon's position (here in the Gemini constellation) is only slightly different, its phase is clearly not the same.

The difference between *sidereal* and *synodic* lunar months is explained in Figure 2.11. In the Copernican system, the Moon is orbiting the Earth, while the Earth is orbiting the Sun. During the time the Moon needs to accomplish one full turn around the Earth with respect to distant stars, Earth moves forward on its orbit around the Sun, so that the solar light illuminates the Moon under a different angle, and the observed phase is not the same. The Moon has to "catch up" in order to present the same configuration with respect to the Sun; it takes some extra time, which explains why the synodic month is slightly longer than the sidereal one.

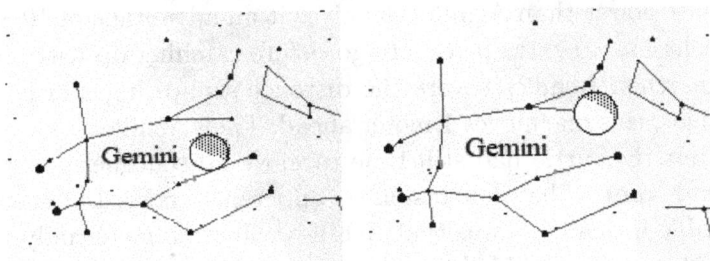

Fig. 2.10   Sidereal month: lapse of time separating two identical positions with respect to the fixed stars (in fact, the same right ascension).

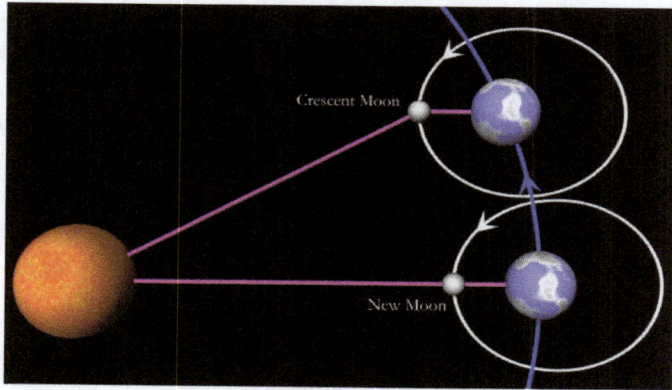

Fig. 2.11   Synodic month: lapse of time separating two identical phases.

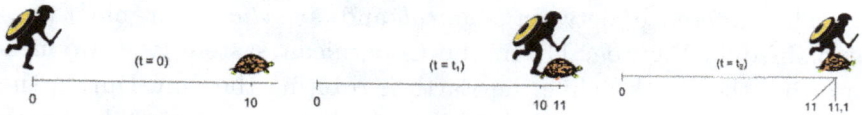

Fig. 2.12   Achilles and the Turtle endless run.

But during the time when the Moon moves to compensate the delay, Earth has moved forward a bit following its annual rotation around the Sun. Such a situation was analyzed by Zeno of Elea (ca. 495–430 B.C.E.), who took as example the contest between Greek hero Achilles and a turtle. Suppose that both start to move forward at the same time, but taking into account turtle's slow pace, supposedly ten times slower than Achilles, he gives it an advantage of 10 meters, so that he has 10 extra meters to go before catching up with the turtle. But when Achilles covers the distance separating him from the challenger, this one moves 1 meter ahead. Then Achilles goes 1 meter forth, but the turtle had still time to move 10 centimeters ahead of him, and so on. Thee conclusion — apparently false, but seemingly irrefutable logically — was that Achilles will remain eternally behind the turtle, in spite of his higher speed (see Figure 2.12).

The false result of Zeno's reasoning comes from the fact that he ignored the time that consecutive moves would take, each being

ten times shorter, therefore lasting ten times less than the previous one. But at the time Zeno and his followers were discussing the
nature of motion, they did not know how to handle infinite series of
fractions decreasing in geometrical progression.

[* We have to prove the following formula which gives the finite
limit of infinite sum of geometric series.

$$\Sigma_{n=0}^{\infty} q^n = \frac{1}{1-q}, \quad (|\,q\,|< 1). \tag{2.5}$$

Proof by mathematical induction:

$$\Sigma_{n=0}^{1} q^n = q^0 + q^1 = 1 + q = \frac{1-q^2}{1-q}. \tag{2.6}$$

Next step:

$$\Sigma_{n=0}^{2} q^n = 1 + q + q^2 = \frac{1-q^2}{1-q} + q^2\frac{1-q}{1-q}$$

$$= \frac{1-q^2+(q^2-q^3)}{1-q} = \frac{1-q^3}{1-q}. \tag{2.7}$$

There seems to be a common pattern in both cases, corroborated by
the next example, easy to calculate: in fact, we also have

$$\Sigma_{n=0}^{3} q^n = 1 + q + q^2 + q^3 = \frac{1-q^4}{1-q}. \tag{2.8}$$

which suggests that the universal formula should be

$$\Sigma_{n=0}^{N} q^n = \frac{1-q^{N+1}}{1-q}. \tag{2.9}$$

But this still does not provide the proof that the formula (2.9) is
valid for any value of N. The rigorous proof, using the method called
mathematical induction, consists in deriving the formula (2.9) from
the formula supposed valid for the previous value $N-1$. Let us
suppose that we have

$$\Sigma_{n=0}^{N-1} q^n = \frac{1-q^N}{1-q}. \tag{2.10}$$

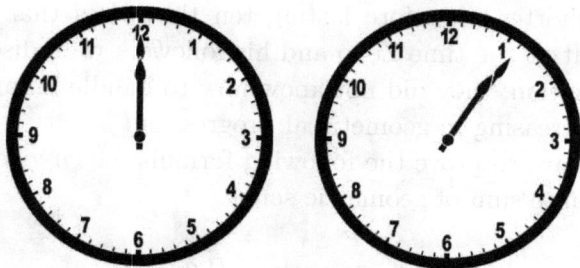

Fig. 2.13   Left: clock's hands meet at noon; right: next time clocks hands meet at 1 h 5 minutes and 27 seconds later. In twelve hours the hands meet eleven times.

Adding up the next term q^N gives

$$\Sigma_{n=0}^{N} q^n = \Sigma_{n=0}^{N-1} q^n + q^N = \frac{1-q^N}{1-q} + \frac{q^N(1-q)}{1-q} = \frac{1-q^{N+1}}{1-q}.$$

$$(2.11)$$

As we saw, the formula holds for $N = 1$, therefore by induction, it holds for any value of N. *]

Now we can find out how long does it take before clock's hands meet again after a given first meeting, e.g. at noon, like in Figure 2.13.

The short hand (hours) moves 12 times slower than the minute (longer) hand. When the longer hand (minutes) continues to turn making one full circle in one hour, the short hand goes slowly, covering only 1/12 part of full circle, which corresponds to 5 minutes on the clock. The longer hand will use five minutes more to catch the short one, but after 5 minutes the short hand is still slightly ahead, because it could cover 1/12 part of 5 minutes. To get the exact answer, we must sum up all the infinitesimal contributions, which can be expressed in the formula for the infinite sum of geometric progression with ratio $q = \frac{1}{12}$. We get in this way the following result:

$$T = \sum_{n=0}^{\infty} \left(\frac{1}{12}\right) = 1 + \left(\frac{1}{12}\right) + \left(\frac{1}{12}\right)^2 + \left(\frac{1}{12}\right)^3 + \cdots$$

$$= \frac{1}{1-\frac{1}{12}} = \frac{12}{11}.$$

$$(2.12)$$

The exact time between two encounters is therefore $\frac{12}{11}$ hours $=$ 1.090909... hour $= 1$ hour 5 minutes 27 seconds.

Let us rewrite the above result with notations designing the three periods by corresponding letters:

$$T = T_1 \frac{1}{1 - \frac{1}{12}} = T_1 \frac{1}{1 - \frac{T_1}{T_2}} \tag{2.13}$$

where $T_1 = 1$ hour $T_2 = 12$ hours, $T =$ time between two consecutive encounters. Let us rewrite the last equation using the corresponding angular velocities:

$$\omega_1 = \frac{2\pi}{T_1}, \quad \omega_2 = \frac{2\pi}{T_2}, \quad \Omega = \frac{2\pi}{T}, \tag{2.14}$$

therefore we can replace:

$$T_1 = \frac{2\pi}{\omega_1}, \quad T_2 = \frac{2\pi}{\omega_2}, \quad T = \frac{2\pi}{\Omega}. \tag{2.15}$$

Substituting into (2.13) and simplifying by the common factor 2π on both sides, we get

$$\frac{1}{\Omega} = \frac{1}{\omega_1} \frac{1}{1 - \frac{\omega_2}{\omega_1}} = \frac{1}{\omega_1 - \omega_2}. \tag{2.16}$$

The inverses of these quantities are also equal, which means that

$$\Omega = \omega_1 - \omega_2, \tag{2.17}$$

which means that the relative angular velocity is the difference of two angular velocities. This applies to the case when the two objects turn in the same direction, like the two hands of a clock, or like Earth and Moon in their common travel around the Sun. Were the two angular velocities equal, the relative angular velocity would become null; in fact, if Moon turned around the Earth with the same angular velocity as the Earth turns around the Sun, it would present eternally the same phase.

Let us apply the formula (2.17) to the case of two lunar months. According to the clock analogy, the apparent synodic angular velocity equals the difference between the *absolute*, or sidereal angular velocity with respect to the distant stars, and the angular velocity of

Earth in its annual motion around the Sun. This means that in the formula (2.17) Ω is the synodic angular velocity, ω_2 is the angular velocity of Earth, and ω_1 the sidereal angular velocity of the Moon.

In equation (2.17) we may assume any two variables as known, and the remaining third variable as the quantity to be determined. In order to make the calculus more amusing, let us suppose that the two lunar angular velocities are known, and what we want to discover is the sidereal period of Earth's rotation around the Sun. Therefore, we have to solve for ω_2, getting from (2.17) $\omega_2 = \omega_1 - \Omega$. Both lunar periods are known to a very good precision; therefore, by definition,

$$
\begin{aligned}
\omega_1 &= \frac{2\pi}{27.322 \,\text{days}} = 0.22997 \,\frac{\text{rad}}{\text{day}}, \\
\Omega &= \frac{2\pi}{29.531 \,\text{days}} = 0.21277 = \frac{\text{rad}}{\text{day}}.
\end{aligned}
\tag{2.18}
$$

The difference is equal to $0.22997 - 0.21277 = 0.0172 \,\text{rad/day}$. The last step is to find the time of one full rotation of Earth around the Sun, which is $T_E = \frac{2\pi}{0.01720} = 365.3$ days $= 1$ year. Not so bad, taking into account the approximation used (e.g. in this calculation the numerical value of π was taken to be 3.1416). The exact value of T_E in days according to the Gregorian calendar is 365.2425.

2.5 The planets

In ancient times only five planets were known, the other ones, discovered later (Uranus in 1781 by German born English astronomer Frederick William Herschel, Neptune in 1843 by French astronomer Urbain Le Verrier)[4] can be seen only with a good telescope. In the past, the five planets had different names in different languages, and were associated with different divinities, but nowadays their Latin names, derived from the names of Greek and Roman gods, are universally used by atronomers worldwide. The very name *planet* is derived

---

[4]In 1843 Le Verrier and John Adams independently predicted a new planet, but it was first observed knowingly by Johann Galle and Heinrich d'Arrest on Sept. 23, 1843. It had already been observed unknowingly by several other people, to begin with Galilei, but also Michel Lalande and James Challis.

from ancient Greek word *planetos* meaning "a wanderer". This is because unlike the stars whose mutual positions in the sky are fixed, planets constantly change their whereabouts traveling along complicated trajectories in the vicinity of the ecliptic, sometimes completely disappearing from the sight when they come close to the Sun.

Planetary motions are complicated, but not chaotic: already in ancient times the periodic character of their wandering was observed, with periods determined quite accurately. The positions of the planets with respect to the fixed stars are determined by their right ascension (R.A.) and declination (DEC). Consequently, their positions relative to the Sun can be obtained by substracting or adding their respective celestial equatorial coordinates. The most characteristic planetary positions with respect to the Sun deserved special names. Thus, the first important concept is the *elongation*, which is the angle between the directions to Sun and to planet as seen from the Earth. If right ascensions of a planet and the Sun coincide, the elongation of the planet is said to be 0°. This situation is called *conjunction*. Planets can be also in conjunction with the Moon and with other planets. For two inferior planets, Mercury and Venus, two conjunctions are possible, in front or behind the Sun, while for superior planets, Mars, Jupiter and Saturn, only one conjunction is possible.

When the elongation reaches 180°, the planet is in *opposition*. This is possible only for superior planets. Finally, when the elongation is equal to 90°, the planet is in *quadrature*. This is also possible only for superior planets, and of course, the Moon. The characteristic planetary positions are displayed in Figure 2.14.

It was also noticed that important difference exists between two *inferior* planets, Mercury and Venus, and three *superior* planets, Mars, Jupiter and Saturn. Today these terms correspond to the radii of planetary orbits as compared with the average radius of Earth's orbit around the Sun: Mercury and Venus are orbiting closer than Earth, while Mars, Jupiter and Saturn much farther away. However, these denominations were given long time before the Copernican system was known and adopted by the astronomers. This is because even in the geocentric paradigm the hierarchy of celestial spheres supposed to bear the Moon, Sun and five planets, was established by Ptolemy

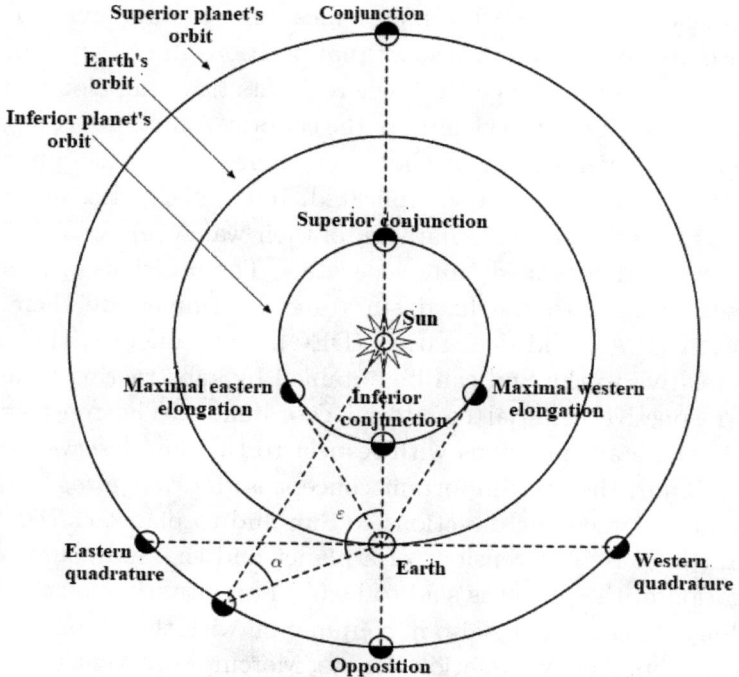

Fig. 2.14 The most important planetary configurations observed periodically, with their astronomical denominations.

according to the following order:

$$Moon - Mercury - Venus - Sun - Mars - Jupiter - Saturn$$

The fact that the Moon is closer to us than the Sun was understood very early due to the eclipses. The observational basis for distinction between inferior and superior planets was clear and simple. Two interior planets, Mercury and Venus, could not be observed further than between 18° to 28° from the Sun in the case of Mercury, and between 45° and 47° from the Sun in the case of Venus. Mercury can be seen only shortly before sunrise or shortly after sunset; similarly for Venus: much longer, although never deeper in the night. During their celestial wandering Mercury and Venus periodically disappear from sight being too close to the Sun, which by its daytime brilliance makes them invisible, like all other stars during the day.

On the contrary, the three superior planets can occupy any position along the Zodiac and can be seen in *opposition*, i.e. at the point of the sky diametrally opposed to Sun's position. In what follows, we give a summary of most important features of five planets accessible to the naked eye observations, as they were perceived by ancient astronomers.

• **Mercury:** Among the five planets known in the Antiquity, Mercury is the fastest - this is why ancient Romans, from whom the actual name of this planet is inherited, identified it with their swift-footed messenger god Mercury, who was also the god of travelers, commerce and thieves. According to ancient Greco-Roman myth, he had a winged hat and sandals enabling him to fly.

Mercury's motion with respect to the Sun displays periodicity of 116 days, which is the time between two consecutive maximal eastern (or western) elongations. It is also the closest to the Sun - independently of the adopted model of planetary motion, geocentric or heliocentric - just because its maximal elongations vary between 17.9° (*perihelic elongation*) and 28.7° (*aphelic elongation*). When the western elongation is at its maximum of about 28°, the consecutive eastern elongation which occurs about 58 days later is only about 17°, and vice versa. And many a time maximal elongations cover the range between those values. In other words, at perihelion Mercury's distance from the Sun is only about two-thirds (or 66%) of its distance at aphelion. Of all planets in our Solar System, Mercury's orbit has the highest eccentricity.

As it happens for all planets, from time to time Mercury's motion with respect to the stars becomes retrograde, when its motion with respect to the stars reverses its direction. The heliacal setting occurs when Mercury makes its last appearance in the evening sky and becomes unobservable the next day as it comes too close to the Sun and disappears in the sunshine; heliacal rise occurs when Mercury makes its first appearance in the morning sky while pulling away from the sunlight. Mercury is invisible at its inferior conjunction, but sometimes visible close to the superior one.

Mercury's orbit is tilted by 7° with respect to the ecliptic. For observers based at 40° latitude North this planet never culminates

higher than 10° above the horizon. When Mercury is close to its
western elongation it can be seen only in the morning, before dawn;
when it is close to its eastern elongation, it can be seen only at dusk,
shortly after sunset. Its brightness varies between magnitude −0.2 at
a perihelic elongation down to +0.3 at aphelic elongation.

• **Venus:** The "morning star" (which periodically becomes the
"evening star"), Venus is the second brightest object in nocturnal
sky, after the Moon, reaching stellar magnitude −4.2. There are
some (unconfirmed) reports that when Venus' brightest phase coincides with new Moon, when the nights are particularly dark, it is
even possible to see very faint shadows cast by its light. This white
shining planet is one of the most beautiful sights in the sky; no wonder that ancient Greeks gave it the name of goddess of beauty and
love, Aphrodite. The Romans followed them, although before Greek
influence became dominant, they called this bright planet "Lucifer"
(bringing the light). Ancient Egyptians identified morning Venus
with goddess *Isis*, and Babylonians with their main goddess *Ishtar*.

Venus is visible in the morning or in the evening most of the
time, but about every 10 months it moves so close to the Sun that it
disappears from the sight, becoming invisible for a couple of weeks.
For a long time, ancient Egyptians believed that the "morning star"
and the "evning star" were two distinct celestial objects; while the
Babylonians and ancient Greeks before Pythagoras recognized that
both were in fact the same planet.

Other ancient peoples have identified the "morning star" and the
"evening star" being the same celestial object, despite its periodical
disappearance in the "underworld". Among ancient Central American peoples e.g. the Maya and the Aztecs, the heliacal rising of Venus
was attended with great awe and apprehension. The Mesoamericans
also managed to fix the length of its complete cycle as 584 days. They
also noted that the ratio between this period and the length of the
year is given by simple fraction; 584 : 365 = 8 : 5.

As it became clear later on, this coincindence is due to the resonance between Venus and Earth revolutions around the Sun. As a
result, Venus seen from the Earth, is periodically in the same position
with respect not only to the Sun, but also to the stars. This happens

because five periods of Venus take eight years: as a matter of fact, $584 \times 5 = 2920$ days, and $2920 : 365 = 8$ years, which explains the "round" fraction $5 : 8$.

The "Venus pentagram" is the diagram on the Zodiac that Venus makes in eight years. After that period, Venus, the Sun, the Earth and the stars are back in the same relative positions. More precisely, if we take a particular point in a synodic revolution of Venus, for instance the greatest eastern elongation, and note which zodiacal constellation it is in, after an interval of 584 days, greatest eastern elongation will occur again, this time almost seven constellations away. These events repeat themselves (roughly) every nineteen months. As a result, in eight years the eastern elongation point will return close to the place from which it started - only about two degrees less - creating a pentagram inscribed inside the Zodiac, as shown in Figure 2.15.

Both Mercury and Venus were called "inner" of "lower" planets due to their proximity with the Sun, resulting in the impossibility of being observed in the middle of the night, far from the Sun's actual position. In the Ptolemaic geocentric system Mercury turned closer to the Earth than Venus, and both were closer to it than the Sun - so that even in that system the phenomenon of *transit* were conceivable, while no transit could be observed for the remaining three "outer" planets.

This is why the order of the inner planets seems to be inverted as compared with the Copernican heliocentric system, where Mercury

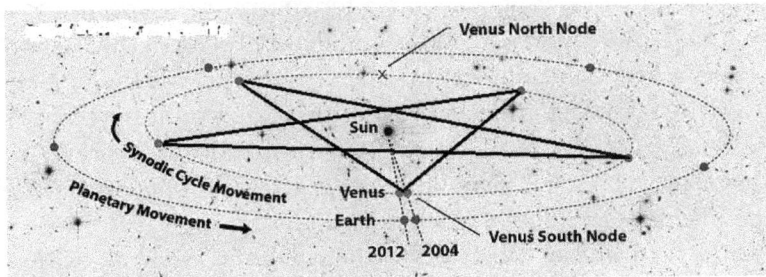

Fig. 2.15 The Venus pentagram. Lower conjunctions of Venus occur every 584 days forming a perfect pentagram.

orbits the Sun closer than Venus does. Still, Venus comes closer to the Earth than Mercury. In Tycho Brahe's mixed geo-heliocentric system Mercury and Venus orbit the Sun, which turns around the Earth.

Venus reaches two points in its orbit where the Sun-Earth-Venus angle lets us see it at its brightest. These events are known as greatest brilliancy, They occur when Venus lies at an elongation of 39°, approximately 36 days before and after inferior conjunction. At these two points the illuminated fraction of Venus and its distance from Earth combine together to maximize the amount of light we detect from our sister planet.

• **Mars:** Often mentioned as the "Red Planet" due to its distinctive red or orange color, Mars is our second closest neighbor planet after Venus. It is also the first of "superior" planets, which can be observed in any elongaton from the Sun, up to the opposition occurring at 180°. At such moments Mars becomes particularly bright, arriving at a −2.9 magnitude, making it brighter than any star including Sirius, but not matching Venus at its maximal brightness. Consecutive oppositions of Mars, when it is visible in the sky at a point exactly opposite the Sun's position at the same moment, occur every 780 days, or two years and 50 days. Mars culminates then at midnight, being opposite to the Sun.

In conjunction, which occurs like the oppositions, every 780 days, Mars passes very close to the Sun in the sky. At closest approach, Mars would appear at a separation of less that one degree from the Sun, making it totally unobservable for several weeks while it is lost in the Sun's glare. Even if Mars could be observed then, it would appear as a very faint object, practically invisible to the naked eye.

Over the following weeks and months, Mars would re-emerge west to the Sun, gradually becoming visible for ever longer periods in the pre-dawn sky. About a year later, it will reach opposition, when it will be visible for virtually the whole night.

Figure 2.16 shows when and where consecutive oppositions of Mars will occur in the coming 40 years, the last one having been observed at the end of July 2018, and the next one in October 2020.

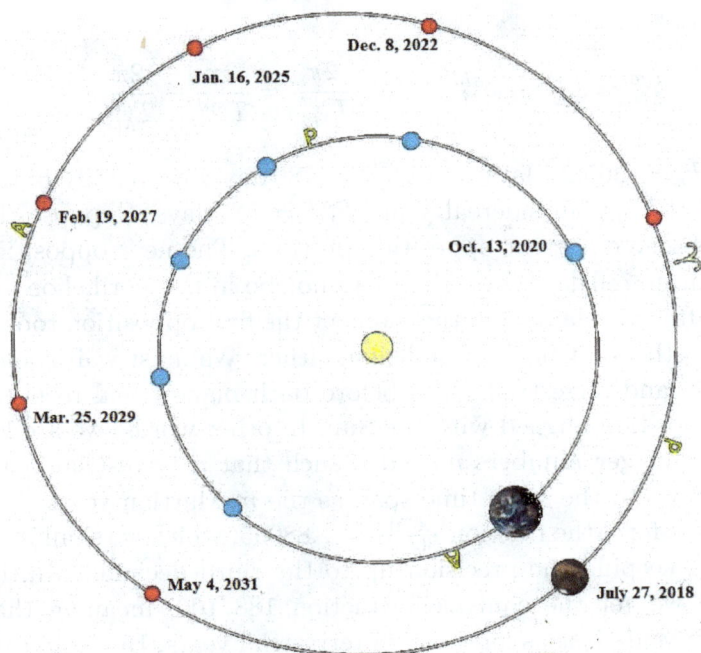

Fig. 2.16 Martian oppositions in XXI-st century.

During certain oppositions Mars is particularly bright attaining magnitude -2.9, which makes it the brightest object in the sky after the Sun, the Moon and Venus. Such events, lasting several weeks, are called *great oppositions*. They occur when during opposition Mars is at perihelion and Earth at aphelion of their respective elliptic orbits. Because when Mars is at the closest to the Sun point of its orbit, it receives a maximal amount of light, and when at the same time Earth is in aphelion, the distance to Mars is minimal, therefore the visual brighness is maximal.

Let us find out how long it takes for next great opposition to happen. To do this, we have to find out how long a Martian sidereal year lasts. Because Earth and Mars are orbiting the Sun in the same direction, the apparent angular velocity of Mars as seen from Earth is the difference between Earth's and Mars' sidereal (absolute) angular velocities:

$$\omega_M^{syn} = \omega_E^{sid} - \omega_M^{sid}, \quad \text{or} \quad \frac{2\pi}{T_M^{syn}} = \frac{2\pi}{T_E^{sid}} - \frac{2\pi}{T_M^{sid}}. \tag{2.19}$$

With $T_E = 365.24$ days and $T_M^{syn} = 780$ days, we readily obtain the length of Marsian sidereal year: $T_M^{sid} = 687$ days. Suppose that we have observed a great opposition of Mars. The next opposition will occur after 780 days, but Mars will not be in the perihelion - it was there after 670 days from the moment the first opposition took place. And Earth won't be in its aphelion either. We must wait some more Martian and terrestrial years before both planets will recover their initial position aligned with the Sun. In other words, we are looking for two integer numbers m and n such that n terrestrial years will correspond to the same time span as the m Martian years.

Let us form the fraction $\frac{670}{365.24} = 1.88095$, which we shall round up to 1.88, keeping the precision up to the third decimal. Multiplying by 100 we get the equivalent fraction 188/100, meaning that 100 martian years last as long as 188 terrestrial years. But who can wait for 188 years? In fact, we do not need precision that high; if Mars and Earth are close to their positions on respective orbits up to $1°$, Martian opposition could still be considered as a "great" one.

In order to get reasonable approximation, we can apply the Diophantian development of improper (greater than 1) fractions as follows:

$$\frac{188}{100} = \frac{47}{25} = 1 + \frac{22}{25} = 1 + \frac{1}{\frac{25}{22}} = 1 + \frac{1}{1 + \frac{3}{22}} = 1 + \frac{1}{1 + \frac{1}{\frac{22}{3}}} = 1 + \frac{1}{1 + \frac{1}{7 + \frac{1}{3}}}$$

If we stop the development before the last step, truncating it and neglecting the last fraction 1/3, we get a much simpler result involving smaller integers:

$$1 + \frac{1}{1 + \frac{1}{7}} = 1 + \frac{1}{\frac{8}{7}} = 1 + \frac{7}{8} = \frac{15}{8}. \tag{2.20}$$

At its closest approach (July 30–31, 2018) Mars came to within 57.59 million km of Earth.

Nearly 60 000 years ago (at the opposition that occurred on September 24, 57 617 B.C.E.) Mars was only 55.72 million km distant. The opposition of Mars on August 28, 2003, which brought Mars to 55.76 million km of Earth, was Mars' closest approach since then. The 2003 record for closeness will not be broken again until August 29, 2287.

• **Jupiter:** Named after the chief god of Greco-Roman pantheon, Jupiter is certainly the most majestuous of all planets. Although its brightness does not match Venus at its maximum, in contrast it keeps the luminosity almost constant: its apparent magnitude varies slightly, from −2.7 to −2.94, which is more than one magnitude brighter than Sirius (magnitude −1.46), the brightest star in the sky.

Jupiter's synodic period equals 398.8 days, which is the time between its two consecutive oppositions. This means that after one year Jupiter has moved westwards about 33.07°, so there is still some time left to catch up until the next opposition happens. The time difference is 398.8 days, which means that when one year passed after Jupiter's opposition, we have to wait still for $398.80 - 365.24 = 33.56$ days. From this we can easily find out how many years it takes to complete the cycle, i.e. after how many years Jupiter's opposition will happen in the same place with respect to the fixed stars. We have to divide the duration of Jupiter's synodic year by its difference from the solar year:

$$\frac{398.8 \,\text{days}}{(398.8 - 365.24) \,\text{days}} = \frac{398.8}{33.56} = 11.88 \,\text{years}. \qquad (2.21)$$

This is the sidereal period of Jupiter.

• **Saturn:** The slowest moving planet was named after the Roman god corresponding to the ancient Greek *Kronos*, the keeper of time and son of Uranus (the Sky) and Gaia (Earth). Its magnitude varies from faintest +1.17 up to brightest −0.55, when it becomes brighter than any star except Sirius and Canopus. Its progression among the stars is very slow indeed: the synodic period between two consecutive oppositions equals 378 days. As in the case of Jupiter, we can easily find out Saturn's sidereal period which amounts to 29.5 years.

As other planets, Saturn undergoes a periodical occultation by the Moon; however, due to the particular inclination of its orbit with respect to the lunar one, these occultations present an interesting time pattern. They occur in cycles: once in five years lunar occultation occurs every month during a year, then during the next five years no Saturn occultations can be observed, after which occultations resume during another year.

The periodicity of retrograde motions did not escape ancient Greek astronomers. Ptolemy in his *magnum opus* known also as *Almagest*, gives the recurrent numbers of retrograde motions versus numbers of planetary periods, taking care to find sufficiently long observation periods ensuring integer numbers for both types of data Table 2.5.

The numbers of retrograde motions were observed and accounted for during many years, so that first two columns in the above table were perfectly known. For two inferior planets the third column contains the *sums* of first two items, i.e number of years N_Y plus the total number of retrograde motions observed, N_R, producing the number of orbits $N_O = N_T + N_R$. For three superior planets the number in the third column is the *difference* between the first two, $N_0 = N_Y - N_R$.

Table 2.4   Durations of planetary retrograde motions.

Planet	Symbol	Retro (days)	Part of T_{syn}	Total arc
Mercury	☿	23	20%	12°
Venus	♀	42	7%	16°
Mars	♂	73	9%	15°
Jupiter	♃	121	30%	10°
Saturn	♄	138	36%	7°

Table 2.5   The numbers of retrograde motions over long periods of time.

Planet	Symbol	Years	Retrogrades	Orbits
Mercury	☿	46	145	191
Venus	♀	8	5	13
Mars	♂	79	37	42
Jupiter	♃	71	65	6
Saturn	♄	59	57	2

Dividing the number of years by the number of retrograde motion cycles we get the *synodic period* of a planet. So, we have for the consecutive planets:

Mercury : $\dfrac{46\,\text{years}}{145} \times 365.24\,\text{days} = 0.317 \times 365.24 = 115.87\,\text{days}$,

Venus : $\dfrac{8\,\text{years}}{5} \times 365.24\,\text{days} = 1.6 \times 365.24 = 584.38\,\text{days}$,

Mars : $\dfrac{79\,\text{years}}{37} \times 365.24\,\text{days} = 2.135 \times 365.24 = 779.84\,\text{days}$,

Jupiter : $\dfrac{71\,\text{years}}{65} \times 365.24\,\text{days} = 1.092 \times 365.24 = 398.95\,\text{days}$,

Saturn : $\dfrac{59\,\text{years}}{57} \times 365.24\,\text{days} = 1.035 \times 365.24 = 378.06\,\text{days}$.

2.6   The Milky Way

The Milky Way is a wide irregular band of faint white light which extends itself across the sky. If a plane containing it were drawn in space, it would be at 60° with the ecliptic plane. It is most easily observed during new moon, when our satellite does not submerge the sky with its brilliance. In the northern hemisphere the most luminous part of Milky Way can be seen near the southern part of the local horizon, and in southern hemisphere it is visible directly overhead, particularly well close to the June solstice. Our eys are unable to distinguish individual stars composing it; they can be seen only with a telescope. The first man who saw that it is just a swarm of stars was Galileo in 1611.

Various cultures and civilizations gave different names to this hazy band of light. The Chinese and their daughter Korean and Japanese cultures called it "The Heavenly Silver River".

The Milky Way, or Galaxy ($\Gamma\alpha\lambda\alpha\xi\iota\varsigma\ \kappa\upsilon\kappa\lambda o\varsigma$ in Greek), owes its name to the myth describing how Zeus had an affair with mortal woman Alcmene, who gave birth to a baby boy Hercules. Zeus was so fond the infant, that he wanted to endow him with immortality. Zeus placed the baby on his wife Hera's breast when she was asleep, to make the baby drink divine milk. When Hera woke up and discovered that she was breastfeeding an unknown child, she pushed the baby away, spilling some of her milk which produced the Milky Way in the sky.

Before the advent of the naturalistic vision of the universe Ancient Greek philosophers expressed different views on the Milky Way's nature.

When Portuguese navigators under the command of Vasco da Gama crossed the equator seeking the southern way to India, they discovered the southern part of the sky, unknown to the people of the Old World living essentially in Northern Hemisphere. But it was another great navigator, Fernão de Magalhaes (Magellan) who described two celestial objects looking like clouds; today they bear his name. After the invention of telescopes it appeared that two Magellanic clouds are composed of myriads of stars, just like the Galaxy. Today we know that they are dwarf galaxies which are gravitationally bound as satellites to our Galaxy.

Our Milky Way has a close neighbor, the Andromeda Galaxy, also known under its more prosaic name $M31$ (Number 31 of Mercier's catalog). Although it is more than twice bigger than our Milky Way, it can hardly be noticed by a naked eye, even during a moonless night. But if it could be seen, it would be an oval spiral structure occupying slightly more than 3.5° across, 7 times Moon's diameter. This means that it is relatively close: just about 16 times its own diameter.

2.7 Comets and meteors

Comets and meteorites were observed since pre-historical times, and were often interpreted as omens announcing forthcoming disasters. How strong this conviction was can be seen in a book published in 1848, *"A Chronological View of the World, exhibiting the Leading Events of Universal History, with an account of the Appearance of Comets and a complete view of the Fall of meteoritic stones in all Ages"* [Haskel (1848)]. One of the most famous coincidences of the appearance of a comet with an important historic event happened in the Middle Ages. The famous Bayeux tapestry was produced in late eleventh century, to commemorate the story of the Norman conquest of England by William Duke of Normandy and his victory over Harold, last Anglo-Saxon King of England. William won the battle of Hastings, at which Harold was killed, in 1066 C.E. One of the several Englishmen in the tapestry looks up at a big star with a large horizontal tail. The Latin inscription above them says "ISTI MIRANT STELLA" - "these (people) admire the star". The name "comet" is derived from ancient Greek word *komé* meaning "long hair". In Latin, *coma* also means "long hair" or "braid", and corresponds well to the usual appearance of comets as stars with a tail.

Usually a comet becomes visible as a faint tiny spot of magnitude +3. When it comes nearer to the Sun, a luminescent tail appears, and the head of the comet also increases its brightness. The tail, which sometimes can cover as much as 15° in the sky, always points in the direction out from the Sun.

In contrast with the more or less periodic behavior of stars, the Sun, the Moon and the planets, comets seemed to pop out of nowhere and disappear after a few days or weeks. Some meteor showers appear quite regularly, but with varying intensity; bigger meteorites that reach Earth's surface are totally random, both in time and location. It is understandable that ancient naturalists considered both comets and meteorites to be some kind of atmospheric phenomenon. This point of view was expressed by Aristotle, and since then was widely admitted by astronomers and geographers.

The greatest comets were so bright that they could be seen during daytime, their magnitudes reaching −15. Here is the list of several exceptional comets: some of them were recorded in ancient times, other were seen in the last few centuries:

- in February 1402, with magnitude probably −3;
- in 1556, magnitude −2;
- in 1577, observed by Tycho Brahe and probably seen by young Kepler, magnitude −3;
- in 1680, magnitude −6;
- in 1682, magnitude −4;

- in 1883, magnitude −7;
- in 1882, the "Great September Comet", at its highest brilliance reaching the magnitude −15, brighter than the full Moon;
- in 1910, Great January comet, visible during the day, magnitude −5.5 (brighter than Venus);
- in 1927, Skjellerup-Maristaing comet, reaching a −6-magnitude, visible even at 5° distance from the Sun;
- in 1965, Ikeya-Seki comet, 10 times brighter than full Moon, magnitude −10.

Comparing his observations of the 1577 comet, Tycho Brahe (1546–1601) saw first that its trajectory was beyond the Moon's orbit, then proved that before disappearing, it went beyond the Mars' orbit. Johannes Kepler believed that comets move along rectilinear trajectories.

The nature of comets was clarified by Newton and Halley after a happy coincidence, two great comets having appeared one after another, in 1680 and in 1682. The first one, called *the Kirch comet* after its discoverer, the German astronomer G. Kirch (1639–1710), appeared in November 1680, with its tail growing until it disappeared in the Sun's graze. Shortly after it reappeared on the other side of the Sun, and initially it was taken to be a new comet by the Royal Astronomer John Flamsteed (1646–1719), but very rapidly identified as the same comet that appeared in November; now, in January 1681, its tail was much bigger, and pointing outwards from the Sun. After a month the comet gradually lost its tail and brilliance, and disappeared from sight when it arrived beyond Mars' orbit.

Newton used the observational data to reconstruct the elliptic shape of the comet's trajectory, thus confirming his theory of universal gravity and the uniqueness of the inverse square law. Today this comet is also referred to as *Newton's comet*. In his famous book *Principia*, Newtons expressed his opinion concerning the nature of comets, which has become a commonplace since than - but the priority certainly belongs to him. Here is what he wrote:

"...universally, the greatest and most fulgent tails always arise from comets immediately after their passing by the neighborhood of the sun.

Therefore the heat received by the comet conduces to the greatness of the tail; from whence, I think I may infer, that the tail is nothing else but a very fine vapour, which the head or nucleus of the comet emits by its heat",

which is confirmed nowadays by modern observational methods, including spectroscopy and even one direct inspection the (*Rosetta-Philae* spacecraft made the first soft landing on the Churyumov-Gerasimenko comet in 2014).

A *meteoroid* is one of countless small bodies orbiting the Sun, mostly in the asteroid belt between Mars and Jupiter, or elsewhere. Their orbits are often unstable and prone to variations under gravitational influence of planets.

A *meteorite* is a meteoroid that hits the ground not entirely vaporized while transiting the atmosphere at high speed.

A *meteor* is a small meteoroid, often the size of grain of sand or a very small stone that entirely vaporizes in Earth's atmosphere being visible as "shooting stars". Small particles swarming around comets' cores or gathering in stable potential wells called *Lagrange points* on the Earth's orbit or beyond. Our planet crosses periodically one of those swarms, and each time a shower of meteors can be observed. The most notorious swarms are the *Leonids*, the remnants of the Temple-Tuttle comet, appearing once every 33 years; the *Perseids*, peaking on August 12 each year, the remnants of the Swift-Tuttle comet.

It seems ironical that the only tangible messengers of the Cosmos accessible for scientific scrutiny, including their physical and chemical properties, were steadily denied cosmic origin. That meteors were interpreted as an atmospheric phenomenon is not strange — after all, they really vaporize in higher atmospheric strata. But that meteorites were continuously believed to be of volcanic origin is less easy to explain. Many ancient men of science repeatedly expressed this view, including Aristotle whose authority remained unquestionable for more than fifteen centuries. Even after the triumph of the heliocentric view of the world and the advent of modern astronomy the belief that the cosmic void, home of planets and their satellites cannot contain any useless garbage, was shared by the astronomers until the end of 18th century. The acknowledgement of extra-terrestrial origin of meteorites came only in the beginning of the 19th century.

On July 24, 1790, at 9 p.m. a huge meteorite fell in the vicinity of Barbotan, a town in south-western France. The fall was seen by thousands of people. The Major of the city gathered declarations of about 300 witnesses, and sent the report to the Royal Academy of France in Paris. The members of the distinguished institution did not believe in the possibility of stones falling from the skies. The renowned chemist Claude Berthollet wrote on that occasion the following: *"How sad that a municipality produces a protocol containing popular fairy tales pretending that it really happened, while no physical reason for such an event can be given, nor by any reasonable cause"*.

But on April 26, 1803 a huge meteorite fell in the vicinity of L'Aigle, a French town in Normandy. In contrast with what happened in 1790, the Minister of the Interior Chaptal ordered Jean-Baptiste Biot (1774–1862), then a young member of the Academy of Sciences, to investigate the veracity of the event on the spot. Biot spent 10 days gathering information, asking dozens of witnesses and ensuring that they did not know each other and never met before or after the event. It was like a genuine criminal police investigation, and when Biot returned to Paris, he succeeded to convince his colleagues that the meteorites are indeed of extra-terrestrial origin.

The scientist who deserves the title of "father of meteoritic" science was Ernst Chladni (1756–1827), known mostly from his original experiments with vibrating plates covered with fine dust, displaying curious patterns of acoustic resonance. His interest in meteorites was aroused by the falling fireball event observed in 1791 in Göttingen.

Peter Simon Pallas (1741–1811), a German geographer and explorer of Russia, brought to St. Petersburg a 680 kilogram lump of metal found in 1772 near the Siberian town Krasnoyarsk. It was examined by many scientists, who concluded that its chemical composition and crystallographic structure did not correspond to any rocks found on Earth, but was similar to several other strange findings apparently fallen from space, according to witnesses. Later on Chladni called the meteorites of this type *"Pallas iron"*.

Planetary scientists distinguish between three main types of meteorites:

i) Stony meteorites: which consist mostly of silicate minerals. The two main subtypes are the *chondrites*, which represent the oldest remnants of times when solar system was formed, and *achondrites*, which are parts of asteroids and presumably of Martian or Lunar origin.

ii) stony-iron meteorites: which have nearly equal amounts of metal and silicate crystals.

iii) iron meteorites: which are almost completely made of metal.

Many meteorite samples of all kinds are dispayed in astronomical observatories and museums of natural history across the world.

Chapter 3

The eclipses

The Moon and its orbit - Solar eclipses - the Saros - Solar eclipse's duration - Lunar eclipses - Duration time - Occultations

3.1 A satellite or a sister planet?

Our Moon is in many ways an exceptional astronomical object, unique in the Solar System. The two most salient features are its large size and mass as compared with Earth, and the very close orbital distance: Moon's radius is just 3.67 times smaller than the Earth's, its mass is 81 times smaller, and the distance is roughly equal to 60 Earth's radii. Figure 3.1 shows the two globes side by side.

Table 3.1 displays relative masses and real distances to their mother planets of several big satellites in our solar system, illustrating the exceptional parameters of our unique celestial companion.

No wonder that some astronomers more than once referred to the Earth-Moon system as "a double planet" rather than "a planet with a satellite". Geographers have agreed on universal criterion distinguishing between the main river and its affluents (tributaries): it is not the length, neither the discharge at the confluence, but the velocity of flow in the two rivers measured at the confluence point. This is why Missouri is a tributary of Mississippi, although its total length measured from the source is longer; but it is Mississippi's water flow which is more rapid. Similar situation occurs with the Russian rivers Volga and its tributary Kama, whose discharge at the confluence is higher; but it is the Volga which has the more rapid flow.

Fig. 3.1   Earth and Moon on scale. The distance between the two *is not to scale!*

Table 3.1   Masses and orbital radii of satellites in the solar system.

Planet	Satellite	Relative mass	Distance in km
Earth	Moon	0.0123	384 000
Jupiter	Ganymede	0.00008	1 070 400
Jupiter	Europa	0.000025	671 100
Jupiter	Io	0.000046	421 800
Saturn	Titan	0.00023	1 270 145
Saturn	Enceladus	0.00015	800 000
Uranus	Titania	0.00008	900 000
Neptune	Triton	0.00015	900 000

Several criterions can be introduced in order to define a "double planet" when a satellite is comparable in mass and volume with its mother planet, which is indeed the case of our Moon. But where should we set the limit on the mass ratio? Obviously, when the masses of two companion planets are almost equal, they would revolve around their common center of mass, which would pursue its way on an elliptical orbit around the Sun. In general, if the masses of the two planets are M and m, with $M > m$ and their radii are R and r, and the average distance between their centers D, then the center of mass C will be found at the distance $\frac{m}{M+m} D$ from the center of the greater mass M and at the distance $\frac{M}{M+m} D$ from the center of mass

m (for simplicity, we consider a circular orbit). In the case when the mass M is high enough, and the radius R big enough, the following relation will be satisfied:

$$\frac{m}{M+m}D < R. \tag{3.1}$$

The center of mass will be inside the bigger planet, in which case the smaller one shall deserve the name of a satellite. This is the case of the Earth-Moon system:

$$M = 81\,\text{m}, \quad D = 384\,000\,\text{km}, \quad R = 6\,370\,\text{km},$$
$$\frac{mD}{M+m} = \frac{D}{82} = 4\,683\,\text{km}, \tag{3.2}$$

which is about two-thirds of Earth's radius. This means that the center of mass of the Earth-Moon system is inside our planet, about $1687\,\text{km}$ under the surface. This qualifies the Moon as a satellite.

But there is an alternative criterion, which can be formulated as follows: a satellite (in our case the Moon) revolving around a planet which itself revolves around a massive central star (our Sun) undergoes two simultaneous gravitational accelerations, one coming from its heavier companion (in our case the Earth), and the other from the common great center of attraction (the Sun). In a case when the attraction of the Earth is felt stronger than the attraction towards the Sun, such a body should be called a satellite; in contrast, when the Sun's attraction overcomes the attraction of the planet, the second body can be called a "sister planet", and the pair deserves to be called a "double planet".

In principle, to decide to which category our Moon belongs, it is enough to compare two formulae for gravitational acceleration:

$$a_S = \frac{M_S G}{D_E^2} \quad \text{versus} \quad a_E = \frac{M_E G}{D_M^2}, \tag{3.3}$$

where a_S and a_E are accelerations of the Moon resulting from gravitational influence, respectively, of Sun and Earth, M_S and M_E Sun's and Earth's masses, D_E and D_M are average distances to the Sun and to the Earth — we can safely set D_E as the average radius

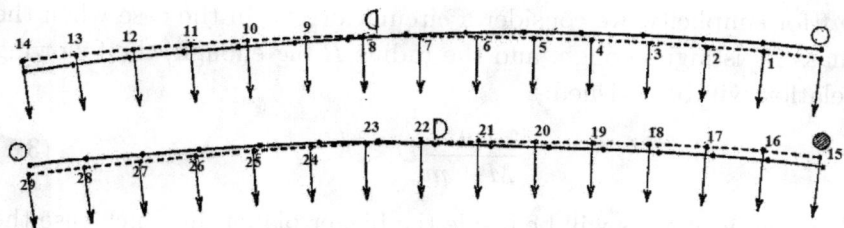

Fig. 3.2 The orbits of the Earth (hatched line) and the Moon (full line) around the Sun; the upper part corresponds to the first 14 days of lunation, the lower part to the next 14 days, from 15 till 29. The radius of Earth's orbit here is equal to 50 cm, and the maximal distance between Earth and Moon is less than 2 mm. Moon's orbit is always concave.

of Earth's orbit, because the Moon's distance to the Sun oscillates around it with negligible amplitude, so small is the radius of Moon's orbit (384 000 km) in comparison with the radius of Earth's orbit (150 000 000 km). The ratio of the two is $2.56 \cdot 10^{-3}$, which can be considered negligible in the present case.

But how to make the estimate without knowing the masses and the value of Newton's gravitational constant G? The answer can be given using pure geometrical properties of both orbits. Supposing for simplicity that they are circular, it is enough to know their radii, D_E and D_M. In its complex motion around the Earth and around the Sun, the Moon is half of the time closer to the Sun than the Earth, and half of the time farther. As seen from the Sun, both heavenly bodies follow quasi-circular orbits, with perturbations so small that both represent convex curves, as shown in Figure 3.2.

The convexity of Moon's orbit as seen from the Sun means that at any point of the orbit Sun's gravitational attraction acting on the Moon is stronger than the attraction of the Earth. This qualifies our Moon as a satellite of the Sun in the first place, rather than a satellite of our planet. If the Earth suddenly happen to disappear, our Moon would continue its way around the Sun, only on a perfectly elliptical orbit.

Let us have a closer look at the two orbits; the geometrical proof of convexity of lunar orbit around the Sun is quite easy, as shown in Figure 3.3.

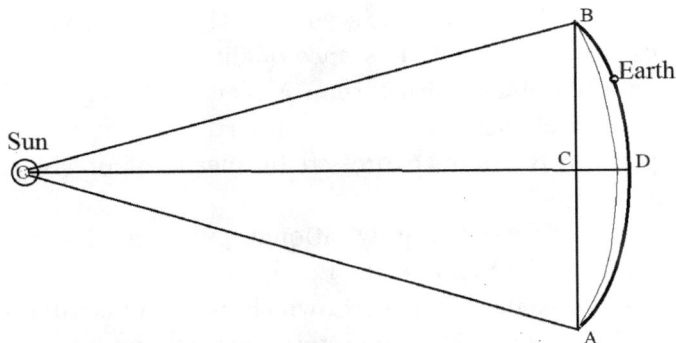

Fig. 3.3 The orbit of Earth, with a sector corresponding to two weeks during which the Moon is on the side closer to the Sun (a week before and a week after the New Moon). The distance CD between Earth's orbit and the chord is more than 2 million km, which can accomodate Moon-Earth distance more than 5 times. The Moon's orbit is a convex curve. (The drawing is not to scale, and the curvature is exaggerated).

We wish to evaluate the width of the space between Earth's orbit (the segment ADB) and the chord ACB. The angles $\angle ASD$ and $\angle DAC$ are equal, and the common value (in radians) is $\alpha = \frac{2\pi}{52}$, because it corresponds to one week between the new Moon and first quarter, and there are 52 weeks in a year. The angle α being relatively small, we can assume that the lengths of the segments AC and AD are practically equal, and the triangles ASC and CAD are similar. The value of α in radians is $2\pi/52 = 0.121$, which gives $\alpha = 6.92°$ in degrees (1 radian $= 57.29°$). Using the similarity between the triangles ACD and SCA we have:

$$\frac{CD}{AC} = \frac{AC}{SA} \rightarrow CD = \frac{(AC)^2}{SA} = \left(\frac{AC}{SA}\right)^2 SA. \qquad (3.4)$$

Here SA is the radius of Earth's orbit, $1.49 \cdot 10^8$ km; the ratio $(AC)/(SA)$ is the sine of the angle α, which in absolute value is very close to the value of α in radians, i.e. $\alpha \simeq \sin \alpha = 0.121$. Inserting this value into (3.4), we get $CD = 2.175 \cdot 10^8$ km, which is more than two millions kilometers, more than 5 times the radius of Moon's orbit around the Earth. This means that the lunar orbit is closer to the circle of Earth's orbit around the Sun than to the chord ACB, confirming that Moon's path as seen from the Sun is convex at each point of the orbit. This circumstance qualifies the Earth-Moon system as a double planet rather than a planet with a satellite. But after all, such a classification is not important; what really counts, are the quantitative data, expressed by means of precise numbers and formulae.

A comparison between gravitational pulls the Moon receives respectively from the Sun and the Earth can be made using the simplified version of Kepler's third law, which in case of circular orbits is reduced to the equality between gravitational attraction and the centrifugal inertial force. This formula enables us to compare the two forces without knowing the values of masses and the gravitational constant, having at hand only the distances and periods. For circular orbits of the Earth around the Sun and of the Moon around the Earth we can write, respectively, for the two accelerations,

$$
a_S = \frac{GM_S}{D_E^2} = \Omega_E^2 D_E = \frac{4\pi^2}{T_E^2} D_E,
$$
$$
a_E = \frac{GM_E}{D_M^2} = \Omega_M^2 D_M = \frac{4\pi^2}{T_M^2} D_M
$$

$$(3.5)$$

where $D_E = 1.49 \cdot 10^8$ km is the radius of the Earth's orbit, D_M is the radius of the Moon's orbit, M_S is the solar mass, M_E the mass of the Earth, G the gravitational constant, and Ω_E and Ω_M the angular velocities of the Earth and the Moon, T_E and T_M the respective sidereal periods, with $T_E = 1\,\text{year} = 365.24\,\text{days}$, $T_M = 1$ sidereal month $= 27.35\,\text{days}$. Taking the ratio a_S/a_E between gravitational pulls acting on the Moon from Sun and Earth respectively, we get

$$
\frac{a_S}{a_E} = \frac{D_E}{D_M} \left(\frac{27.35}{365.24} \right)^2 = 2.18,
$$

$$(3.6)$$

confirming the preponderance of solar attraction over Earth's gravitational influence on the Moon.

3.2　How the Moon moves

The Moon is the principal object in the sky at night, and its motion has been observed and analyzed since Antiquity. The best and most reliable observers were the Babylonians, whose scribes meticulously noted its motions during many centuries. The Moon is large enough to enable quite precise observations and measurements without telescope. Ancient astronomers have left many observations that are valid until today, although the precision of modern measuring devices is incomparably higher.

The Moon is also the only astronomical object in our solar system whose role as Earth's satellite did not change in spite of the consecutive worldview modifications: whether it was the geocentric Ptolemaic system, or a heliocentric one, proposed for the first time by Aristarchus of Samos and revived by Copernicus, or a mixed one proposed later by Tycho Brahe — in all these models the Moon was orbiting the Earth, first with all other celestial objects, or only with chosen planets and the Sun — by now it is the only natural celestial object whose orbit has the Earth as center. Another unchanged observation is that our Moon is the closest of all celestial bodies. This was understood mostly due to the existence of solar eclipses and *occultations*, when Moon's disc passes in front of a planet.

In Figure 3.4 the essential elements of the Moon's orbit are displayed. During one revolution, the Moon comes very close to the Sun (then it is said to be in *conjunction*, and cannot be seen — this is called the new Moon) or opposite to the Sun (in *opposition*), and is perceived as full Moon. In both cases the Moon and Sun are said to

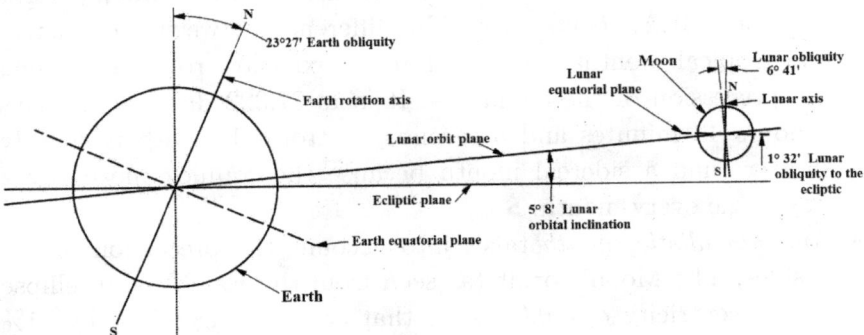

Fig. 3.4 Earth and Moon: Their orbits and rotation axes. Both planets' sizes are to scale, and all the angles are exact, but distance between Earth and Moon is NOT to scale.

be in *syzygy*, when the Moon, Sun and Earth are aligned. Conjunction is equivalent to the Moon's *elongation* 0°, while the opposition corresponds to the elongation of 180°. When the Moon is in its first or third quarter, i.e. when exactly half of its disc is visible, its is said to be in *quadrature*, which corresponds to elongations of 90° and 270° respectively.

The particularities of the lunar orbit combined with the Earth's own axial rotation, precession and annual revolution around the Sun result in a complicated periodicity pattern of the Moon's motion in the sky. Let us recall the definitions of various lunar periods, each of which has its specific astronomical role.

- A *sidereal month* marks the time between two identical positions Moon takes with respect to the fixed stars. It lasts 27.322 days, or 27 days 7 hours 43 minutes and 12 seconds.
- A *synodic month* is the time between two *lunations*, e.g. between the consecutive new moons. It lasts 29.531 days, or 29 days 12 hours 44 minutes and 3 seconds. It is longer than the sideral month because during one sidereal month Earth has moved on its orbit by about 27 degrees, and it takes a couple of days more for Moon to catch up with respect to the Earth-Sun direction. It is useful to know that 13.37 sidereal months are roughly equal to 12.37 synodic months.
- A *tropical month* is the time span between Moon's passage through vernal equinox, i.e. between two moments when Moon's right ascension (R.A.) equals 0 h. The difference between the sidereal and tropical months is due to Earth's axis slow rotation causing the precession of the equinoxes. It lasts 27.322 days, or 27 days 7 hours 43 minutes and 5 seconds. A tropical month is slightly shorter than a sidereal month because the equinox moves by a tiny angle every month.
- An *anomalistic month* takes into account the precession of the apsides. The Moon's orbit (as seen from the Earth) is an ellipse with eccentricity $e = 0.0549$, so that at the perigee it is by 11% closer to us that at the apogee. The semi-major axis of the elliptic orbit is called the *line of apsides*. The line of apsides slowly

rotates. The precession of the apsides has the period of 3233 days, or 8.85 years, in the same direction as the orbital motion of the Moon. Therefore, it takes slightly more than one sidereal month to come to the same apsis (perigee or apogee) after one revolution. This is why the anomalistic month lasts 27.554551 days, or 27 days 13 hours 18 minutes and 33 seconds.

- *A draconic month.* Under the influence of the Sun, due to the inclination by 5.145° of lunar orbit with respect to the ecliptic, the plane of lunar orbit slowly rotates. As a consequence, the line of nodes, which is the intersection of lunar orbital plane with the ecliptic, completes a full turn of 360° in about 6798 days (18.6 years). As a consequence, the period between two consecutive passages of the Moon through the same node lasts 27.212220 days, or 27 days 5 hours 5 minutes and 36 seconds, which defines the duration of a draconic month.

3.3 Why do the eclipses happen?

Solar and lunar eclipses fascinated people of all civilizations since the most ancient times. Our ancestors were mostly impressed by the *full solar eclipses*, when for a few minutes the solar disc disappears completely and a short "night" takes place, with dark sky and stars becoming visible at noon. The surprise is so great that even animals often show symptoms of panic and bewilderment. In various mythologies solar eclipses were explained by the malicious action of a celestial dragon swallowing the Sun, and shortly after spitting it our because it was too hot.

The most spectacular recent full solar eclipse visible in Europe and Middle East occurred in 1999. Its principal stages are exposed in Figure 3.6.

Our Moon's orbit is an ellipse whose eccentricity is not very high, its exact value is $e = 0.0549$, the *perigee* distance equals 362 600 km and the *apogee* distance is 405 400 km. The difference is of almost 12%, which is due to the fact that the eccentricity counts twice: the distance between the focus of the ellipse and the geometrical center of the ellipse is given by ea, where a is the major semi-axis; therefore,

Fig. 3.5   The solar and the lunar eclipse.

Fig. 3.6   Total solar eclipse visible in Europe in 1999.

this distance is added or substracted from the major semi-axis value when the Moon is in the apogee or in the perigee, respectively. This distance difference is perceptible, as shown in Figure 3.7.

In fact, the observation of regular variations in the size of the full Moon led to the conclusion that its orbit must not be circular. In the Ptolemeian system the eccentricity of lunar orbit was taken into account by adding a specific *epicycle*, a small circle along which the Moon turns while the center of the epicycle turns around the Earth.

If the Moon's orbit were coplanar with Earth's orbit, in other words, if it were contained in the *ecliptic plane*, both solar and lunar eclipses would occur regularly every synodic month: every new Moon would result in the obstruction of solar radiation by our satellite, and full Moon would inevitably pass through Earth's shadow. But the

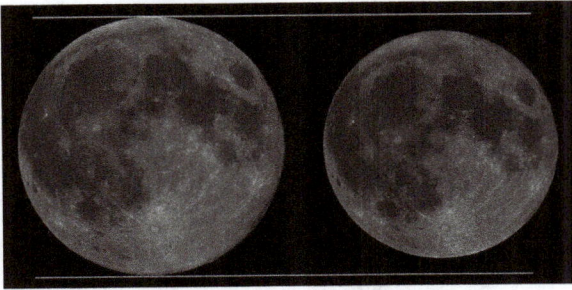

Fig. 3.7 The full Moon in perigee (left) and in apogee (right).

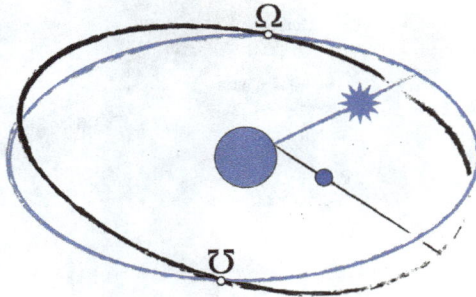

Fig. 3.8 The ascending and descending nodes of lunar orbit.

actual Moon's orbit is contained in the plane which is inclined with respect to the ecliptic, the angle between the two planes being equal to 5°15′ on the average (it varies slowly between 4°58′ and 5°18′). Both solar and lunar angular sizes are close to 32′, roughly half a degree, and can easily miss each other even when they pass through the same celestial longitude. The eclipses can occur only when both Sun and Moon cross one of the *nodes*, which are the points where the lunar orbit cuts the ecliptic plane, as shown in Figure 3.8. Nevertheless, lunar eclipses occur more frequently, because the angular size of Earth's shadow in the vicinity of lunar orbit, i.e. at the average distance of 384 499 km.

Full solar eclipses can happen only when the new moon occurs when our satellite crosses one of the nodes, because only then both Sun and Moon are aligned as seen from the Earth. Again, if the nodes

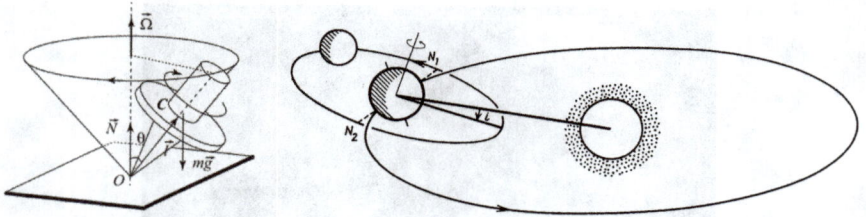

Fig. 3.9   Precession of a rapidly spinning top (left) and precession of Moon's orbit (right). The angle i is exaggerated to make the image more perceptible.

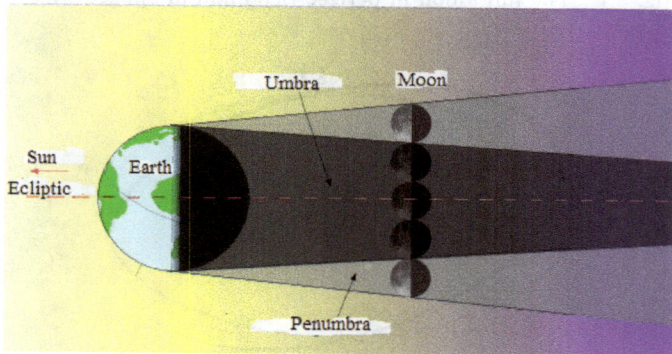

Fig. 3.10   The lunar eclipse scheme.

were always found in the same direction with respect to the distant stars, the calculus of the periodicity of solar and lunar eclipses would be quite simple. But the nodes themselves are moving slowly, due to the *precession* of the Moon's orbit. This phenomenon is caused by the gravitational pull of the Sun, whose direction is not coplanar with the orbit. This means that there is an extra component, perpendicular to the orbital plane, which acts in the same way as the terrestrial gravity acts on a rapidly spinning top whose rotation axis is not exactly vertical. This extra component provokes the *precession* of the top's proper rotation axis around the vertical axis. The angular velocity of the precession is much lower than spinning top's proper angular velocity; this is also the case with Moon's orbit: its precession frequency is equal to one full cycle in more than eighteen years (more precisely, in 18.6 years, or $19°21'$ per year).

3.4 How often do eclipses occur?

Let us first evaluate how often solar eclipses do occur. They are obviously less frequent than lunar ones, because the shadow Earth casts on the Moon is about 2.6 larger than the Moon itself, so the probability that our satellite will pass through Earth's shadow is much higher than the probablility that the Moon will cross the node at exactly the same time when the Sun appears in the same direction.

The *draconic month* is defined as the period of time between two consecutive passages of the Moon through the same orbital node (ascending or descending one). The draconic month lasts $T_{drac} = 27$ days 5 hours 3 minutes 35.8 seconds, or 27.2123 days transcribed in decimals, The *synodic month*, which is the period between two identical phases, lasts longer: $T_{syn} = 29$ days 12 hours 18 minutes and 36 seconds, or 29.531 days if expressed in decimals.

Suppose that having observed an eclipse we want to know how soon the next one will take place. This means that we want to compute the natural numbers of synodic and draconic months, denoted respectively by S and D, satisfying the following simple equality:

$$29.5306 \times S = 27.2123 \times D \qquad (3.7)$$

The simplest and the most obvious solution would be to choose the following integers:

$$S = 272\,123, \quad D = 295\,306, \qquad (3.8)$$

where S and D are the numbers of respective months. But this solution corresponds to about 809 years, and a forecast of next solar eclipse occurring after more than eight centuries does not seem to have any practical value — human beings do not live that long! On the other side, do we really need to know the exact timing of the next full solar eclipse up to a second? This was the precision used while establishing equation (3.7). An average full solar eclipse lasts at least two hours, so the prediction up to half an hour seems acceptable. Therefore, there is no harm if one uses approximations. In ancient times people did not have the decimal system with as many digits after the comma as wanted; they knew only fractions. But a clever use of *infinite fractions* was introduced by the Greek mathematician

Diophantus; this system of notation is also known under the name of *Diophantine fractions*. Here is how this system works in the present case.

First of all, we form the "improper" fraction, which is bigger than 1:

$$\frac{295306}{272123} = 1\frac{23183}{272123}.$$ (3.9)

Then we re-write the fractional part that appears after the integer 1 by inversing it, so as to get another improper fraction in the denominator[1]:

$$1 + \frac{23183}{272123} = 1 + \frac{1}{\frac{272123}{23183}} = 1 + \frac{1}{11 + \frac{17110}{23123}} = 1 + \frac{1}{11 + \frac{1}{\frac{23183}{17110}}}.$$ (3.10)

Continuing in the same manner, we arrive at the fifth step at the following continuous fraction:

$$\frac{295306}{272123} = 1 + \frac{1}{11 + \frac{1}{1 + \frac{1}{2 + \frac{1}{1 + \frac{1}{4}}}}}.$$ (3.11)

Now we can stop the development in consecutive fractions at any stage of approximation. For example, the rudest approximation would consist in neglecting anything but the integer part of the fraction (3.9), i.e. just 1. This would mean that we neglect the difference between the draconic and the synodic months, which would in turn lead to the conclusion that the next full solar eclipse should occur one lunar month after the first one. As we know, this is not what is observed — the "zeroth approximation" does not work. The next approximation consists in keeping the fraction $\frac{1}{11}$ that appears at the first step of the development. This would give us $1 + \frac{1}{11} = \frac{12}{11}$. Such an approximation compares twelve draconic months with eleven synodic months, treating them as (roughly) equal. The error committed

---

[1]See the discussion of Mars in Chapter 2.

is still too great, and in fact, we do not observe another full solar eclipse occurring less than a year (eleven synodic months) after the first one.

Let us evaluate the margin of error resulting from this approximation. Twelve draconic months last $27.2123 \times 12 = 326.5426$ days, while eleven synodic months last $29.5306 \times 11 = 324.9246$ days. The difference is more than one day and a half, more exactly 38 hours and 54 minutes. During one day (24 hours) the Moon moves westwards by $13°10'$ with respect to the stars, and by $12°11'$ with respect to the Sun. Both Sun's and Moon's angular diameters are roughly equal to *half a degree*, i.e. about $32'$ on the average, so they can be found anywhere in such a vast strip. This means that we should sharpen our estimate so as to reduce the width of the accessible strip to the width comparable with Sun's and Moon's diameter, and by the same token, comparable with the time span of an average solar eclipse.

The full set of consecutive approximations is given by the following series of fractions:

$$1; \ \frac{12}{11}; \ \frac{13}{12}; \ \frac{38}{35}; \ \frac{51}{47}; \ \frac{242}{223}; \ \frac{1019}{939}; \ \text{etc.} \qquad (3.12)$$

The first four approximations happen to be still too rough, but the fifth one, $242/223$, offers a compromise: the difference is less than eight hours, while the average full eclipse lasts more than two hours, so the probability that at least a partial solar eclipse will occur becomes quite high. Indeed, this is the case. The periods corresponding to 242 draconic months or 223 synodic months are close enough, both equal to 18.6 years. This period, known already to the Babylonians and Egyptians, is called *the Saros*.

3.5 How long does a solar eclipse last?

Let us evaluate as precisely as possible the time duration of a full solar eclipse, including the total time during which the Moon is obstructing at least some part of the solar disc, and minimal and maximal durations of total eclipse, when no solar light can reach the observer, creating a momentaneous nightfall. On the Northern Hemisphere the orbital motion of the Moon is from West towards East, the Moon turning around the Earth in the same direction as the Earth itself turns around the Sun. The diurnal motion of our planet is more rapid, and concerns both Sun and Moon in the same manner — they rise in the East and set down in the West. But the Moon moves eastwards by about 12° every day, until completing the full cycle (lunation); the 360° of the full circle should be divided by the number of days in a synodal month:

$$\Delta\phi = \frac{360}{29.53} = 12°11' \text{ per day.}$$

The figure below represents two situations we are interested in. On the right, the total time, from the first contact of the Moon with solar disc until the last contact, when the Moon ceases to obstruct the slightest part of the Sun. On the left, the time of total solar eclipse. It is clear that if the two luminaries had exactly the same angular size, the duration of a total solar eclipse would be infinitesimally short — as soon as the Moon disc covers the totality of Sun, it also leaves it, letting some rays appear again. But neither the Earth's orbit around the Sun, nor the Moon's orbit around the Earth are circular. Both are elliptic; the Earth comes closer to the Sun at perihelion and farther in aphelion, and the Moon is closest to the Earth at perigee, and farthest at the apogee. The differences of distance lead to differences in apparent size of the discs:

The Sun: from 31'30" in aphelion to 32'30" in perihelion;

The Moon: from 29'24" in apogee to 33'30" in perigee.

Among the continuum of various combinations, the extremal cases are either when Earth is at perihelion and Moon is in apogee, or contrarily, when Earth is at aphelion and Moon at the perigee. In the first case the solar disc is bigger than the lunar one, and the Moon cannot cover the Sun totally, which results in an *annular eclipse*. In the second case the Moon's apparent angular diameter is two arcminutes larger than the Sun, so that total eclipse may last a few minutes. Figure 3.11 shows the first and last contacts, as well as the stage of total obstruction of solar glare.

Fig. 3.11 The solar eclipse duration: Left: from first to last contact, Right: the total eclipse phase.

The Moon's synodic angular velocity is

$$\omega_M = \frac{360°}{29.53} = 12°11' \text{ per day, or } \quad \frac{731'}{24} \simeq 30'28'' \text{ per hour.} \tag{3.13}$$

With this speed the disc of the Moon has to cover the distance of $2 \times 32' = 64' = 1°4'$, i.e. twice its own diameter, between the first and last contacts with solar disc. Dividing distance by speed we get the time it would take:

$$\frac{64'}{30'28''/\text{hour}} = \frac{3840''}{1828''/\text{hour}} = 2.1\,\text{hours} = 2\,\text{hours}\,6\,\text{minutes.} \tag{3.14}$$

To establish the short duration of the total eclipse phase we should perform the same calculation for just the tiny difference between slightly bigger angular size of the Moon and the angular size of the Sun, which at best can attain $2'$. A calculation similar to (3.14) with $2'$ replacing $30'28''$ yields

$$\frac{2'}{30'28''/\text{hour}} = \frac{120''}{1828''/\text{hour}} = 0.0328\,\text{hours} = 3.94\,\text{min,} \tag{3.15}$$

i.e. close to four minutes. This estimate is made with the tacit hypothesis that the eclipse occurs close to the noon time, when both Sun and Moon are high in the sky, and the cone of Moon's shadow hits Earth's surface at an angle not too far from 90°.

But this is not the end of the story. The real duration is always longer due to Earth's rotation. The Moon's umbra, whose diameter on Earth's surface is maximally 260 km is running very quickly, its average speed reproducing the Moon's own orbital speed (the synodic one, because it is taken with respect to coordinate system centered on Sun), which is (using the average distance Earth-Moon, in the approximation of quasi-circular orbit)

$$v_{syn} = \frac{2\pi \cdot 385000\,\text{km}}{T_{syn}} = \frac{2412749\,\text{km}}{29.53\,\text{days}} \simeq 0.948\,\text{km/sec}. \qquad (3.16)$$

Dividing the diameter of the lunar umbra by its speed on Earth's surface, we get

$$\frac{260\,\text{km}}{0.948\,\text{km/sec}} = 275\,\text{sec} \simeq 4\,\text{min}\,34\,\text{sec}, \qquad (3.17)$$

i.e. four minutes and a half, close to previous estimate (3.15). Usually the umbra is smaller (ca. 160 to 200 km), making the total duration of the full eclipse shorter, usually between 2 and 3 minutes.

However, these calculations were made without taking into account the diurnal rotation of Earth around its axis. In fact, the Moon's shadow follows the path parallel to its orbit's plane, close (within 5.5° obliquity) to the ecliptic, forming an angle 23°26′ with the equator or any local parallel. The Earth rotates with an angular velocity of 360° per 24 hours, from West towards the East; as a result, a terrestrial observer is chasing the *umbra* with speed which depends on his local latitude. The linear speed on the equator is about 463 km/sec; at the latitude 50° it is still quite high, $463 \times \cos 50° = 298\,\text{m/sec}$. This speed has to be substracted from the speed of lunar shadow advancing on the surface of Earth, but not entirely, because they do not follow the same direction: the rotational speed is directed along the local parallel, whereas the Moon's shadow follows a path roughly parallel to the ecliptic. The angle between these two directions is roughly equal to the obliquity of Earth's axis, which is 23°24′. The cosine of this angle is $\cos 23.4° = 0.91$, so that the projection of the linear speed due to Earth's rotation on the speed of Moon's *umbra* is 422 m/sec on the equator, and 271 m/sec at the latitude 50°.

Let us evaluate the maximal possible duration of a full eclipse stage, which can be observed with all optimal conditions combined: the Moon being at perigee and Earth in aphelium to make the difference of angular dimensions as great as possible in favor of the

Moon by $2'$; moreover, the full eclipse occurring near the equator, and close to noon when the shadow is cast almost vertically. Now, with Moon being in the perigee, we should make a correction to the value of its velocity, and consequently, the speed of its shadow on Earth surface, due to Kepler's second law. The speed of the Moon's shadow on the surface of Earth is the same as its orbital speed, which is at perigee is at its maximum. The orbit of our satellite is almost circular, but its eccentricity $e_M = 0.055$ is significant enough to be taken into account.

For an almost circular orbit one can suppose that the velocity is everywhere orthogonal to the radius vector, so that one can interpret the areal velocity as a simple product of radius vector and orbital velocity $r^2(d\varphi/dt) = r \cdot v$. Then Kepler's second law will reduce to

$$r_1 v_1 = r_2 v_2 = \text{Constant, so that} \quad v_2 = v_1 \frac{r_1}{r_2}.$$

Setting the average Earth-Moon distance as $r_1 = 385\,000$ km, the perigee at $r_2 = 363\,000$ km and $v_1 = 948$ m/sec, we obtain $v_2 = 1004$ m/sec.

The maximal relative speed of the umbra is equal now to $1004 - 422 = 582$ m/sec., and this value should be used for the evaluation of maximal time duration of total eclipse phase. Dividing the diameter of the Moon's umbra by its speed as seen by terrestrial observer, we get the final result:

$$\frac{260\,\text{km}}{582\,\text{m/sec}} = \frac{260\,000}{582} = 447\,\text{seconds} = 7\,\text{min}\ 27\,\text{sec}, \qquad (3.18)$$

i.e. 7.5 minutes, in agreement with observational data gathered over many centuries — the longest full eclipses ever observed lasted indeed slightly longer than seven minutes.

The duration of the total eclipse phase becomes shorter when it occurs at higher latitudes. For example, if a total solar eclipse with the same optimal conditions (Moon in perigee, Earth in aphelium) is observed somewhere at latitude $50°$, then the projection on the ecliptic direction of linear velocity due to Earth's spin is only 271 m/sec., and the relative speed of umbra is $1004 - 271 = 733$ m/sec. Inserting this value instead of 582 m/sec in formula (3.18) we get

$$\frac{260\,\text{km}}{733\,\text{m/sec}} = \frac{260000}{733} = 355\,\text{seconds} = 5\,\text{min}\,55\,\text{sec},$$

Such optimal coincidences are very rare; usually total eclipse phase lasts from 2 to 4 minutes.

3.6 Measuring the distance to the Moon by parallax

Solar eclipses provide us also with a natural astronomical tool for measuring the actual distance to the Moon if we can compare observations performed at the same time from two sufficiently distant sites. The best result can be obtained when the straight line joining two localities forms a right angle with the path of the Moon's shadow. The last spectacular solar eclipse observable practically everywhere in Europe happened on August 11, 1999. The path of the Moon's umbra is shown in Figure 3.12. Total occlusion could be observed first in Cornwall, then in Normandy, North of Paris. In London and Paris only a partial eclipse was visible, although the shining portion of solar disc was really tiny, as shown on the right in Figure 3.12.

The path of total shadow went south of London, this is why the lunar disc was seen a bit lower than the solar one, leaving a narrow crescent on the top. Observers looking at the same transit of the

Fig. 3.12   The path of the Moon's umbra on August 11, 1999, and the solar eclipse as seen from London, England (left) and from Paris, France (right).

lunar disc in front of the Sun saw a tiny crescent below. The total angular width of both crescents covered about 3′, i.e. about one tenth of solar diameter.

On the map (Figure 3.12) we drew a triangle with three sides: Paris, London and third summit C, near to the city of Caen. The distance from London to Paris is known to be 345 km in straight line. With some elementary trigonometry we can estimate the length of the side London-Caen, perpendicular to the path of the shadow, as being equal to 320 km. This is the basis of the triangle whose two other VERY long sides meet at the Moon's center M (not visible on the map); the angle between these sides pointing at the Moon is the parallax angle 3′. Denoting the distance London-Caen by d_{LC} we can easily find its length in kilometres:

$$d_{LM} = \frac{320\,\text{km}}{3'/1\,\text{radian}} = \frac{320\,\text{km} \cdot 57.29°}{3'}$$

$$= 320 \cdot 1146 = 366\,720\,\text{km}, \tag{3.19}$$

which gives a fair estimate of the distance Earth-Moon at perigee.

The method of parallax can be used to find the distance from Earth to the Sun observing a very rare astronomical event called the transit of Venus, which is similar to solar eclipse, the planet Venus playing the role of obstructing disc — which looks like a fly crossing a plate, so small its diameter is as seen from Earth — less than 1 arcminute compared with 32′ of solar disc. The analysis of Venus' transit is the subject of Chapter 15.

3.7 Lunar eclipses

Due to the greater angular size of Earth's shadow, lunar eclipses not only occur more frequently than the solar ones, but their duration is much longer. As we know, the shadow Earth casts at the distance of the lunar orbit covers a disc whose radius is about 2.65 times the apparent Moon's radius. This corresponds to the angular size of 1°25′. This is the Earth's *umbra*, the region in which the Moon does not get any light, because there the Sun is totally hidden by the Earth.

As it was conjectured by Aristarchus, the *penumbra* region surrounds the umbra disc, and its width is exactly equal to one lunar diameter. The total angular size of penumbra and umbra inside corresponds to 4.65 lunar diameters, which is equivalent to 2°24 as seen from Earth. As a result, due to the obliquity 5°30′ of the lunar orbit, the Moon can pass over or below Earth's shadow, or enter the penumbra partly, which would be almost imperceptible by the naked eye, or enter the penumbra and only partly the umbra. Finally, it can pass through the penumbra and then cross the umbra, entering the dark zone entirely.

The corresponding types of lunar eclipses are called *Penumbral, Partial* and *Total.* A total eclipse of the Moon is the most spectacular, because when the Moon enters completely in the shadow, it does not disappear, but continues to be seen as a faint, intensely red disc. This phenomenon is due to solar rays refracted by the Earth's atmosphere. As we have learned in Chapter 1, our atmosphere disperses mostly the blue part of the spectrum, while the red part passes more easily — this makes our sky blue and the Sun red during the sunset.

Figure 3.13 shows all types of paths of the Moon through the Earth's shadow, i.e. all types of lunar eclipses. As can be clearly seen from the figure, one can distinguish between seven stages of lunar eclipse. Here is the scenario of full eclipse of the Moon; for partial eclipses several stages will be missing, e.g. sometimes the Moon's disc enters only the penumbral region, and even that only partly; such "truncated eclipses" are hardly even noticed. But when the Moon passes entirely inside the umbra, all seven stages can be observed. They are as follows:

1. Penumbral eclipse begins when the penumbral part of Earth's shadow starts covering the Moon. This phase is not easily seen by the naked eye.
2. Partial eclipse begins: Earth's umbra starts covering the Moon, making the eclipse more visible.
3. Total eclipse begins: The Earth's umbra completely covers the Moon and the Moon is red, brown, or yellow in color, depending on atmospheric conditions on Earth.

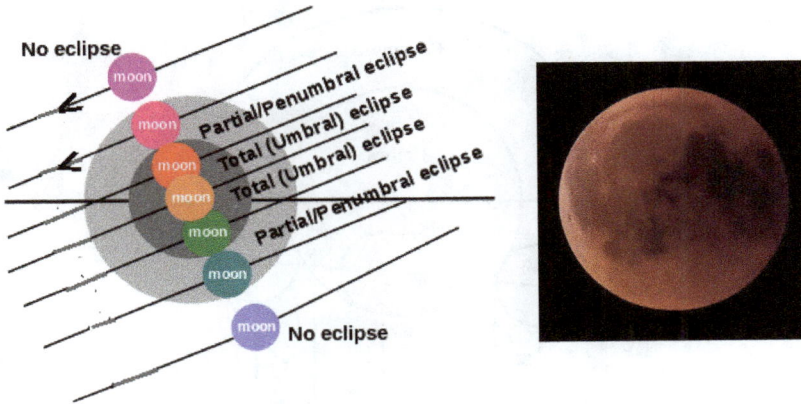

Fig. 3.13   Left: Possible types of lunar eclipses. Right: Red Moon color during total eclipse.

4. Maximum eclipse: This is the middle of the total eclipse.
5. Total eclipse ends: At this stage, the Earth's umbra starts moving away from the Moon's surface.
6. Partial eclipse ends: the Earth's umbra completely leaves the Moon's surface.
7. Penumbral eclipse ends: At this point, the eclipse ends and the Earth's shadow completely moves away from the Moon.

All stages of the full lunar eclipse that occurred on September 6, 1979 can be identified in Figure 3.14. Aristarchus used lunar eclipse to evaluate the real diameter of the Moon comparing it with the size of Earth's shadow. His estimate of the size of this shadow to be twice as big as the apparent angular size of the Moon, concluding that Moon's diameter must be about three times smaller than Earth's diameter; in fact his estimate was contained between 0.32 and 0.40 Earth's diameters. The error was big, but the order of magnitude was right.

In Aristarchus' times there were no clocks able to mark minutes or seconds, so the duration of full lunar eclipse could be measured only up to a quarter of an hour or so. But nowadays we can use time measurements to determine Moon's dimensions with greater

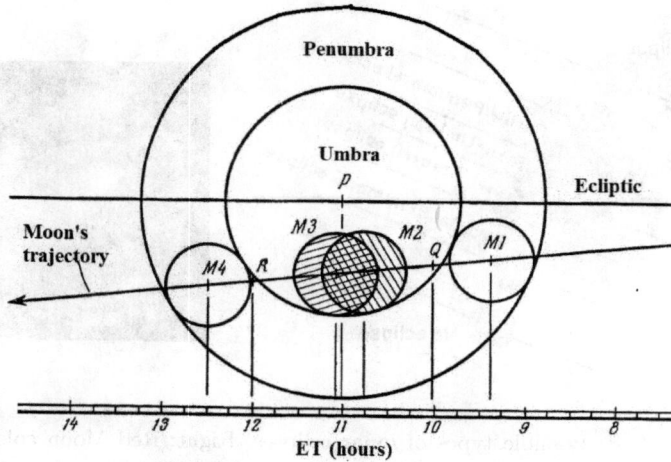

Fig. 3.14   The lunar eclipse of September 6, 1979, as seen in Eastern Europe.

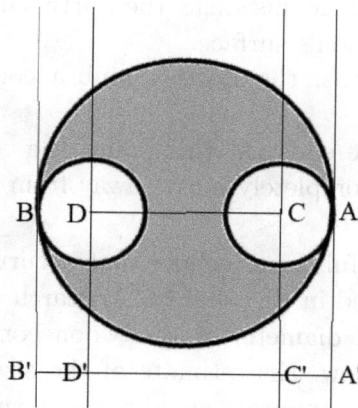

Fig. 3.15   The exact timing of full stage of lunar eclipse gives clue to Moon's real dimension as compared to Earth's radius.

precision. The idea is to fix the exact time interval of total *central* lunar eclipse, i.e. stages $3, 4$ and 5, as shown in Figure 3.15.

From the beginning till the end of total eclipse, when the Moon is completely inside the umbra, its center has to cover the segment CD of unknown angular size x. Let us denote the apparent angular diameter of the Moon by d. Its average value is 31 arcminutes, so that according to the figure, the angular diameter of Earth's shadow $AB = y$ is $y = x + d = x + 31'$. The time duration of a total lunar eclipse depends on whether Moon is at the apogee or at the perigee, or in between, because accoding to Kepler's second law, its pace is more or less rapid. The maximal time will be when the Moon is at the apogee. The current value for the longest duration of a total lunar eclipse is about 100 minutes. We can find the unknown value y comparing the time the Moon needs to pass from C to D with the angular distance Moon covers during an hour, i.e. in 60 minutes. To do this properly, we have to use the Moon's synodic period, because what we observe is the position of the Moon with respect to the Sun, and not with respect to the stars. Also, contrary to the case of solar eclipses, the Earth's axial rotation does not influence the observed time of lunar eclipse.

The synodic month lasts 29.53 days which is 708.72 hours. Dividing $360°$ by the number of hours we get the result: on the average, during one hour Moon travels $0.508° = 30.48'$. We have to solve the following simple equation:

$$\frac{x}{100} = \frac{y - 31'}{100} = \frac{30.48'}{60}, \quad 6(y - 31') = 10 \cdot 30.48',$$

$$y = 81.8' \simeq 1°22'.$$

(3.20)

This is the angular diameter of total shadow (umbra) cone at distance at which the Moon crosses it during the total eclipse stage. As compared with the Moon's angular diameter, it is $82'/31' = 2.65$ times greater. As Aristarchus rightly conjectured, we have to add an extra lunar diameter to get the ratio between the Earth's and Moon's diameters, which leads to the final estimate 3.65.

Now it is easy to find the Moon's diameter in kilometers and the distance separatiing us from our satellite. With actual data the mean Earth's diameter is 12 742 km; dividing this by 3.65 we obtain for the Moon's diameter 3490 km. (This is slightly above the official value 3474 km found in modern atsronomical textbooks, but the error is less than half of percent, and we cannot expect better precision using only average values for the Moon's angular orbital velocity and its angular diameter.)

The last equation serves to find the distance D at which the 3490 km large Moon's diameter is seen with an angular size of $31' = 0.52°$. We have:

$$\frac{d}{D} = \frac{3490 \text{ km}}{D \text{ km}} = \frac{31'}{\text{radian}} = \frac{0.52°}{57.3°}, \qquad (3.21)$$

from which we get

$$D = \frac{57.3}{0.52} d = 110.2 \cdot 3490 \text{ km} = 384\,571 \text{ km}. \qquad (3.22)$$

This is the distance to the Moon from the observer who is staying on the Earth's surface. To get the average radius of the Moon's orbit one should add the Earth's radius, i.e. 6378 km, which gives for the distance between the centers of the Earth and Moon $384\,571 + 6378 = 390\,949$ km, which is a bit more than the average. This is probably due to the fact that the maximal full eclipse duration of 100 minutes was taken when the eclipse occurred while the Moon was close to the apogee, when its velocity is minimal. Due to the ellipticity of its orbit, the Moon can come as close as 363 229 km at the perigee and as far as 405 400 km at the apogee.

3.8 Occultations

The term *occultation* designates an event that occurs when one astronomical object is hidden by another one that passes between it and the observer on Earth. Solar eclipses are occultations of the Sun by the Moon, while lunar eclipses can be defined as occultations of the Sun by the Earth, as seen from the Moon. The Moon often blocks, or occults, stars from our view. And human beings have been watching such occultations at least since Aristotle recorded the Moon covering Mars on April 4, 357 B.C.

Occultations of planets by the Moon are quite rare, while the occultations of stars are quite frequently observed. For example, the star Aldebaran disappeared behind the lunar disc on January 19, 2015. The *International Occultation Timing Association* keeps a detailed survey of past and future star occultations by the Moon.

The most interesting cases are the so-called *grazing occultations* when the Moon's visible edge — sometimes just a high mountain — passes in front of a planet or a star. A star can wink in and out of sight within a few seconds, so small is the part of the Moon hiding the star from our sight. The timing data are so precise that they can substantially improve our knowledge of positions and motions of stars, or reveal the double stars so close to each other that they are below the resolution of telescopes: this happens when a double wink is observed.

There is one important shortcoming of grazing occultations: they can be observed exclusively from a very narrow stripe of our globe, sometimes only a few kilometers wide. Depending on the case, the edge can mean just $d = 100\,\text{km}$, which correponds to the parallax of about 53 arcseconds, slightly less than one arcminute, when we divide d by the distance to the Moon, $384\,000\,\text{km}$. Therefore the width of the strip on Earth from which a grazing occultation can be observed is also no more than 100 kilometers.

Occultations of a planet or star by another planet are extremely rare due to planets' tiny angular sizes.

Chapter 4

Ancient astronomers and mathematicians

On ancient world views - Astronomy and astrology - Egyptian science - Babylonian advancements - Early Greek science - Eudoxus' first model of Solar System - Aristotle's cosmography - Aristarchus' first heliocentric system - Hipparchus' stellar atlas - Ptolemy's geocentic system - Science in Medieval Islam - Great Muslim astronomers

4.1 Preamble

Exploring the history of astronomical observations and discoveries is like traveling backwards in time. As seen from modern perspective, the views of ancient people on Earth and Sky, on stars and heavenly bodies, on the origin of the world and living creatures may seem sometimes primitive or childish; but we have to keep in mind how limited their observation possibilities were, relying only on the naked eye, without any magnifying or amplifying device. The surrounding world seemed full of unknown parts and unexpected dangers.

One striking feature common to many civilizations, no matter how distant from us in time and space, is that science and religion were intimately intertwined, as a matter of fact, the learned people were charged with determining the time of religious services and agricultural events alike, and predicting the future from omens in the sky. For a long time astronomy and astrology were one for ancient Egyptians, Babylonians and Phoenicians. The world they lived in was filled with gods and spirits; not only humans and animals had souls, but everything was animated and alive, the Sun and the Moon, the stars, the sea, the winds, great rivers that provided water for agriculture — everything was identified with divinities. Of which some were benevolent, others obnoxious, depending on their whims, so that humans had to observe them constantly, to foresee their mood, to influence it by prayers, and to behave accordingly.

The views of ancient peoples were anthropomorphic: not only the faith that the world was created at some point in the past was very common, but it was often believed that it owed its birth to divine parents, like a human being. Marriages between gods and goddesses gave birth to other divine creatures. According to Egyptian mythology, the ancestor of all gods was a self-created god named *Atum*, meaning "perfection". He emerged from primordial waters and became the father of two gods, *Shu*, the god of air and vital breath, and *Tefnut*, the goddess of heat. Together they begot *Geb* (or *Seb*), the god of Earth, and *Nut*, the goddess of Sky, bearing stars on her body. Figure 4.1 shows ancient Egyptian representation of Earth (Geb) and Sky (Nut) supported by Air (Shu). Similar myths about world's origin were common in ancient Akkad and Babylon kingdoms; they were exposed in a Babylonian creation epos named *"Enuma Elish"* (from its first words "When on High"), written on seven clay tablets

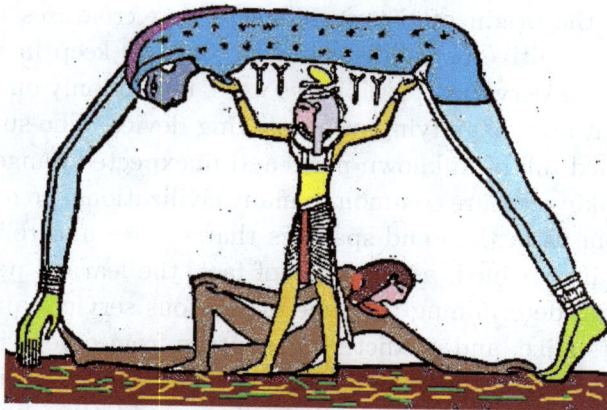

Fig. 4.1   One of the countless copies of ancient Egyptian creation myth.

found in the ancient king Assurbanipal's library in Niniveh (in contemporary Iraq). According to this story, the world was created by impregnation of the goddess of salt sea water and chaos, *Tiamat*, who mated with the god of fresh water *Apsu*. Human beings were created by the young god named *Marduk*, who won the fight against *Tiamat*.

Echoes of these most ancient documented beliefs can be found in *"Genesis"*, the first book of Hebrew bible, as well as in ancient Greek myths describing the origin of the world.

Great caution should be taken while speaking of ancient and very ancient scientific views and achievements. This is because we very often extrapolate our present possibilities of inquiery and understanding the nature towards the remote past. To take the simplest and most common example of misinterpretation of certain historically confirmed data, let us consider the claim that "Ancient Egyptians were able to evaluate the number π, which determines the ratio between the half perimeter of a circle to its radius, up to the fourth decimal".

In reality, ancient Egyptians did not use decimal fractions; they used just the fraction $\frac{22}{7}$ which provides an astonishingly good approximation of what we denote now by "π", an irrational number whose representation in the decimal system amounts to an infinite series of digits following the point separating the integer part (in this particular case the digit "3") from its non-integer part. Here are the first ten digits of this infinite series of numbers: $\pi = 3.1415926578....$

It turns out that the simple fraction used by the Egyptians, once expressed in the decimal system, gives a surprisingly good approximation:

$$\frac{22}{7} = 3.1428571428... \tag{4.1}$$

(Note the group of numbers 142857 that will repeat itself *ad infinitum*, like any rational number (fraction) encoded in the decimal system.)

To evaluate the precision with which the fraction 22/7 approaches the real number π, it is enough to write down the first four digits of both approximations, up to the place where they start to diverge:

$$\pi \simeq 3.142 \quad \text{and} \quad \frac{22}{7} \simeq 3.143 \text{ respectively.} \qquad (4.2)$$

The difference is only 0.001, which corresponds to a relative error of $\frac{0.001}{3.142} \simeq 3.4 \times 10^{-4}$, remarkably small. For a wheel of the size of 1 meter that would mean a deformation less than a millimeter, hardly perceptible in practical use.

Another example of a hasty backward-in-time extrapolation is praising the pre-Columbian Maya and Aztec civilizations as excellent astronomers, giving as one of the most striking proofs their evaluation of the synodic period of Venus. As in the case of the quasi-perfect knowledge of the value of π by ancient Egyptians, a similar degree of precision was attained in the determination of the period after which the planet Venus appears as the morning star after being seen as the evening star, and vice versa. The exact value known today is 584.57 days, drawn from very precise astronomical observations involving best telescopes and time measuring devices.

Without any of these, with the naked eye, the Amerindians could arrive at the same result, just noticing, after many years of observations, that Venus makes a full turn five times in eight years. The time span of the year was known to them due to ingenious stone structures enabling the observation of Sun's annual motion with precision up to a day.

It turns out that if we use that simple relationship, i.e. in our understanding, the fraction $\frac{8}{5}$ multiplying the 365 days that represent one sidereal year, we obtain the period of 584 days — a result even more astonishing than ancient Egyptians' knowledge of the value of π! In fact, the precision is of the order of $0.57/584 \simeq 0.001$, i.e. less than one part in thousand. In real time it would correspond to about 14 minutes, certainly too short a span to be meaningful and discernable with naked eye observations.

All this being said, we should still admire the perspicacity and intelligence of ancient people who were able to follow the complicated motions and varying positions of celestial objects and note the most important periodic laws without any of modern optical devices — just with naked eye and the simplest settings serving to fix directions. In the following sections we shall succintly describe what the oldest civilizations were able to achieve in mathematics and astronomy.

4.2 Astronomy and astrology

The idea that the five "wandering" celestial bodies must greatly influence all life on Earth was common to practically all ancient civilizations: Sumerians and Babylonians, Egyptians, Chinese and Indian civilizations, Greeks and Romans alike. Stars and planets were often identified with gods, and their motions and changing appearance, especially differences in luminosity, were interpreted as omens to be taken into account by the rulers in order to take appropriate decisions. Solar and lunar eclipses, as well as comets appearing from time to time, were also believed to be an expression of divine whims or signs of displeasure. Ancient rulers relied on the advice of diviners, which made these knowlegeable people very influential. Especially that they knew how to predict real phenomena important in everyday life, like the beginning of river floodings and seasonal weather changes.

For a very long time astronomy and astrology were so intimately intertwined that it was difficult to discern between the two. Both names are derived from ancient Greek and contain the root "astro-", which comes from the word "astron" meaning "star". Now, the second part comes either from "nomos" meaning "law" or "rule", and appears in the names of other sciences, like economy, taxonomy, or from "logos", meaning "reason" or "plan", and by extension "wisdom" and "understanding". This ending is found in the names of many sciences such as biology, geology, pharmacology, immunology, etc.

As of today, economy which uses mathematical tools is much closer to what one would define as science than ecology which is akin to philosophical stance more than an exact science. Isidore of Seville (ca. 560–636 C.E.), Archbishop and the last scholar of the ancient world, was the first to draw a clear distinction between the two practices. Nevertheless the final divide between astronomy and astrology occurred in European science quite lately, during the enlightenment that started in the 17th century. In principle, the scientific exploration of motions and physical properties of celestial bodies deserved to be named "astrology", and guessing, which supposed influence of stars and planets on human fate (the so-called *judicial astrology*), could have been called "astronomy". But now we have been accustomed to the contrary.

Babylonian astrology served exclusively the kings and the state affairs. When the Greeks inherited Babylonian and Egyptian science, they adopted the science of numerical analysis of celestial motions together with the supposedly predictive power of various combinations and positions of the Sun, Moon and planets with respect to the stars. The Greeks democratized Babylonian astrology extending the influence of stars and planets on individuals. This gave rise to the great popularity of *horoscopes*, which were believed to have important predictive value for determining the character and future events of any individual. A horoscope (from Greek "ora" meaning "time" and "skopos" meaning "watcher" or "observer") is a detailed report of exact positions of the Sun, Moon and five planets with respect to the Zodiac constellations at the moment of someone's birth, this practice being called the *natal astrology*.

It is amazing that in spite of tremendous development of modern astronomical knowledge astrology is still popular, even among people whom one would not consider as backward or superstituous. Many journals and weekly magazines publish horoscopes regularly on their last pages; specialized astrological journals flourish in many countries. An individual horoscope should take into account not only the day, but also local or sidereal time of birth of a person in order to determine the exact celestial positions of the Sun, Moon and planets. Notwithstanding the fact that after 2300 years while the positions of the Sun during a year were observed and fixed, the slow precession of Earth's axis resulted in the shift of the equator with respect to the Zodiac by about 15°, so that at the time of vernal equinox Sun is in the constellation of Pisces and not Aries as by the time when the Babylonians had discovered Sun's yearly motion with respect to the stars and defined the twelve constellations of the Zodiac.

The belief that certain psychological or physical features are shared by people born under a given Zodiac sign is still popular. Usual weekly or monthly predictions for each sign are vague enough to impose a unique interpretation. More detailed horoscopes take into account not only one of the twelve Zodiac signs containing a person's birthday, but also positions of the Moon and five planets at the same time. It is easy to see that this gives rise to $12^7 = 35\,831\,808$ combinations, i.e. more than thirty-five million. If one adds as an extra parameter the day of the week, one will get $7 \times 12^7 = 250\,822\,656$ combinations — pretty enough to personalize a horoscope.[1]

4.3 Ancient Egypt

Ancient Egypt was one of the oldest civilizations produced by humanity, going back to 3000 years B.C.E. when the first written texts did appear. After about 300 B.C.E. when Alexander the Great conquered Egypt, hellenistic culture prevailed, opening an entirely new period, which will be discussed in the chapter devoted to Greek science. Here we give a brief account of the ancient Egyptians' mathematical and astronomical skills.

• Egyptian mathematics

Ancient Egyptians used a positional decimal system, with separate symbols for 1, 10, 100, 1000, etc. Addition and substraction amounted to couting how many symbols of each kind were in both numbers. During later periods they were replaced by hieratic numerals, nine different symbols for the first nine numbers, then nine different symbols for first nine multiples of 10, and so on.

For multiplications of bigger numbers tables of doubles were used, reducing all such operations to subsequent additions. For example, to multiply 27 by 13, the following preliminary table was set up:

$$27 \times 1 = 27, \quad 27 \times 2 = 54, \quad 27 \times 4 = 108,$$

$$27 \times 8 = 216, \quad 27 \times 16 = 432, \ldots$$

---

[1]Nevertheless the hundreds of people born on the same hour of the same day all have the same birth-horoscope, but very different lives.

then the multiplier was represented as sum of numbers appearing in the table, here $13 = 8 + 4 + 1$, so that the desired answer is

$$27 \times 13 = 27 \times (8 + 4 + 1) = 216 + 108 + 27 = 351. \qquad (4.3)$$

Tables of multiplications helped scribes to make the calculus of great numbers easier; the idea of a square root was also known to them, but without general method of finding it; the trial and error method prevailed.

More complicated fractions were reduced to sums of primitive ones, with numerator equal to 1. A simple and ingenious method of "closest guess" was used by egyptian scribes. For example, $\frac{2}{5}$ was represented as sum of two *unit parts*, or primitive fractions, $\frac{2}{5} = \frac{1}{3} + \frac{1}{15}$. The first fraction plays the role of initial rough approximation, the second one is found to complete the count. The next term could be added or substracted, depending on circumstances. For example, take the fraction $\frac{11}{23}$: the closest simple fraction is $\frac{1}{2} = \frac{11}{22} > \frac{11}{23}$, so one can expect the following representation:

$$\frac{11}{23} = \frac{1}{2} - \frac{1}{x} \text{ so that } \frac{1}{x} = \frac{1}{2} - \frac{11}{23} = \frac{1}{46} \text{ and } \frac{11}{23} = \frac{1}{2} - \frac{1}{46}.$$
$$(4.4)$$

Linear equations with fractional coefficients were solved by the method called "false assumption", popular also in ancient Hindu, Chinese, and Muslim and European civilizations until the Renaissance in Europe. As an example, we can cite a problem found in a document dating from the Middle Kingdom, about 1650 B.C.E., actually conserved in the British Museum under the name of "Rhind papyrus", which contains dozens of arithmetical and geometrical problems for the training of future scribes. The question reads: "A quantity and its seventh part equals 19. What is it?". The solving strategy is as follows: the first guess is the denominator of the fraction which is 7; but 7 and its seventh part is $7 + 1 = 8$ which is not 19. The next step is to divide 19 by 8, represent the result as sum of entire number and primitive (with numerator 1) fractions, and then multiply it by 7. The answer is

$$7 \times \left(\frac{19}{8}\right) = 7 \times \left(2 + \frac{3}{8}\right) = 14 + \frac{21}{8} = 16 + \frac{5}{8}, \qquad (4.5)$$

which in modern version would be formulated as solving for unknown x the equation $x + x/7 = 19$, equivalent with $8x/7 = 19$ whose solution is $x = 16\frac{5}{8}$.

Ancient Egyptians needed not only arithmetics for accountancy, but also geometry for their architecture and land measures. They knew the ratio between the circumference of a circle and its radius, which was approximated by simple fraction 22/7. Compared to what we define now as $\pi = 3.1415926...$ we can see that their aproximation was good enough for architectural and other practical needs (see equation (4.2)).

• Egyptian astronomy

Ancient Egyptians were neither a sailing, nor nomadic people; theirs was a sedentary agricultural civilization, depending greatly on fertile floods of the great river Nile. This is why their timing was of primordial importance, because it defined sowing and harvesting periods. The flood of the Nile was preceded by the heliacal rising of Sirius, called *Sothis* by the Egyptians. Egyptian priests noticed that the Sun is drifting across the sky. When it is found in a certain region of the sky, its graze prevents us from seeing any stars in its vicinity. But as the days go by, the Sun changes position, allowing the stars it was concealing to be visible again. The heliacal rising of a star occurs at the first day when that star becomes visible again in the east during dawn, just before sunrise.

There were special places outside the temples where two priests would sit down facing each other. One was the observer, another played the role of observational device, sitting right in the east direction as seen be the observer. When rising Sirius touched the priest's ear, they knew that it was time to announce the imminent flood. In ancient times it happened in the beginning of July. Nowadays the heliacal rising of Sirius as observed in Egypt has shifted to August due to the precession of the equinoxes.

The ancient Egyptian calendar year was 365 days long: twelve months 30 days long plus extra five festive days. The leap years were unknown; however, they noticed that after four years the heliacal rising of Sirius occurred one day later. The full cycle would last $365 \times 4 = 1460$ years (more exactly, 1461 years). This long period was called "the Sothic cycle".

Egyptian priests also used the rising of other stars to mark the passage of time during the night. These "stellar clocks" enabled them to divide the night into twelve hours.

4.4 Ancient Babylon

Babylon was one of the most advanced civilizations in ancient Mesopotamia. It was heir to Akkadian and Sumerian cultures, which developed the first rudiments of astronomy and mathematics. The Babylonians, under the Achaemenid (550–312 B.C.E.) and Seleucid (312 B.C.E.–63 C.E.) Empires, advanced far beyond.

• Babylonian mathematics

Sumerians were probably the first who invented a positional system of expressing huge numbers as sums of successive powers of 60. Extremely simple and efficient, it used only two signs: < for tens, and † for units. For example, number such as 23 would be noted << † † †, but when it was bigger than 60, say 85, it would become a sum of two terms, $85 = 60 + 25$, of which the first one would be noted by a single unit † in place for multiples of 60, then the rest would follow after a short space: $85 = † << † † † † †$. Very big numbers required higher powers of 60. According to statistical data, the United Kingdom population was about 63 182 000 in 2011. The two numbers would be written by Babylonian scribe as follows;

$$63\,182\,000 = † † †† <<<<< †† <<< <<< † † † << \qquad (4.6)$$

meaning

$$4 \times (60)^4 + 52 \times (60)^3 + 30 \times (60)^2 + 33 \times 60 + 20. \qquad (4.7)$$

The year 2011 is expressed as $<<< † † † <<< † = 33 \times 60 + 31 = 1980 + 31 = 2011$.

The arithmetic system based on powers of 60 included also fractions, expressed in negative powers of 60. Any fractional number was reduced to an integer divided by one of the powers of 60, usually no higher than $(60)^3 = 216\,000$, which in terms of the decimal system corresponds to the precision down to 10^{-5}. The best example is given by the value of the square root of 2, expressed in one of Babylonian clay tablets as

$$\sqrt{2} = \frac{1}{1} + \frac{24}{60} + \frac{51}{(60)^2} + \frac{10}{(60)^3} \simeq 1.414213, \qquad (4.8)$$

as compared with the value of the square root of two expressed in decimal system, $\sqrt{2} \simeq 1.4142136$. The difference can be seen only after the sixth decimal.

Any fraction would be first expressed, after multiplication of numerator and denominator by the same factor, as a similar one, but with denominator being close to a number that could be transformed by further multiplication into a power of 60. Then the numerator would be split in parts corresponding to lower negative powers of 60 (including $60^0 = 1$, i.e. the integer part for improper fractions, like in the case of square root of two shown in (4.8) and the remaining part representing the remaining negative powers. In practical applications Babylonian mathematicians accepted errors of the order of 1 to 2 percent. Take for example the fraction $\frac{3}{19}$. A skillful Babylonian mathematician would first of all multiply both numerator and denominator by 21, and then proceed to the closest approximation permitting an easy expression in powers of 60:

$$\frac{3}{19} = \frac{3 \times 21}{19 \times 21} = \frac{63}{399} \simeq \frac{63}{400}. \qquad (4.9)$$

The error resulting from replacing 399 by 400 is one-fourth of one percent, which is quite acceptable. Now it is easy to go from 400 to the closest power of 60: indeed, one has $9 \times 400 = 3600 = (60)^2$. This enables us to write

$$\frac{3}{19} \simeq \frac{63}{400} = \frac{9 \times 63}{3600} = \frac{567}{3600} = \frac{540 + 27}{3600} = \frac{9}{60} + \frac{27}{(60)^2} \simeq 0.1575, \qquad (4.10)$$

not bad at all when compared with the value of the same fraction obtained by any computer and expressed in decimal system: $3/19 = 0.1578947....$ The difference between this result and its Babylonian counterpart appear only at the fourth decimal place after zero.

• Babylonian astronomy

The Babylonians were certainly the finest astronomers of the ancient world. They paid some tribute to earlier cultures of Mesopotamy, Akkadians and Sumerians, but it is not clear to what extent. Astronomy and astrology, which at that time meant one and the same thing, were the domain of scribes-priests referred to later by ancient Greeks as *Chaldeians*. They left an overwhelming imprint on many cultures and civilizations that followed; some of their findings are in use until now, as the division of circle into 360 parts (degrees), each of these parts divided into 60 minutes, the division of day and night into 12 hours each, the division of the Zodiac into twelve constellations 30 degrees wide each. Archeologists and linguists were able to reconstruct Babylonian's atsonomical knowledge due to the vast collections of clay tablets containing an enormous amount of information in the form of astronomical diaries and the ephemerides, i.e. exact times of passages of heavenly bodies through the local meridian.

The development of Babylonian astronomy can be divided into two periods, the old one was practiced from 1830 B.C.E. when the reign of the first Babylonian dynasty started, till 626 B.C.E. when the New Babylonian Empire replaced the old one. The new Babylonian astronomy developed in the period after 626 B.C.E. until the conquest of Babylon by the Greek armies of Alexander the Great. Greek astronomy developed rapidly after the Babylonian treasures of knowledge were assimilated by curious minds.

Babylonians were interested in stars and planets for two distinct reasons. The first was practical, connected with everyday life, the necessity of measuring time and the establishment of a reliable calendar. The second was the belief in the divine nature of heavenly objects identified as gods, and the necessity of interpreting various omens in the sky, which were supposed to help the rulers to make correct decisions. Astrological skills of the Chaldeians, were proverbial in the ancient world, even after the Roman conquest. The horoscopes in today's popular magazines reproduce with little modifications the ancient Greek recipes transmitted from the Babylonians who invented them more than twenty-five centuries ago, when their kings would not make any serious decision without consulting the astrological situation first.

The most remarkable astronomical heritage of ancient Babylon is found in well conserved libraries of clay tablets containing an enormous amount of astronomical information, records of eclipses; heliacal risings, ephemerides of planets and brightest stars. What we could learn from the tablets, the most ancient of which are known under the names "Enuma Enlil" and "Mul.apin", is that the Babylonians were the first to acknowledge the periodicity of celestial phenomena, like equinoxes and solstices; they also discovered that the Morning Star and the Evening Star were in fact the same celestial body.

Contributions made by the Chaldeian astronomers during the later period include the confirmation of eclipse cycles and Saros cycles; they also observed that the Sun's motion along the ecliptic was not uniform, though remaining unaware of the reason of this behavior. It is known today that this is due to the Earth moving along an elliptic orbit around the Sun, moving swifter when it is nearer to the Sun at perihelion, and slower when it is at aphelion, at a slightly greater distance from it.

Apparently, the Babylonians were not interested in the physical nature of celestial phenomena, interpreting the luminaries and planets as gods or as divine omens. But their analysis of temporal relations was extremely precise for the simple observational methods that were at their disposal. Their records of solar and lunar eclipses were used centuries later by the Greeks. The Babylonian lunar calendar was based on synodic months, but they were aware of other periodicities of the Moon's motions, including the *anomalistic month* i.e. returning of the apsis (the perigeum-apogeum line) to the same position with respect to the stars, and the *draconic month* which is the time span between lunar visits of the same node. From this, they derived the *Saros* period, lasting about 18.6 years, separating two consecutive solar eclipses. The three lunar periodicities discovered by the Babylonians were:

$$223\,\text{synodic} = 239\,\text{anomalistic} = 242\,\text{draconic months}$$

Most of the astronomical knowledge of the Babylonians has been assimilated by such great Greek astronomers as Hipparchus and Ptolemy.

4.5 The science of ancient Greece

Ancient Greece is an unparalleled cultural and civilizational phenomenon, which became the source of inspiration for both Christian and Islamic science. In contrast with the science of ancient Mesopotamia and Egypt, it was not the exclusive activity of a special caste of scribes, but was developed by people with curious minds belonging to the upper class, but without any specific religious function — just curious minds exploring the surrounding world. Although the Greeks were aware of the mathematical and astronomical knowledge of the Babylonians and Egyptians, their attitude towards astronomy was radically different: instead of treating the Sun, the Moon and the planets only as divinities, they assumed that these were physical objects, whose size, distance and physical nature could be investigated. In other words, the Greeks replaced the question "who is it?" by the question "what is it?". By doing so, they cleared the path towards modern astronomy as we understand it today. It is worhwhile to learn how the Greek philosophy evolved during a time span almost ten centuries long.

Ancient Greek philosophy, including its continuation in the Roman Empire, can be divided in four distinct periods:

- The pre-Socratic period, from the 7th till the 5th century B.C.E.;
- The classical, or Socratic period, from the 5th till the end of the 4th century B.C.E.;
- The Hellenistic period, from the end of the 4th century B.C.E. till the 2nd century C.E.;
- The Roman period, from 2nd till the 5th century C.E.

This chronological division in principal development stages of Greek philosophy is by no means unique, but it gives a fair account of its most salient particularities in each of the aforementioned periods.

Although Greek philosophy was involved in understanding human nature as well as the notions of beauty, justice and social behavior, here we are interested in what can be called "natural philosophy", explaining the ways and phenomena of the surrounding world.

Among the sources of Greek philosophy of nature one can cite mythology, primitive beliefs in the divine nature of natural forces, anthropomorphic representations of constellations, stars and planets, all this gradually abandoned in favor of a vision of the Universe gradually freed from myths and superstitions. A characteristic feature of Greek philosophy is its search for unifying principles underlying the apparent variety of natural phenomena, and building coherent philosophical systems of thought.

4.5.1   *Before Aristotle*

• **The Miletus School**. The oldest Greek philosophical school appeared in the Ionian city of Miletus (situated in Anatolian Peninsula on the shore of Aegeian Sea, now in Turkey). Its founder, Thales of Miletus ($\Theta\alpha\lambda\eta\varsigma$, 625 ~ 547 B.C.E.) who was considered by ancient Greeks as one of their *Seven Sages*, as the father of scientific approach to the description of natural phenomena and perhaps as the first person deserving the title of "philosopher of nature".

Thales became famous for his alleged prediction of the solar eclipse of 585 B.C.E., and for his ability to evaluate dimensions of objects at a distance, by comparing their shadows with the shadow of a stick of known length. None of his writings remains, all we know of him was transmitted in the works of Plato and Aristotle.

One of Thales' main ideas was the belief that all material phenomena take their origin in one common element. He attributed the role of such fundamental substance to "wet nature", or water. According to Thales, all other materials were created from water, and after some time return back to water. The universe was endowed with soul of its own, and full of gods. Primordial water had also a divine nature.

In Thales' scheme of the Universe Earth was a huge disc floating in the middle of huge sea, due to its natural buoyancy, like a piece of wood. The Sun, the Moon and the stars turned above in the air.

• After Thales' death his allegedly preferred pupil **Anaximander** (610 – 546 B.C.E.) took the lead, but no more than a couple of years. He is supposed to be the author of a treaty "On Nature" of which no traces survived, but his ideas were reported in writings of Aristotle and other Greek and Roman philosophers. Anaximander can be called the first astronomer, because he applied mathematical proportions to describe the sky. He had also an original view on the material world, which he supposed to be composed of *apeiron*, a substance more primitive than the four elements. The constant and eternal motion of primitive matter results in creation of observed variety of elements and their compositions.

Anaximander was also the first philosopher proposing that life on Earth started in water. The first living creatures resembled fish, then went out to colonize the land. Man evolved from animals, in particular from fishes, but from the very beginning he was different from other animals.

• **Anaximenes of Miletus** (588 – 525 B.C.E.) was the latest representative of Milesian School. As his predecessors, he tried to find out what was the matter underlying all substances known, and he chose air as the primordial source of all being. He argued that in contrast to water, air is unlimited in space. Water can condensate from the air, and solids can condensate from water. Fire is just rarefied and heated air. The soul was also a variety of air. And the air, according to Anaximenes permeated the entire Universe. Also the Earth, the Sun, the Moon and planets had their origin in the air, and were created from it by condensation. Anaximenes was also the first philosopher who explained solar eclipses by supposing that they happen when the Moon passes between the Earth and Sun.

As his predecessors, Anaximenes believed in multiple worlds, all of them created from the air. His was also the first attempt to explain the nature of celestial bodies like the Sun, Moon, planets and stars. He believed that they were derived from Earth. The most distant stars were supposedly created from the rarefied air prevailing far from the Earth.

• **Heraclitus of Ephesus** (540 – ??? B.C.E) was an original thinker remembered for his short aphorism "$\pi\alpha\nu\tau\alpha\,\rho\epsilon\iota$" — "Everything flows", meaning that the world is in constant motion and a process of transformation. Another famous saying of his states that "No man ever steps in the same river twice". No wonder that Heraclitus chose fire as primordial element from which all other forms of matter were constantly created. The world is not static, but an eternal fight of oppositions.

• **Pythagoras** (ca. 570–490 B.C.E.) was born on the island of Samos. Pythagoras moved in 532 or 531 B.C.E. to Croton in Southern Italy, which at that time was a Greek colony, part of *Magna Graecia*, where he founded a society of close followers later known as Pythagoreans, based on rigor and self-discipline, as well as on a relentless quest for perfection. He was admired by the next generations perhaps more for his religious and political views than for pure scientific achievements, although his name is associated with the famous geometric theorem and first scientific approach to musical scales. One of the surviving discoveries was the incommensurability of the musical scale based on perfect ratios of chord lengths (today we know that it was related to the corresponding frequencies). Pythagorean tuning was based on the observation that a strained chord with half of its length gives rise to the sound that is an octave of the same chord with basic length; and when the length is reduced in proportion 2 : 3 (two thirds), the sound perceived is the perfect fifth (like the notes Do-Sol, for example). Continuing the process, the new reduction of length (maintaining the same tension strength) by 2 : 3, i.e. 4 : 9 of the initial length would produce the sound re'. In order not to go beyond one octave from the initial tune, let us double the length, which will produce the ratio 8 : 9, corresponding to the second (Re), just above the initial Do. Continuing in this manner, we could reproduce the full scale — but the intervals will never exactly coincide, because no stack of 2 : 3 can be expressed as a stack of 1 : 2.

In modern terms, the Pythagoreans discovered that $\log_2(\frac{2}{3})$ cannot be expressed as any finite fraction.

The most original Pythagorean ideas are 1) the belief in reincarnation of souls, 2) the belief in the underlying harmony of the Universe, 3) the belief that numbers and mathematics rule the world, 4) an original view of the Cosmos, with the Earth, Sun and Moon rotating around an invisible eternal fire, elaborated by one of Pythagoras' follower Philolaus (470 – 390 B.C.E.). In politics, which was the field of his predilection, Pythagoras believed that human society should be ruled by aristocratic minority of knowlegeable and educated people, opposed to the uneducated class of plebeians. His followers succeeded to grab power in the city of Croton and ruled there for some time, until the plebeians of Croton helped by their allies from neighbor cities organized a riot and put an end to Pythagorean supremacy, killing many of the philosopher's followers and banishing the survivors.

Pythagorean philosophy influenced Plato and Aristotle. In particular the belief in the underlying harmony of the Universe and the idea that behind all physical phenomena one can find mathematical explanation.

• **Parmenides of Elea** (540–470 B.C.E.). His most original ideas can be found in fragments of his poem that survived; they were also reported by Plato. At a first glance Parmenides' was just playing with words, and his conclusions may appear to us as rather naïve, nevertheless the questions he formulated were so profound that they haunt philosophers, astronomers and physicists until today: is the Universe finite or infinite in time and space? Is a vacuum, i.e. space deprived of any form of matter possible?

Parmenides defined the Universe as the *Being*, i.e. "all that exists", in opposition to "Non-being", meaning probably "empty void", or "nothingness". The first assessment was: "Being exists; Non-being does not exist". From this premise Parmenides derived the *uniqueness* of Being (today we would say that there is only one Universe). The logical proof was as follows: let us suppose that there are two Beings. If they are connected with something that exists, that something is by definition a part of Being, which contains the two Beings and the part that plays the role of a bridge. Or, if they are distinct, there is a non-being in between. But Non-being does not exist, therefore they are connected, forming one bigger Being.

In the same manner Parmenides argued that Being is continuous in time, and eternal — it can have neither a beginning, nor an end.

● **Empedocles of Acragas** (494–434 B.C.E.) was born in Acragas (Sicily) and died at the age of sixty. He was the son of a certain Meton, and was from an important and wealthy local aristocratic family.

As is the case of his predecessors, Empedocles' work survives only in fragments, but in a far greater number than any of the other Presocratics. These fragments are mostly quotations found in other authors such as Aristotle and Plutarch, who mention a poem written in hexameter, traditionally entitled *On Nature*. Only some of its fragments are conserved until the present.

Empedocles was influenced by other Greek philosophers who preceded him, among whom Pythagoras, Anaximander and Anaximenes, and above all, Parmenides whose work was certainly familiar to him, and from whom he took the inspiration to write in hexameter verse, and whose physical system he adopts but seeks to rectify.

● **Anaxagoras** (ca. 500 − 428 B.C.E.) was born in one of the peripheral Greek colonies in Asia Minor, still under Persian rule, but as a young man he moved to Athens, where he became a friend and associate of Pericles. Anaxagoras followed Parmenides accepting the principle of an eternal Universe, explaining apparent generation and destruction of material objects by replacing them with mixture and separation of ingredients which he considered to be the essential "seeds" of all that exists, called by him *homoeomeries* combining and agglomerating under the impulse of the universal Mind (*"noos"* in Greek). He supposed that human or animal bones are composed of tiny bones, the stones of tiny stones, leaves of plants of tiny leaves — a kind of atomistic view of matter.

Anaxagoras was probably the first Greek philosopher who understood correctly the nature of solar and lunar eclipses. He also professed the theory according to which all heavenly bodies are not of divine, but of terrestrial nature: the Sun is an enormous ball of hot iron, the Moon and the planets are huge or small rocks ejected from the Earth; they shine because they become hot due to the velocity causing resistance of the air. According to several ancient sources, Anaxagoras also gave explanations for the light of the Milky Way, the formation of comets, the inclination of the heavens and the solstices.

Anaxagoras views on the Cosmos seemed so extravagant to his contemporaries, that soon after Pericles' death, having no political protector, he was accused of blasphemy and banned from the city.

• **Democritus of Abdera** (ca. 460 − 370 B.C.E.) was a disciple of Leukippos (ca. 480 − 420 B.C.E.) of whom little is known, but who is credited by Aristotle to be the true inventor of *atomism*. According to his views, all matter is composed of tiny undivisible ("a-tomos" means "impossible to be cut down" in Greek) particles moving in the void. When their motion is disordered, they behave like the air or fire, when they come closer they form liquids, and when they come to stand still in order, they form solids. This idea, postponed by most of the later philosophers in favour of the continuum, was seriously considered again by Gassendi (1596–1655) only in the 17$^\text{th}$ century.

4.6   Greek philosophy after Socrates

• **Socrates** (in Greek, Σωκρατης, 469 − 399 B.C.E.), one of the greatest ancient philosophers, has not left any written text, probably by principle. He spent most of the time discussing with his fellow citizens in the streets of Athens, or teaching young people forming a circle of enthusiastic followers of his philosophy. Convinced that the philosophers' main goal was the quest for truth, he invented an original way of seeking it, often called the *Socratic method*, consisting in asking questions, gradually driving his opponent in discussion to arrive to a common conclusion. Socrates considered himself not as a bearer of truth, but rather as obstetrician, helping the truth pre-existing in human mind to come out. No wonder that he won the nickname "Gadfly" among the Athenians.

Socrates' teachings, his questioning the established truths (his favorite saying was "One thing only I know, and that is that I know nothing") and above all, his popularity among young men who gathered to learn from his wisdom, were gradually frowned upon by his fellow citizens. The Athenian comedy author Aristophanes (∼455 − 386 B.C.E.) attacked Socrates and his followers nastily in his play "Clouds", which ends with a thinly veiled appeal for murder.

In 399 B.C.E. Socrates was put to trial, and a death sentence was pronounced, motivated by the opinion that his teachings corrupted the youth drawing it away from the worship of Gods, as well as the accusation of personal impiety. Although his pupils proposed to organize his flight from prison, Socrates said that such a behavior would give reason to his persecutors; on the contrary, he wanted to prove his rightness and not escape the sentence. He died in prison after having drunk the poisonous extract of hemlock, the current Athenian method of execution.

Perhaps the most valuable of Socrates' advices to humanity, besides his exemplary moral stance, was his famous phrase "Γνῶθι σεαυτόν", or "Know yourself". How strong Socrates' influence was can be confirmed by a similar sentence pronounced by the Swiss psychologist C.G. Jung almost twenty-five centuries later: "*Who looks outside, dreams. Who looks inside, awakens*". And — last but not least — it might be that Socrates would have been forgotten had he not have among his pupils and followers such an outstanding human being as Plato.

• **Plato** (in Greek Πλατων, ca. 428 − 347 B.C.E.) is the central figure not only in Greek philosophy — we owe him all we know about Socrates and his teachings, and much more of his own — but it can be safely said, as British philisopher A.N. Whitehead put it once, alluding to the wealth of general ideas scattered through Plato's texts — that the entire European philosophical tradition that followed, "consists of a series of footnotes to Plato". Although such statement is a deliberate literary hyperbola, the fact is that Plato's *Dialogues* have set the foundations of Western philosophy, defining in an everlasting manner most of the basic philosophical concepts and their mutual relationships: the difference between opinion and knowledge, between myth and reality, pure form and matter.

Besides Socrates, Plato was strongly influenced by Pythagoras, from whom he took the belief in *metempsychosis*, or transmigration of souls, as well as the concept of form as distinct from matter, and that the physical world is an imitation of an eternal and ideal mathematical world, hidden to our senses, but open to our mental investigation. He introduced the distinction between pure and applied mathematics, calling "arithmetic", what is now called number theory and "logistics", now called arithmetic. In the dialogue *Timaeus* Plato associated each of the four classical elements (earth, air, water, and fire) with a regular solid (cube, octahedron, icosahedron, and tetrahedron respectively) due to their shape, the so-called Platonic solids. The fifth regular solid, the dodecahedron, was supposed to be the "fifth element" which made up the heavens. The Universe was supposed to be of perfectly spherical shape, with Earth, also spherical, in its center. The Sun and Moon, as well as the five planets, had their own spheres rotating with different angular velocities. Plato's views were developed further by one of his disciples, Eudoxus of Cnidus.

Plato was interested mostly in the human condition, social relations, government and morals; natural sciences were also discussed, but not as much as politics, justice and society. He reported Socrates' views on political systems in *The Republic*, one of the longest dialogues. Four basic types of government are defined; the *aristocracy* (rule by the best), the *oligarchy* (rule by the few), the *democracy* (rule by the people), and finally the *tyranny* (rule by one person, a "tyrant"). These names are still in use.

Plato held the belief that knowledge was not purely the result of inner reflection but instead, could be sought through exchange of ideas, and therefore, transmitted to others. It was because of this belief that Plato founded in Athens ca. 387 B.C.E. his famous Academy. It was not a formal school or college in the modern sense;

Fig. 4.2   Left: Plato. Center: Plato and Aristotle. Fragment of Raphael's fresco painted in 1509–1511 on the walls of one of the most beautiful chapels of Vatican. Elderly Plato points his finger to the sky, while Aristotle shows Earth with his hand. Right: Aristotle.

it was rather an informal group of intellectuals who shared a common interest in studying philosophy, mathematics, and astronomy.

Aristotle studied there for twenty years before founding his own school, the Lyceum. The Academy persisted throughout the Hellenistic period as a skeptical school, until coming to an end after the death of Philo of Larissa in 83 B.C.E.

• **Aristotle** (384 − 322 B.C.E.) (in Greek Aριστοτελης) was born in Stagira, in Northern Greece, then the Kingdom of Macedonia. At the age of twenty he left for Athens, where he studied in Plato's Academy during 20 years. It was reported that Plato said once, comparing another of his pupils, Xenocrates, with Aristotle, that "The one needed a spur, the other a bridle." Plato was admirative of Aristotle's boundless energy and unsatiable curiosity; Aristotle respected Plato enormously and gratefully recognized his master's great influence.

During the twenty years spent in the Academy, Aristotle's philosophical views constantly evolved. Under great influence of Plato's ideas, he became more critical with time. Although he subscribed to Plato's belief in an immortal soul to whom the body is a temporary dwelling, he rejected Plato's theory of Forms, which claimed that the utmost reality is stored in ideal forms, of which the objects we see and touch are imperfect imitations: a dog is a dog because it imitates the "ideal form" of a dog. With all due respect towards Plato, Aristotle kept developing his own philosophy, summarizing his attitude in his famous statement:

Amicus Plato, sed magis amica Veritas (in ancient Greek: Φιλοσ μεν Πλατων, φιλοτε′ρα δε αληθηια)

After Plato's death in 348 B.C.E. Aristotle left Athens for Assus, a city on the northwestern coast of Anatolia, where Hermias, one of his fellow students in Plato's Academy, was ruler. After Hermias' demise by the king of Persia and death, Arisotle left for the island of Lesbos, where with the help of his students he carried zoological and marine biology research, summarized later in several books.

In 343 B.C.E. Aristotle was invited by the Macedonian King Philip II to become a tutor for his son Alexander, then 13 years old. After a few years Alexander became king and soon was conquering Asia, and Aristotle returned to Athens. He established his own school in a gymnasium known as the "Lyceum", endowed with substantial library. A group of brilliant students gathered in the Lyceum. They were called *"peripatetics"* from the name of the cloister (*peripatos* in Greek) in which they held their discussions while walking around.

During his Lyceum period Aristotle wrote the most important part of his philosophical works, whose systematization and division into separate disciplines was one of his greatest innovations valid until the modern times. This can be put in contrast to Plato's legacy, whose "Dialogs" constantly mix together different domains of human knowledge.

When Alexander the Great died in 323, democratic Athens became much less friendly towards Macedonians. Saying that he did not wish the city that had executed Socrates "to sin twice against philosophy", Aristotle fled to Chalcis, where he died the following year.

4.6.1 *Aristotle's logic*

The work of Aristotle covered all domains of science and philosophy of his time, and its influence is beyond any comparison; his writings were edited, studied and commented with equal devotion by Greeks, Romans, then the Arab and European wise men, philosophers and scientists. Considered the *nec plus ultra* and translated in all major languages they became the basis of universally accepted knowledge. Paraphrasing Whitehead's quip on Plato, one can say that both medieval European and Islamic science were a series of footnotes to Aristotle's works. Aristotle's authority was so great that it took many centuries before some of his views started to be criticized and often proven to be false.

Aristotle's contribution to universal human knowledge is so vast and rich, that even a short summary relating the totality of his writings would take too much space to be presented here. We shall restrain our narrative to his views on physics and astronomy. The most important volumes of his tremendous work are: *"Metaphysics"*, *"Physics"*, *"Politics"*, *"Poetics"* and the *"Organon"*, the last one containing the foundations of Aristotelian logic, which remains the basis of our thinking since then. According to Aristotle logic is not a separate science, but the necessary tool for practicing any science. It can be defined as the "science of thinking".

Aristotelian logic stemmed from Pythagoreans' and Sophists' roots, then from the teachings of Socrates and Plato, but Aristotle's merit was giving logic a unique formal and precise frame; he also introduced the clear distinction between general propositions, hypotheses, axioms, corollaries and logical conclusions.

The propositions were divided into four categories: *affirmative true, affirmative false, negative true* and *negative false*. The relations between propositions are subject to the following three laws of formal logic:

— The law of identity: logical conclusions are reliable only under the condition that all denominations have the same commonly agreed sense; which essentially means that everything is itself, and it cannot be something else.
— The law of no contradiction: out of two propostitions concerning the same subject and mutually contradicting each other, one must be necessarily false. In other words, no statement can be true and false at the same time.

— The law of excluded middle: something either exists, or does not exist; a statement must be either true, or false. "Tertium non datur" — there is no third possibility.

These were the fundamental axioms regulating all statements; the next development consisted in the introduction of *categorical syllogism*, or combining different statements and producing new ones, just like in arithmetics we can combine (by adding or multiplying) two numbers in order to produce a third one. In its simplest form, a syllogism should start with two statements called *premises*, one *general*, another *specific*; by applying the general one to the specific one, a conclusion is drawn. A well known example is: a) All men are mortal (general premise); b) Socrates is a man (specific premise); c) Therefore Socrates is mortal.

These are just a few examples of Aristotle's approach to rational thinking. His laws of logic became the foundation of scientific reasoning for millenia. Logical coherence has become the most praised quality of scientific explanation of reality. This worship of logic was inherited from Socrates and Plato. Of course observing surrounding reality was also important, but very often logical coherence would prevail, leading sometimes to erroneous statements. However, in many cases we should rather admire the perspicacity of Aristotle's thought, and how close he happens to come to many of our present-day views. Let us examine the very basis of Aristotelian physics.

4.6.2 *Aristotle's physics*

According to Aristotle, all material objects can appear, move, evolve and disappear during their existence, but the Universe itself is eternal. The ultimate goal is to make the *form* prevail over *matter*, and to achieve perfection of life in all its manifestations.

The material world is composed of *four elements*, which are: fire, air, water and earth. The elements are endowed with two types of opposite qualities: "lightness" and "heaviness", and they can be also "dry" or "wet". The light elements display he natural tendency to go up, and the heavy elements tend to go down. This natural motion of elements results in Earth's spherical shape. Physical bodies are

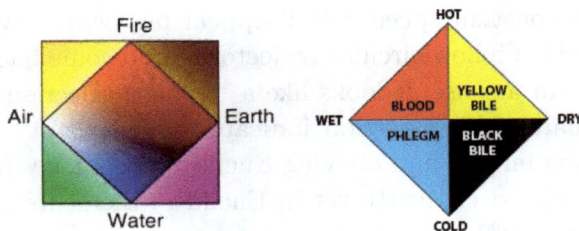

Fig. 4.3   The four elements of which all matter on Earth is composed. Parallelly, four fluids called "humors" circulate in human body, according to Hippocrates (460 − 370 B.C.E.).

endowed with some internal force, or "spirit" that pulls or pushes them towards similar ones: thus the heaviest metals and high density ores are found deep under the ground, water naturally tends to join the ocean, if some amount of air is contained in water, upon heating it liberates itself and joins the atmosphere. The flames of fire go up attracted by the eternal fire above the skies. All physical bodies are just different mixes of four elements, in various proportions. The four elements are found in the "sublunar world", while the heavens are filled with the fifth element (from which the word *quintessence* is derived).

• On the nature of motion: — All motions are composed of two elementary types, rectilinear motion and rotational motion. To corroborate this axiom, one is invited to observe the trajectory of an arrow shot at some angle upwards. According to Aristotle, the motion of the arrow is composed of three phases: ascending one, along the straight line following the initial shooting angle, followed by a segment of a circle at the apogeum of the trajectory, and the final descending stage along another straight line.

Then comes the next postulate which sets a neat distinction between the motions observed on the Earth and the motions of heavenly bodies. While both rectilinear and circular motions are observed on Earth, only circular motions are followed by heavenly bodies. To prove this, Aristotle uses seemingly flawless logical conjecture: Let us suppose that heavenly bodies can follow either circular, or rectilinear motion. After certain time all bodies which follow rectilinear motion with constant speed will disappear far away, leaving behind only those which follow circular trajectories and come back infinitely many times. In a sense, it looks like a "natural selection" principle. Seemingly naive, this assertion indicates that Aristotle considered space as being infinite and obeying Euclidean geometry (although it was not fomulated explicitly yet in Euclid's "Elements", written at least 30 years later).

• On free fall: — A simplistic version of the law of free fall attributed to Aristotle consists of affirmation that "heavy bodies fall faster than the light ones", a statement corroborated by the example of a lead ball and a feather dropped simultaneously from the same altitude; as everybody can verify, the lead ball will fall down very rapidly, while the feather will arrive on the floor much later. It took more than 1900 years before Stevin in 1580 and Galilei in 1586 discovered actual laws of free fall, with constant acceleration independent of mass of falling object, if one can neglect the effects of friction of air.

The full statement concerning the free fall made by Aristotle is the following: the velocity of freely falling material body is proportional to its density, and inversely proportional to the density of the surrounding medium. As a corollary, Aristotle concluded that a vacuum cannot exist, because falling bodies without being surrounded by any material substance would acquire an infinite velocity, which was considered impossible.

Let us summarize up the main postulates and axioms of Aristotelean physics:

• **1**. All bodies tend to approach their natural position with respect to the center of the Earth. For example, stones fall down, while sparks rush upwards from fire.

• **2**. The universal gravity acts either towards the center of the Earth, or outwards, depending on objects and their characteristic features. This is why fishes swim in water, and birds fly in the air.

• **3**. No motion can exist without natural cause. When an external force different from gravity acts on a body, it starts to move along a straight line with constant velocity.

• **4**. Velocity of free fall is proportional to the density of the falling object and inversely proportional to the density of the surrounding medium. The same stone falls rapidly in air, and slowly in water.

• **5**. There is no vacuum in nature, because the bodies would fall down with infinite velocities.

• **6**. Space is filled with matter. The outer space beyond the Earth's atmosphere is filled with aether, or the "fifth element" (later on called *quinta essentia* in Latin), different from the four elements of which ordinary matter is composed.

• **7**. The Universe is spherical, finite and perfect. There is nothing beyond it, and the questioning the whereabouts of the Universe makes no sense.

• **8**. Matter can not be composed of atoms, because if such were the case, there would be vacuum between them.

• **9**. Also all celestial bodies, including the Sun, the Moon and the planets, are made of aether, which means that they are radically different from material bodies we meet in our everyday life.

• **10**. The Cosmos is eternal and not subject to changes. The Sun, the Moon and the planets are perfect spheres which keep their form forever.

• **11**. All celestial bodies move with constant velocity along circular orbits.

4.6.3 *The post-Athenian period*

Aristotle's fame went far beyond Athens and Greece. The king of neighboring Macedonia, Philip sent his young son Alexander to learn philosophy and science in Aristotle's academy. Alexander became the greatest conqueror of all times, and founder of many cities, among which Alexandria in Egypt became the new center of Greek culture and science, illuminating the entire Hellenistic world. The center of gravity of Greek science moved from Athens and Southern Italy to Alexandria. But even before that, one could clearly distinguish two trends in ancient Greek science, the first one speculative, flourishing before hellenistic expansion, peaking in Athens with Plato and Aristotle, the second more practically oriented, which came to maturity in Alexandria and in hellenistic parts of the Roman Empire.

The most prominent representative of Greek science with practical inkling was undoubtedly Archimedes of Syracuse (287 − 212 B.C.E), one of the greatest mathematicians and engineers of the Antiquity. In mathematics, he determined the number π by *exhaustion method*, inscribing and overscribing regular polygons in or on a circle, thus getting better approximations for circle's cincumference and surface area as the number of sides was growing. The final estimate made by Archimedes was $\frac{223}{71} \leq \pi \leq \frac{22}{7}$, which means the precision of the order of $2.5 \cdot 10^{-4}$, a very fine result.

He is credited with inventing the Archimedes screw, used until today for pumping water over short distances. In astronomy, he is supposed to have made two movable spheres: a star globe and the other a device (the details of which have been lost) for mechanically representing the motions of the Sun, the Moon, and the planets.

Archimedes made indirectly an important contribution to astronomy, because in his small treatise *The sand-reckoner* he gave the most detailed surviving description of the heliocentric system of Aristarchus of Samos (ca. 310−230 B.C.E.). The same book contains an account of an ingenious procedure that Archimedes used to determine the Sun's apparent diameter by observation with an instrument he invented for that purpose.

Another great post-Aristotelean scientist is Euclid (ca. 310 − 250 B.C.E.), who lived in Alexandria during the reign of Ptolemy Soter I. Little is known of his life, including the exact dates of birth and death, but his treatise *The Elements* was conserved in Greek, and translated later in Arabic in the Middle Ages, and then back into Latin in late medieval Europe. Euclids subsequent influence on Islamic and European science can be compared only with Aristotle's.

4.7 Ancient Greek astronomy

• **Eudoxus of Cnidus** was undoubtedly the first of the great Greek astronomers. Born in 408 B.C.E. in Cnidus, an ancient Greek city facing the island of Rhodos in Asia Minor (now in Turkey). He studied mathematics with the Pythagoreans in Sicily, philosophy in Plato's Academy in Athens and astronomy while visiting Heliopolis in Egypt. After completing his studies, he left Athens to found his own school in Cyzicus, another Greek town in north-west Asia Minor, on the shore of sea of Marmara, and had many students there. Later on Eudoxus returned to Cnidus where he built his own astronomical observatory, where he saw Canopus, the second brightest star after Sirius. He also constructed a sundial which has been discovered by archeologists in the 19th century.

Then he came back to Athens, with several of his followers, and taught in a school that was somewhat a rival to Plato's academy. Unfortunately none of his writings survived, but his astronomical observations have been reported by Hipparchus, who states that Eudoxus collected all his observations made in Cnidus and in Heliopolis in two main books, *Mirror* and *Phenomena*. Eudoxus' observations turned out to be precious two hundred years later, paving the way for discovery of precession phenomenon.

Some of Eudoxus' mathematical findings were reported by Euclid in his *Elements*. The most important was the proof that irrational lengths can be compared with rational ones, and the very idea of existence of such entities, like the square rooth of 2, which cannot be expressed by any ratio of finite integers, no matter how great they might be. The exact proof by contradiction was given by Euclid. In geometry, Eudoxus used the *method of exhaustion*, which is based on successive approximations by inscribing or circumscribing simpler form inside or outside a given body in order to find its area or volume. Euclid found that the volume of a pyramid is one-third the volume of the prism having the same base and equal height, and the volume of a cone is one-third the volume of the cylinder having the same base and height.

But the most important of Eudoxus' achievements that played a fundamental role in the development of astronomy was his model of planetary motion. It was not only highly appreciated by Aristotle, but became the basis of Aristotle's cosmography exposed in his *Metaphysics*. Eudoxus' idea of rotating spheres to one of which a planet was attached served as a model for subsequent cosmographical models, including the most sophisticated one developed by Ptolemy many centuries later.

Eudoxus explained erratic planetary behavior, including the retrograde motion, by combination of several circular motions. A planet was supposed to be attached to a rotating sphere, which in turn was carried by a bigger sphere whose center coincided with the center of smaller sphere, but whose rotation axis was tilted with respect to the first one. If necessary, a third sphere was introduced, with the same center, but with another independent axis of rotation, and another angular velocity. This is why this model was also related to as "homocentric".

In Figure 4.4 one can see the result of composition of two independent rotations of concentric spheres. With clever choice of angle between the two axes one can obtain a trajectory having the form of a double loop, imitating the retrograde motion. Then the two spheres were attached to a third concentric sphere turning around the Earth with desired angular velocity, thus completing the imitation of the observed planetary motion.

- **Aristarchus of Samos** (310 – 270 B.C.E.) was born on the island of Samos. He probably studied in Alexandria and in Athens, under the third director of the Lycaeum founded by Aristotle. Only one of Aristarchus' treatises survived, *On the Sizes and Distances of the Sun and Moon*, containing the description of geometrical methods and interpretation of solar and lunar eclipses leading to the possiblity of measurement of distances to the two luminaries. How he proceeded will be explained in the next chapter.

Aristarchus was the first astronomer who — following mostly aesthetical arguments of Philolaus (ca. 470 – 390 B.C.E.) — challenged

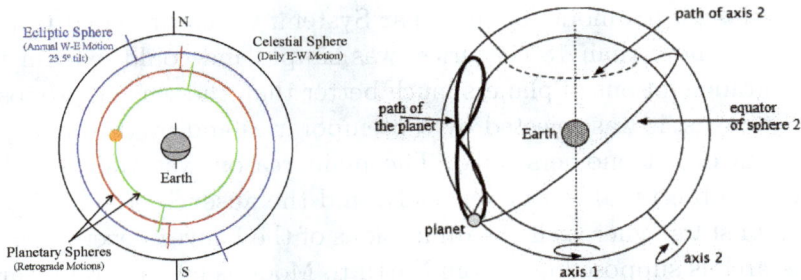

Fig. 4.4 Left: The scheme of nested spheres in Eudoxus' model of planetary motion. Right: Retrograde planetary motion obtained by composition of circular motions.

the firmly anchored belief that Earth is the motionless center of the Universe. He was also the first to evaluate the distances separating us from the Moon and Sun. His estimate of the Moon's distance was quite correct: 63 Earth radii (e.r.), but his estimate of the distance to the Sun was erroneous by a factor of 20. Nevertheless it is undoubtedly because Aristarchus found the Sun to be at least 5.6 times greater than the Earth, that he concluded that the Earth is orbiting around the Sun, and not the contrary. The heliocentric system is not mentioned in Aristarchus' treatise, but it is summarily described by Archimedes in his treatise *"The Sand-Reckoner"* as follows: according to Aristarchus, the fixed stars and the Sun are stationary, the Earth revolves in a circular orbit about the Sun, which lies in the middle of the Earth's orbit, and that the sphere of the fixed stars, having the same center as the Sun, is so great in extent that the circle on which he supposes the Earth to be borne has such a proportion to the distance of the fixed stars as the center of the sphere bears to its surface. The last statement was strongly criticized by Archimedes, who took it literally, and protested against "dividing by zero", because the "center" mentioned by Aristarchus has no dimension at all. Most probably Aristarchus was meaning a small central volume, or the Sun itself, not just a point.

Plutarch (ca. 100 C.E.) gives a similar brief account of Aristarchus' hypothesis, stating specifically that the Earth revolves along the ecliptic and that it is at the same time rotating on its axis.

Although his model of the Solar System, which preceded Copernicus by more than 18 centuries, was simple and could explain the retrograde motions of planets much better than the scheme proposed by Eudoxus, it was rejected by contemporaries and by the next generations of astronomers alike. The main reason was the fact that massive objects fall down vertically, and the absence of annual parallax in stars, whereas if the dimensions of the Universe were such as Aristarchus supposed (i.e. from Earth to Moon is 60 e.r., from Earth to Sun is 18 to 20 times the distance to the Moon, and to the sphere of fixed stars is between 400 and 420 e.r.), the stellar parallax (all stars were supposed to be at the same distance, attached to their sphere) would be equal to $(2 \cdot 18\,\text{e.r.})/(420\,\text{e.r.}) \simeq 0.0857$, which in degrees would correspond to $0.0857 \times 57.3° = 4.9°$, which should be visible with the naked eye.

Aristarchus is also credited with inventing the *skaphe*, a sundial consisting of a hemispherical bowl with a needle erected vertically in the middle. Such sundials combine the advantages of vertical and horizontal ones.

- **Apollonius of Perga** ($A\pi o\lambda\lambda\omega\nu\iota o\varsigma$ in Greek) (262 − 190 B.C.E.) was born and lived in Perga (now in Turkey), then a city of the hellenistic Seleucid empire. Not much is known about his life; he visited Alexandria and learned Euclidean geometry there, and later taught geometry, too. He also visited Pergamon, a city with the next great library of the ancient world, and studied there. He was acquainted with the works of Archimedes, which strongly influenced his work.

Apollonius was granted the title of "Great Geometer" by his contemporaries. His great treatise *"Conics"* contained eight books of which four survived in Greek, seven were conserved in Arab translation, and only the eight one was definitely lost. It seems that he was the first to use the names "ellipse", "parabola" and "hyperbola" designing the three types of conical sections. Besides, Apollonius authored other important geometrical treatises, whose originals are unfortunately lost, but whose titles and partly the content were mentioned by Pappus of Alexandria (ca. 290 − 350 C.E.), a Greek mathematician of the latest Alexandrian period. The treatise *"On cutting of ratios"* survived in translation into Arabic.

Apollonius compiled and generalized many results of his predecessors, and developed new methods in the realm of Euclidean geometry. His findings played an important role in the subsequent development of astronomy. Firstly, he introduced the construction of epicyclic curves, obtained by rolling smaller circles along the bigger ones. This invention enabled the astronomers to create models of planetary behavior still respecting the Aristotelean postulate of the exclusively circular character of the motions of celestial bodies. Secondly, he investigated properties of epicyclic motions determining the points of a standstill, when a planet changes the direction of its apparent motion.

In his geometrical investigations Apollonius went far beyond the content of Euclid's *Elements*.

• **Hipparchus of Nicaea** was the greatest astronomer of ancient Greece. Born in Nicaea, he spent at least 20 years of his life working on the island of Rhodos, where he built his own observatory; it seems that he made some of his observations in Alexandria. Hipparchus' scientific legacy is important in astronomy and mathematics alike, and it would take many pages if we wanted to expose it in a more detailed way; here we shall highlight his principal and lasting achievements.

Hipparchus is best known for his discovery of the precession of the equinoxes, which was a direct consequence of another achievement: the first stellar catalog containing 850 stars visible to the naked eye in the Northern Hemisphere. This monumental work was completed in 129 B.C.E. in spite of repeated accusations of impiety made by his contemporaries. The idea of creation of star catalog was stimulated by the Nova star that appeared in 134 B.C.E., proving that not all things remained unchangeable in the sphere of fixed stars. Hipparchus labeled the stars by their celestial latitude and longitude, and reportedly (by Ptolemy) used a system of six magnitudes according to their apparent brightness like we do it today.

His careful and systematic observations were made with a precision of one arcminute. Comparing the latitudes of brightest stars with those obtained by Timocharis of Alexandria about 150 years earlier, Hipparchus noticed systematic differences between former stars' longitudes and the ones he measured, and they were much greater than observational errors could be. It looked like as all longitudes had increased by almost 2° during those 150 years. Stellar longitudes were already measured in equatorial coordinates, starting from the point where celestial equator intersects with the ecliptic at the moment of vernal equinox. In Hipparchus' time this point was situated in the constellation of Aries; today it has moved towards the constellation of Pisces. Hipparchus rightly concluded that equinox points are slowly moving forward, so that the Sun arrives to vernal equinox a bit earlier than if they were fixed with respect to the stars. This is why he called the phenomenon "precession of the equinoxes", or simply *precession.*

Hipparchus proved that a system of movable eccentrics was equivalent with the system of circular motions with deferents and epicycles, showing that both were able to predict Sun's and Moon's positions within the accuracy of 1′. His evaluation of the inclination of the ecliptic with respect to celestial equator fell within less than 5′ error of the correct value. Figure 4.5 below shows Hipparchus' explanation of the variable angular velocity along the orbit of celestial bodies (the Sun, the Moon or a planet) via the introduction of excentricity.

Hipparchus is also credited with mathematical inventions which have laid the foundations of trigonometry. In his astronomical works he constantly needed to translate the angles and circular segments into the lengths of corresponding chords. He produced an exhaustive table of chords corresponding to various angles, which were given in modern terms by the formula $C(\alpha) = 2 \sin \frac{\alpha}{2}$. Historians generally agree that the famous theorem of Ptolemy was originally established by Hipparchus.

Hipparchus did some important work in the field of geography. He was the first to use mathematics to characterize places on the Earth's surface with their latitude and longitude — notions used before exclusively for positioning objects in the sky. He proposed to determine latitudes by comparing the ratios of the longest to shortest

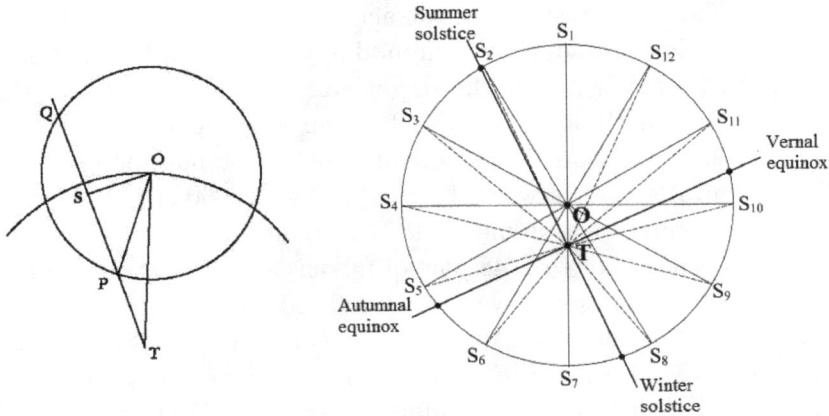

Fig. 4.5 Left: Construction of epicyclic motion by Apollonius; Right: The excentric motion with constant angular velocity creates the illusion of variable angular velocity if seen not from the center.

days and nights in the respective places. He also divided the world in climatic zones as we do it today.

- **Posidonius** (ca. 135–51 B.C.E.) was born in Hellenistic city of Apamea (now in Syria), but spent most of his adult life on the island of Rhodos. He followed the school of Stoics in Athens, but after several philosophical disputes turned to the philosophy of Plato and Aristotle, becoming an ardent follower of the latter. During Posidonius' life the Roman Empire included most of the Mediterrenian basin countries, Gaul and the Iberian Peninsula. Posidonius traveled a lot, and contributed to the geographical knowledge of his time. He observed giant tides on the Atlantic shore of Spain and formulated the hypothesis of the lunar origin of this phenomenon, although with a false explanation: he thought that the Moon was made of a mix of air and fire, its influence on the oceanic water causing it to swell.

Posidonius brought important improvements to astronomy. His estimates of the Moon's size and distance confirmed those of Aristarchus, while his evaluation of the Sun's distance from Earth was probably the closest to the correct one among all ancient astronomers: it was just half of the actual astronomical unit. He calculated the Earth's radius by a method independent of Eratosthenes' approach. He realized that the bright star Canopus cannot be seen on the island of Rhodos, but can be seen at as much as 7° above the horizon in Alexandria. The distance along the meridian between Rhodes and Alexandria was estimated to be of 5000 stadia. According to Posidonius' measurements, the difference in Canopus' observed altitudes corresponds to 1/48 part of full circle, so he concluded that the Earth's circumference was about 240 000 stadia.

If the stadium used by Poseidonius was worth 1/10 of the statute mile, i.e. 162.5 metres, the circumference obtained was 39 000 km, pretty close to the real value. Unfortunately, after Posidonius' death his writings were either distorted or misinterpreted. Strabo noted that the distance between Rhodos and Alexandria was only 3750 stadia, and substituting Greek units by Roman ones, underestimated Earth's circumference by almost 33%, reducing it to 180 000 stadia. This result was reproduced by Ptolemy and remained a quasi official standard.

4.8 Ptolemy and his system

Claudius Ptolemy lived in the Egyptian city of Alexandria from 100 till 170 C.E. or even till 178 according to some Arab commentators and translators of his writings. At that time Egypt was a Roman province ruled by the Greek dynasty bearing the same family name, Ptolemaios. Ptolemy was the last great representative of Greek science, heir to a tradition eight centuries old. Ptolemy lived and worked 400 years after Euclid, 600 years after Pericles and 700 years after Anaximander of Miletus who was the first to suggest the idea of the celestial sphere and to produce a map of the inhabited world known in his time.

Little is known about Ptolemy's life except that he probably spent it all in Alexandria, where he observed two solar eclipses: one in the ninth year of Hadrian's rule (125 C.E.) and in the fourth year of Antoninus' rule (141 C.E.). Besides his *Syntaxis Mathematicus*, known also as *magnum opus* due to its importance for astronomy, and mostly under the name of its arabic translation *Almagest*, Ptolemy was also a renowned geographer in his time. The geocentric system elaborated in *Syntaxis* was universally accepted during Hellenic and Roman antiquity, and later on by medieval Christian and Muslim civilizations alike, and was used to describe and predict motions of celestial bodies by astronomers and astrologists until the Copernican heliocentric system was confirmed by the revolutionary discoveries made by Galilei and Kepler in the beginning of 17th century.

Ptolemy spent twenty years working on his great publication. During the first fifteen years he gathered information on ancient astronomical observations, from Egyptian and Babylonian sources, as well as those of Eudoxus and Hipparchus. In order to produce a coherent model of motions of the Sun, the Moon and the five planets, he made his own observations: during the period from 127 till 141 he saw and described several eclipses, occultations of stars and planets by the Moon, conjunctions of planets with the Moon and with each other; he performed precision measurements of equinoxes and solstices and tracked planets' motions with respect to the fixed stars. In order to better determine the positions of the planets, Ptolemy produced a catalogue containing 1022 stars with their celestial coordinates specified.

All Ptolemy's observations were dated, and can be tracked back using today's modeling tools and mathematical models of the Moon's motion. Due to the incommensurability of planetary and lunar orbits, many conjunctions and occultations cannot occur in the same configuration before millenia. Ptolemy was aware of this when he wrote his volume devoted to astrology, insisting on the fact that most of astronomical events he observed will not repeat themselves in any foreseeable time.

According to the Ptolemaic model of the Universe, which was a creative generalization of previous geocentric models initiated by Eudoxus, developed by Aristotle and improved by Hipparchus, all heavenly bodies were attached to specific celestial spheres, rotating with various angular velocities around appropriate centers, the Earth being placed in the center of the Universe coinciding with the center of the sphere of fixed stars, the highest of all, and the ultimate border of the world.

Following Aristotle's philosophy, all celestial motions were perfect, i.e. not only the shape of the orbits must be circular, but also circular motion must be uniform, which means that angular velocity of such motion should remain constant. This assertion is in contradiction not only with apparent retrograde motions of the planets, but also with the non-uniform motion of the two principal luminaries. It was already known to Babylonian astronomers that the Sun travels along the Zodiac with variable angular velocity: slower in summer, and more rapidly during winter. The period extending from vernal to autumnal equinox is by 5 to 6 days shorter than the period between autumnal and vernal equinoxes, which can be easily checked by counting the number of days in each month.

As a result, the apparent progress of the Sun along the Zodiac is not uniform, which seems to contradict Aristotle's principle of perfect celestial motions. The motion of the Moon was even more irregular, but at least both luminaries advanced always in the same direction, Eastwards, completeing one full circle with respect to the fixed stars in sidereal year (the Sun) or sidereal month (the Moon). The planets posed an even more complicated problem due to their retrograde motion.

The Ptolemaic system responded to the challenge of reconciling the non-uniform and erratic motions of the two luminaries and five planets with the Aristotelian axiom of exclusivity of perfect circular motion in the skies.

The answer given by Ptolemy consisted in superposing several uniform circular motions, and admitting that centers of circular orbits did not necessarily coincide with the center of the Earth. Figure 4.6 displays the set of points and circles that are the basis of the Ptolemaic geocentric system. As a matter of fact, the adjective "geocentric" is only approximate since the center of circular motions did not coincide anymore with the center of the Earth. But the Earth was still suposed to be motionless.

To take the eccentricity of the orbits into account, a circle called the *deferent* was introduced, whose center O was displaced with

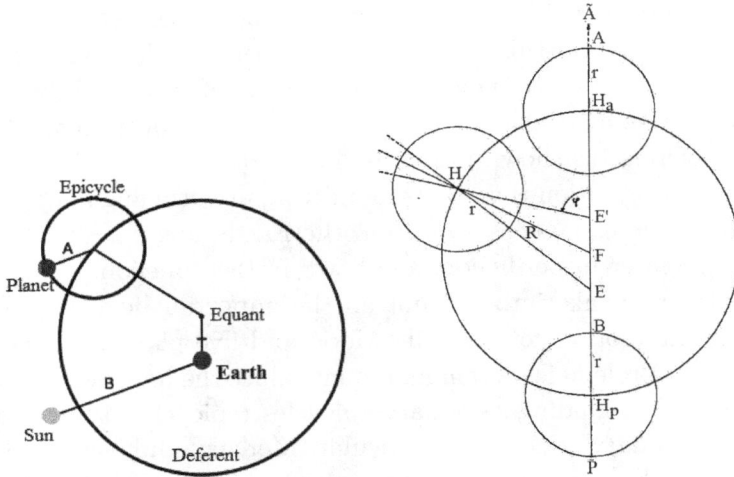

Fig. 4.6 Left: The basic elements of Ptolemaic model: Deferent, Epicycle and Equant. Right: The most important features of planetary motion reproduced: eccentricity (EF), retrograde motion (R), and variation of angular velocity between apogeum (A) and perigeum (B). Earth is at (E), equant is at (E').

respect to the Earth. This created automatically the closest and the most remote points of the orbit, called *perigeum* and *apogeum*, respectively. If the construction were stopped at this stage while maintaining uniform motion along the deferent, no retrograde motion of planets could be observed. To take it into account, an extra circular motion had to be introduced, along a supplementary circle called the *epicycle*, a construction invented a few centuries before by Apollonius of Perga (240 − 190 B.C.E.). An epicycle carried along a concentric circle reproduced retrograde motion and eccentricity at the same time, creating planetary positions closer to or farther from Earth. It was also possible to combine an eccentric deferent with an epicycle.

However, a problem remained with the variation of the apparent angular velocity of celestial bodies, which seemed to be in contradiction with the Aristotelian postulate that only uniform motions are allowed in the heavens. If planets orbit around the Earth along their own circles with uniform motion, but have epicycles, then, at least as seen from the Earth, planets do not travel at a constant angular speed. So high was Aristotle's authority that Ptolemy and his followers introduced another element in order to conform the model with Aristotle's postulate that only uniform circular motions were allowed in heaven. The new element, known later by its Latin name *"equant"*, was a point opposite to Earth with respect to the center of the deferent, as shown in Figure 4.6.

Now the situation is saved by admitting uniform angular speed as seen from the equant; as seen from the Earth, angular speed would become greater in perihelion, and lower in the aphelion.

In order to take into account all the important features characterizing the motions of the Sun, Moon and five planets, in its later version the Ptolemaic system had to introduce the impressive number of 80 circles, including secondary epicycles (epicycles orbiting along bigger, "primary" ones). In particular, Mercury and Venus have to remain on the line between Earth and the Sun, so their epicycles' centers rotate in an opposite direction at a different speed.

Many centuries later a Viennese astronomer Georg Peurbach (1423–1461) reduced this number to 40 without depreciating the agreement with observational data.

The Ptolemaic system was taught at schools and universities all over Europe until the end of the 17th century, when gradually the Copernican system prevailed, especially after Newton's Principia brought the ultimate proof of the Copernican model of our Solar System.

But there are still some places on our planet where the Ptolemaic, and not Copernican system, is not only used, but reproduced as a working mechanical model. These places are called *Planetariums*, and serve as wonderful tools facilitating better astronomical education. The first device of this sort was produced in 1923 in the optical factory founded by German scientific instruments maker Carl Zeiss (1816–1888).

A device projecting on a spherical dome spots of light imitating the visible motions of stars, the Sun and the Moon, and the five planets as seen from the Earth is equivalent to a mechanical realization of the Ptolemaic system. The projector must be placed in the center of the sphere imitating the sky (in practice, only the upper half of the sphere is represented by the dome of the planetarium).

This central point is fixed, like the Earth in the geocentic system. The observed motion of celestial objects is obtained using multiple gears and wheels, like in a clockwork — another feature of Ptolemaic system based on Aristotle's assumption that exclusively circular motions are realized in heaven. And of course only five planets are being displayed, the Earth being represented by the planetarium's center.

4.9 Science of Medieval Islam

At the end of fifth century C.E. the Roman Empire was in shambles. It was divided into the Western and Eastern parts already in 395 C.E., and prone to steadily increasing barbarian invasions ever since. The last Roman emperor Romulus Augustulus was disposed of in 476 C.E. by war Germanic chief Odoaker, who proclaimed himself emperor. The Eastern part of the Empire with its capital in Constantinople gradually transformed into the Greek-speaking Byzantian Empire. All these territories, including Egypt and North Africa up to Spain were Christian since the Emperor Constantinus proclaimed Christianity to become state religion. However, since the fall of the Western Roman Empire, most of these territories were deprived of any organized army and statehood, prone to barbarian invasions, among which the Huns, the Goths and the Vandals were the most terrific.

In the midst of seventh century C.E. a new religion was proclaimed by Muhammad in Arabia. Derived from Jewish and Christian traditions, simplifying them and professing with great conviction the uniqueness of God, it was an important revolution, because it replaced the tribal principle by a more general religious and social unity.

While Europe entered the era of Dark Ages, lasting roughly from 500 C.E. until the 12th century, the Arabs conquered vast territories creating a buoyant Islamic Civilization, which spread from Moorish Spain in the West, through Western North Africa, through Egypt and Mesopotamia, touching India and even the most western parts of China. The only precedent of such a rapid expansion was the Hellenistic empire built by Alexander the Great. Like ancient Greek, and

Latin in the Roman Empire, the Arabic language became universal among Islamic scholars, with Persian and Turkish as subsidiary in the East. Ancient Greek science was assimilated and developed further. Aristotle, Euclid and Ptolemy were translated and multiplied in many copies. Medical science of the ancient Greeks was also honored, and Hippocrates and Galen were translated, too.

The *Golden Age* of Islamic civilization covers the time span from the 9th till the 13th century C.E. in Baghdad under the *Abassid* dynasty, and in Moorish Spain until the 14th century; it lasted a little bit longer in Central Asia, even after falling under the Mongol conquest in the 13th century. During that time of splendor sciences flourished, especially mathematics, chemistry (called *al-khimiya*, or alchemy), and astronomy. A number of scientific terms of arabic origin are in use in Western science until today, and known to everybody: it is enough to cite a few, like *algebra, algorithm, zenith, azimuth, nadir, alcohol, alkali*.

Mathematics and astronomy in Islamic world developed in a larger context of cultural, philosophical, religious and scientific development, due to remarkable thinkers, physicians and polymaths, among whom the greatest ones were the following:

• **Al-Farabi** (..? − 950), known in Christian world under the latinized name *Alpharabius*. He spent most of his life in Baghdad. A prolific writer, he authored comments on Platonic and Aristotelian philosophy, and original works on alchemy, psychology, astronomy and music. His authority was so great in the Islamic world, that he was often called "The Second Teacher" — meaning second after Aristotle. His comments paved the way for the philosophical synthesis by Ibn Sina (Avicenna) and Ibn Rushd (Averroes). In his late years he moved to Damascus, where he died in 950 C.E.

• **Avicenna (Ibn Sina)** (..? − 1037), renowned physician and philosopher. His most appreciated work was the *Canon of Medicine*, a manual of medical science in many volumes, which became the authority for European doctors after it was translated into Latin. Among his lasting contributions were the drug testing method and the discovery of the infectiousness of tuberculosis.

• **Al-Biruni** (..? − 1048), a famous geographer and astronomer, who used Eratosthenes' estimate of circumference of the Earth in his astronomical observations and measuring stellar positions. He also wrote a geographic description of India, a source of new historical, botanical and zoological facts. Unfortunately, never translated into Latin before the modern times.

• **Averroes (Ibn Rushd)** (..? − 1048). Considered as one of the greatest Muslim scientists, he was born in Andalusia, then ruled by the Almoravid dynasty. His philosophical views became popular among some part of European medieval scholars; but his scientific opinions, especially on astronomy and physics, were rather retrograde.

As Bernard Lewis [Lewis (2002)] has put it in his book *What went wrong?*, "for many centuries the world of Islam was in the forefront of human civilization and achievement". However, due to many interior and exterior factors, Islamic science stopped its development around the 15th century; since then, European science took the lead.

4.9.1 *Mathematics in the Islamic world*

Mathematics became one of the most studied sciences in Arab and Islamic world. All of the most important mathematical books of ancient Greece, and above all the Euclid's great book "Elements", were translated into Arabic as early as in the 9th century C.E. by Al-Hajjaj Ibn Yusuf in Baghdad, under the Abbasid Kaliphate. Ptolemy's books on astronomy followed, translated by Al-Hajjaj, too, under the name of "Almagest". and studied diligently ever since. Speaking of "Arabic science" is rather reductive — the Arabic became universal, and the study of Quran compulsory, but many scientists were not Arabs, but often of Persian, Syriac or Egyptian origin.

The earliest acquaintance of the Islamic world with mathematics and astronomy was due to Indian and Persian influence. From India, the Arabs took the decimal system with numerals and zero which we use today. The main Indian mathematical text by Brahmagupta (598 − 670 C.E.) was translated into arabic. It contained, among others, the idea of negative numbers and their algebraic properties. They were called "debts", in opposition to positive numbers called "fortunes"; not only algebraic addition was defined, but also the multiplication, with well-known rules, including the fact that a product of two negative numbers is positive.

One of the most often repeated claims is that Muslims invented algebra. This is largely true, even if initially they took source and inspiration from ancient Greek and Indian mathematics. The word "algebra" comes from the arabic *al-jabr* and means "restoring" or "putting parts together". It was coined by the mathematician and astronomer al Khwarizmi (ca. 780–850 C.E.) who lived in the 9th century in Baghdad, in his treatise *"Ilm al-jabr wa al-muqabala"*, which means "Science of putting parts together and balancing". In modern algebraical terms, the first word referred to the possibility of replacing terms from one side of an equation to another, and adding or substracting the same quantity from both sides. Al-Khwarizmi's name refers to the province of Khorezm (today in Uzbekistan) and was known in the West in its latinized version *Alkorizm* or *Algorithm*, and became the name of a general prescription in mathematics.

Al Khwarizmi's book contained plenty of examples how to solve problems involving commerce, trade, inheritance, marriage and redemption of slaves. The examples did not involve any algebraic symbols yet, using geometrical figures modelling the relations between numbers. But in other Arabic mathematical treatises symbols, numbers and words tended to replace geometrical constructions. This was a major revolution in mathematics, with enormous impact on its further development.

For the mathematicians of ancient Greece a major difference existed between *numbers* and *magnitudes*, the second ones meaning measurable lengths, areas or volumes. Numbers were used as long as they could be expressed as integers or their ratios, e.e. fractions. These quantities were called "rational numbers", in contrast with "irrational ones", which could be constructed geometrically, but not algebraically, like the most famous square root of 2, which is simply the diagonal of a unit square, but no finite fraction exists whose square would be equal to 2.

The new approach to mathematics made possible simple and elegant proofs in place of sophisticated geometric constructions. For example, whereas the well known geometrical theorem saying that the area of a square with sides $a + b$ is equal to the area of a square with side a plus the area of another square with side b plus the area of two rectangles of sides a and b can be proved with no effort by an obvious geometric construction, the proof of another formula, that the area of a rectangle with sides $(a + b)$ and $(a - b)$ (with $a > b$) is equal to the difference between the areas of squares with side a and with side b, i.e. $(a+b) \cdot (a-b) = a^2 - b^2$, needs a more sophisticated, as shown in Figure 4.7.

Among the shortcomings of the purely geometrical approach of the ancient Greeks was the lack of interest in the possibility of negative numbers and arithmetical operations involving negative entities, which were known in ancient China since 300 B.C.E. Islamic mathematicians not only incorporated negative numers in algebraic operations, but also started to treat rational and irrational numbers on the same footing, making no distinction when adding, substraction of multiplying them.

The advantage of the algebraic approach is that it can be easily generalized beyond its usefullness in describing geometrical constructions in two and three dimensions. The formula (4.7) can be generalized to three dimensions geometrically, giving the volume of a cube with side $(a + b)$ as a sum of one cube with side a, another cube with side b, three parallelepipeds with base a^2 and length

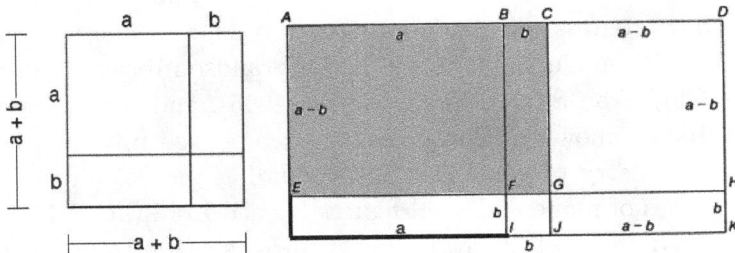

Fig. 4.7   Ancient Greeks' derivation of algebraic formulas $(a+b)^2 = a^2 + 2ab + b^2$ (left) and $(a + b) \cdot (a - b) = a(a + b) - ab - b^2 = a^2 - b^2$ (right).

b, and three parallelepipeds with base b^2 and height a, as follows: $(a + b)^3 = a^3 + b^3 + 3a^2b + 3ab^2$.

But nothing forbids to write down similar formulas for any power of the sum $(a + b)$, e.g. $(a + b)^4 = a^4 + 4a^3b + 10a^2b^2 + 4ab^3 + b^4$, and so on.

Nevertheless most of Muslim authors do not consider negative numbers on the same footing as the positive ones. They knew how to solve linear and quadratic equations, but whenever there was a negative solution, they rejected it as "absurd".

An important step forward due to Muslim mathematicians is the establishment of new arithmetic algorithms: the extension of root-extraction procedures, known to Hindus and Greeks only for square and cube roots, to roots of higher degree and by the extension of the Hindu decimal system for whole numbers to include decimal fractions as computational devices. Omar Khayyam (1148 − 1131 C.E.), Persian mathematician, philosopher, astronomer and poet (remembered more for his beautiful poetry than for mathematical achievements), considered the general problem of extracting roots of any desired degree. He also closely approached the general solution of cubic equations, and elaborated a unifying approach to geometry and algebra.

Islamic algebraists of the 10th century made a substantial progress, generalizing al-Khwarizmi's quadratic polynomials to the algebra of expressions involving arbitrary positive or negative integral powers of the unknown. Several algebraists explicitly stressed the analogy between the rules for working with powers of the unknown in algebra and those for working with powers of 10 in arithmetic.

Similar progress was made in geometry. The Islamic mathematicians Thabit ibn Qurrah (836 − 901), his grandson Ibrahim ibn Sinan (909 − 946), Abu Sahl al-Kuhi (died ca. 995), and Ibn al-Haytham (965 − 1040), known in Europe as Alhazen, solved problems involving the geometry of conic sections, including the calculus of areas and volumes of plane and solid figures formed from them. They also investigated the optical properties of mirrors made from conic sections, which became of crucial importance centuries later, for the construction of optical devices.

4.9.2 *Arab and Islamic astronomy*

During the "Golden Age" of Islam, astronomy was one of the most important sciences. It helped to impose on Muslim religion many requirements which needed serious astronomical skills in order to improve time-keeping and space orientation. Exact timing of five obligatory prayers a day, beginning and the end of fast during the sacred month of Ramadan, linked to the first and last crescent of the Moon, establishing times of islamic holidays — all that created the need for astronomical observations. Mosques had to be oriented towards Mecca in the Arabian Peninsula, and so should a Muslim be oriented during his prayer — this alone needed improved geographical and astronomical knowledge, too. Finally, the vast territories of Islam, with flourishing terrestrial and maritime trade, made necessary the emergence of practical astronomy, improving observational devices like sextants and astrolabiums, to be used on land or in the open sea.

How great the impact of the Arabs was on European astronomy that took the lead after the 16th century can be easily seen in any modern stellar atlas. Here are a few among the best known bright stars whose names are of Arab origin:

- Aldebaran, α Tauri, from arabic *Al-Dabaran*, the "Follower."
- Algol, β Persei, from arabic *Al Ghoul*, the "Demon", or "Scary Monster". Named so probably because its periodical blinking in matter of hours could be easily seen by a naked eye.
- Altair, α Aquilae, from arabic *an-Nisr ut-Ta'ir*, the "Flying Eagle".
- Betelgeuse, α Orionis, from arabic *Yad al Jawza*, the "Hand of Al-Jawza", a mythical character.
- Deneb, α Cygni, from arabic *Dhaneb ud-Djadjab*, "Tail of the Hen".
- Dubhe, α Ursae Majoris, from arabic *Dubb*, "Bear".
- Fomalhaut, α Piscis Austrini, from arabic *Fum al-Hul*, the "Mouth of the Whale".

Arab astronomers continued observations using the techniques inherited from the Greeks and incorporating ancient traditions from

Persia and India. They were often able to improve and enrich the
data collected by Ptolemy in his *Almagest*. Following Hipparchus,
Ptolemy estimated precession of the equinoxes to be of 1° per 100
years, which would correspond to completing the full cycle in 36 000
years. Egyptian astronomer Ibn Yunus (850 − 1009 C.E.) corrected
this estimate, reducing it down to 1° in 70 years, which corresponds
to the full cycle of 25 200 years, very close to the modern value.
This is one of the countless examples of Arab and Islamic astronomical achievements; we shall enumerate below the most remarkable
astronomers and their contributions.

Most eminent Muslim astronomers include Al-Battani, Al-Sufi,
Al-Biruni, Al-Bitruji and Ibn Yunus.

- **Al-Khwarizmi** (ca. 786 − 850 C.E.), of Persian origin, was
appointed by al-Mamoun, the son of Harun-al-Rashid of the Abbasid
dynasty, as the chief astronomer and librarian in in Baghdad.
Al-Khwarizmi's important strides in astronomy and geography
include the measurement of the length of a degree of a meridian in
the plain of Sinjar, improving Eratosthenes' evaluation of the Earth's
circumference, the creation of a world map based on the geography
of Ptolemy, providing more accurate coordinates of approximately
2400 sites in the known world. These results were contained in his
book *Kitab surat al-ard* ("The Image of the Earth").

Another text of al-Khwarizmi that made it into the Western canon
of mathematical studies was a compilation of astronomical tables.
This included a table of sines, and was translated into Latin. He also
produced two treatises on the sundial, on the astrolabe, and one on
the Jewish calendar,

- **Al-Battani** (died in 929) known to Europe as Albategni or
Albatenius was the author of the *Sabian tables* (al-Zij al-Sabi), a
work which had great impact on his Muslim and Christian successors.
His improved tables of the orbits of the sun and the moon comprise
his discovery that the direction of the sun's eccentric as recorded
by Ptolemy was changing. This, in modern astronomy, means that
the Earth's line of apsides is slowly moving. He also worked on the

timing of the new moons, the length of the solar and sideral year, the prediction of eclipses, and the phenomenon of parallax.

Al-Battani was also a pioneer in the field of trigonometry. He was among the first, if not the first to use trigonometric ratios as we know them today. During the same period, Yahya Ibn Abi Mansour had completely revised the Zij of Almagest after meticulous observations and tests producing the famous Al-Zij al Mumtahan (the validated Zij).

• Belonging to the same era, **Abd-al Rahman al-Sufi** (903 − 986) made several observations on the obliquity of the ecliptic and the motion of the sun (or the length of the solar year). He became renowned for his observations and descriptions of the stars, their positions, their brightness and their color, setting out his results constellation by constellation. For each constellation, he provided two drawings, one from the outside of a celestial globe, and the other from the inside (as seen from the sky). Al-Sufi also wrote on the astrolabe, finding numerous additional uses for it (including where one is located, measuring distances and heights...).

• **Ibn Yunus** (d. 1009), in his observational endeavours included, amongst others more than 10 000 entries of the sun's position throughout the years using a large astrolabe of nearly 1.4 m in diameter. His work, in a French edition, was centuries later an inspiration for Laplace in his determination of the "Obliquity of the Ecliptic" and the "Inequalities of Jupiter and Saturn".

Ibn Yunus made observations for nearly thirty years (977 − 1003) using amongst others a large astrolabe of nearly 1.4 m in diameter, determining more than 10 000 entries of the sun's position throughout the years. The famous European astronomer Newcomb also used his observations of eclipses in the motions of the moon.

• **Al-Biruni** (973–1050) used the knowledge of the Earth's circumference for fixing with precision the direction of Mecca from any point of the globe. Al-Biruni wrote in total 150 works, including 35 treatises on pure astronomy, of which only six have survived. In the late 10th century Abu-al-Wafa and the prince Abu Nasr Mansur stated and proved theorems of plane and spherical geometry that could be applied by astronomers and geographers, including the laws of sines and tangents.

Al-Biruni, who was their pupil, became one of the masters in applying these theorems to astronomy and to problems in mathematical geography e.g. the determination of latitudes and longitudes, the distances between cities, and the direction from one city to another.

- **Al-Farghani** was one of Caliph Al-Mamun's astronomers. He wrote on the astrolabe, explaining the mathematical theory behind the instrument and correcting faulty geometrical constructions of the central disc, that were current then. His most famous book *Kitab fi Harakat Al-Samawiyah wa Jaamai Ilm al-Nujum* on cosmography contains thirty chapters including a description of the inhabited part of the Earth, its size, the distances of the heavenly bodies from the Earth and their sizes, as well as other phenomena.

- **Al-Zarqali** (Arzachel) (1029 – 1087) prepared the Toledan Tables and was also a renowned instrument maker who constructed a more sophisticated astrolabe: a *safiha*, accompanied by an explanatory treatise.

- **Jabir Ibn Aflah** (d. 1145) was the first to design a portable celestial sphere to measure and explain the movements of celestial objects. Jabir is specially noted for his work on spherical trigonometry. Al-Bitruji's work *Kitab-al-Hay'ah* was translated by the Sicilian based Michael Scot, and bore considerable influence thereafter.

- **Nur ad-Din al-Bitruji** (ca. 1150 – 1200 C.E.), known in Europe as *Alpetragius* is the author of *Kitab fi al-haya*, "A Book on Cosmology". He lived in Moorish Spain, in Andalusia, at the end of the Islamic Golden Age. He was probably a disciple of astronomer Ibn Tufayl.

The problem faced by al-Bitruji was that faced by all Aristotelians who read Ptolemy's Almagest. Aristotle clearly stated that the planets must move with circular motions and implied that the center of these motions must be identical with the center of the Earth;

he further desired a mechanism to transfer the motion of the prime mover to the planetary spheres. Ptolemy, on the other hand, while preserving the principle of circular motions (on eccentrics and epicycles), placed the centers of these motions elsewhere than at the center of the earth; for Saturn, Jupiter, Mars, Venus, and Mercury he placed the centers of their uniform motions not at the centers of their respective eccentric deferents but at points called equants.

Eudoxus of Cnidus had already shown that it is theoretically possible to explain the two most obvious anomalies in planetary motion retrogression and latitude by means of homocentric spheres. Aristotle, by adding more spheres, converted this system to a mechanical model of the universe (though technical details make it impossible for such a model to yield correct predictions of the retrogressions and latitudes of Mars and Venus). Al-Bitruji followed the suggestion of Ibn Tufayl, as did the latter's other pupil Averroës, and attempted to adjust the Aristotelian solution in such a way that it would correspond to observed reality. The attempt failed owing to the inherent inadequacy of the homocentric system to describe the phenomena.

- **Al-Tusi** (1201 – 1274), was the last great astronomer of Islamic Golden Age, and without exaggeration can be given the title of "Hipparchus of Islamic Astronomy". Of Persian origin, he studied in his native town Tus, then the nearby town Nishapur, educated in philosophy, medicine and mathematics. Already in Nishapur al-Tusi acquired a reputation as an outstanding scholar, and proposed himself to be a member of the Ismaili Court, to the shi'ite ruler And ar-Rahman. Al-Tusi stayed and worked in the Alamut castle until it fell to the troops of Dzhengis Khan's grandson Hulagu. Al-Tusi's reputation was so great that Hulagu named him his own advisor, and took him along to Baghdad, which was conquered by the Mongols in 1258. Hulagu made Maragheh (in northwestern Iran) his capital, and the Observatory of the same name was built there. Under al-Tusi's direction, it has become a renowned center for mathematiical and astronomical studies in the Islamic World. Al-Tusi designed many unique astronomical instruments, including a giant *azimuth quadrant* with radius of 4 meters, and an improved astrolabe.

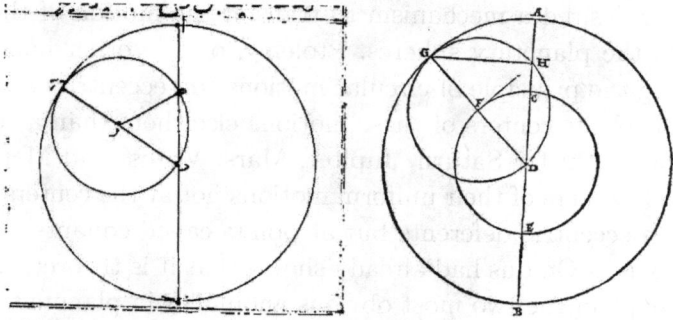

Fig. 4.8   Left: The Tusi couple, from his *Memoir on Astronomy*; Right: The same construction from Copernicus' *De Revolutionibus* 250 years later.

Al-Tusi is mostly known for his geometric decomposition of linear motion into a sum of two circular motions, called "the Tusi couple" (Figure 4.8), shattering the established Aristotelean view on exclusivity of circular motions in the heavens. The same construction was used by Copernicus 250 years later, probably re-invented independently. Al-Tusi published the *Ilkhanic Tables*, relating his atsronomical observations made during 12 years. Using his geometrical findings, he improved substantially the model of lunar motion. He also determined almost exactly the value of precession of the equinoxes, establishing it at 51' per century. Al-Tusi's *Commentary on the Almagest* contained excellent trigonometric tables with values of chords calculated to three sexagesimal places for each half degree of argument.

With all its remarkable achievements, the Arab and Islamic science represent the last chapter of the science of Antiquity, a natural extension of the Babylonian, Indian, and above all Greek scientific heritages of a small world centered on the Earth, whose boundary was the sphere of fixed stars.

Modern science was germinated within the new university framework of Medieval Europe between the 13th and 15th centuries, openly born with the intellectual revolution brought by Copernicus in the 16th century, and then pursued by Descartes, Galileo, Kepler, Leibniz, Newton, and other European luminaries.

Chapter 5

Observers and observatories

Ancient astronomical observations - First observational devices - Nautical astronomical tools - Sextants, quadrants, astrolabes - Medieval Islamic observatories - Telescopes - European observatories: Uraniborg, Danzig, Paris, Greenwich

5.1 Preamble

Initially, astronomical observations were made without any instruments, just with the naked eye. The positions of the rising and setting Sun, or the passing of the brightest stars and planets through the meridian, could be determined using huge stone structures oriented in an appropriate way.

It is commonly thought that the giant circular megalithic structure called *Stonehenge* found in Wiltshire, England, is the oldest astronomical observatory, erected between 3000 and 2400 B.C.E.; however other hypotheses have been made, including the idea that it was a place of worship and healing.

In the ancient world astronomical observations were performed by temple servants and scribes, who used to climb the roofs of temples in Egypt or specially built massive structures called *ziggurats* in Mesopotamia. These buildings hardly deserve the name of astronomical observatories; all that was needed was to ensure an elevated platform from which a total view of the sky is possible, including

Fig. 5.1   The starry sky and the Great Pyramid in Giza, at different night hours, as seen today.

the parts near the horizon. The most important goal was to enable time-keeping.

The starry sky provided a perfect celestial clock supplemented with calendar. Many important ancient structures were oriented in such a way that they could serve as markers on the celestial dial. The three great pyramids in the valley of Kings in Giza were oriented according to four cardinal directions, their sides facing respectively North, East, South and West. In Figure 5.1, the North Pole star is the Polaris ($\alpha\,Ursae\,Minoris$), but at the time when Great Pyramid was erected, it was Thuban ($\alpha\,Draconis$) which indicated the North and was right above the Pyramid as seen from its southern side. Since then, the precession of Earth's axis resulted in the modification of the positions of circumpolar stars to the actual configuration. In about 9000 years from now the star Vega (α Lyrae) will replace the Polaris, taking its place near the North Pole.

The priests in ancient Egypt were in charge of predicting as precisely as possible the next flood of the Nile, crucial for agriculture: peasants had to plant millet on time in order to ensure a better harvest a few months later. The best celestial time marker was the *heliacal rising of Sirius*. The brightest star in the sky, Sirius was periodically invisible when the Sun in its annual journey through the sky was passing close to the *Canis Major* ("Great Dog") constellation to which Sirius belongs. The period during which Sirius cannot be seen starts each year at vernal equinox and lasts about 70 days. The heliacal rising of Sirius occurred in the beginning of June.

In ancient Babylon, the observations were performed by Chaldeians from the top of pyramidal constructions called *ziggurats*. They were able to determine with astonishing precision the celestial coordinates of stars and planets, and follow the Sun's annual and the Moon's monthly motions. As in ancient Egypt, timekeeping was one of their most important functions, as well as interpreting omens from planetary configurations and other noticeable celestial events, especially solar and lunar eclipses, some of which they successfully predicted.

5.2 First observatories

Ancient Greek astronomers performed measurements of stellar and planetary positions using simple devices which were the predecessors of sextant and astrolabe. Hipparchus is credited with invention of first astrolabe (some historians claim that it was rather a version of the armillary sphere), which he used for astronomical observations performed on the island of Rhodos. His collection of astronomical tools was still modest and hardly deserves the name of observatory.

Ptolemy quotes in his *Almagest* a description by Hipparchus of an equatorial ring in Alexandria; a little further he describes two such instruments present in Alexandria in his own time.

Impressive astronomical observatories were built by the Mayans followed by the Aztecs in ancient Mexico. According to recent estimates, they were built in the 9th century C.E. Similar constructions were found in the pre-Columbian culture of the Incas, now in Peru. In a well conserved site of Xochicalco one can admire two exceptional astronomical devices: the stone structure oriented in such a way that the exact time of vernal and autumn equinox could be observed within a precision of one day, and a natural "camera obscura" realized in a cave with special opening in its ceiling, to let Sun rays fall in only during certain period of the year.

However, one of these entrances leads to a cave that was possibly man-made and certainly adapted to allow the Sun to enter. By manipulating a crack leading up to the floor of the Central Plaza above, the people of Xochicalco created an incredibly restrictive zenith tube measuring 8.7 m in length, that allowed the Sun to cast a brilliant beam on the floor when it moved directly overhead on its transit north, on May 14 or 15 and again on its way back south, on July 28 or 29. Surrounded by sophisticated temple alignments, and panoramic views of the horizons, there seems little need to manipulate a cave to identify the zenith, but there are three interesting phenomena that occur only within the cave: firstly, by reducing the shaft entrance to little more than a pinprick, it has been demonstrated that the zenith tube will work as a camera-obscura and allow detailed observation of the Sun and its physical features, such as solar flares and Sun spots; secondly, as the Sun moves west overhead, the round beam of light moves east upon the cave floor, as if being reborn through the dark underworld; thirdly, direct light only enters the cave for 105 days, between April 30 and August 15, which leaves 260 days of darkness, the exact number of days in the sacred-calendar. The cave was apparently plastered and painted in black, yellow and red, the colors of the Sun, which leaves little doubt that the observatory was designed to play an important role in the religious calendar and that it was probably devoted to rituals surrounding the rebirth of the Sun.

5.3   Ancient astronomical devices

Ancient nations of sailors and merchants, Phoenicians and Greeks in the first place, needed devices providing quick and reliable orientation on the open sea, where no characteristic visible markers exist except for the starry sky, the Sun and the Moon.

• **The Kamal**. This was the simplest device measuring angular distances between the horizon and a given star, which in practice meant the Polaris star. The angular distance of Polaris from the horizon determines the latitude of the place — in fact, *is* the latitude. The same procedure was used to measure the altitude of the Sun at noon or the altitude of Polaris star at night. It is hard to imagine a device simpler than this one. It was a rectangular piece of wood threaded through with a string. It was used by Arab and Indian sailors before the invention of the compass and the sextant, of which

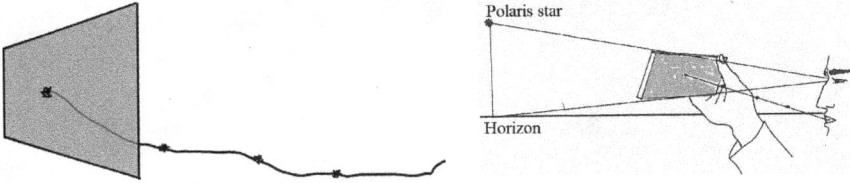

Fig. 5.2   The Kamal, and how it was used.

the kamal was a primitive prototype. The string had a number of knots marking different corresponding latitudes. A navigator would keep one of these knots between his teeth, and the piece of wood in his hand at arm's length. If the knot agreed with the latitude of the place, the upper edge of the kamal would coincide with Polaris star while its lower edge would coincide with the horizon. The right length of the string was found by trial and error, until the proper knot was identified.

• **The Gnomon**. Gnomons were simple vertical sticks, casting its shadow on a flat surface on the ground or on a specially prepared piece of wood playing the role of a dial. Gnomons were used for determining local noon by observing the time of the shortest shadow, and incidentally, the exact local south direction, opposite to the shade at noon. They were often set in the center of a circular plate, calibrated with degrees like a compass, to improve the reading of the shadow's position. On the sea, they could be floated on water if the weather was calm enough.

• **The Jacob's staff**. The *Jacob's staff*, or *Cross-staff* was widely used during the Middle Ages. Its improved conception is attributed to the Jewish scholar and theologist, Rabbi Levi Ben Gershom (Gersonides) (1288–1344). More practical and precise than the kamal, it was no match to the sextant and astrolabe which gradually replaced it.

• **The Sextant**. Sextants served for astronomical observations on land and on sea. Simple devices covering a sixth of a circle, i.e. a 60° sector. They were used primarily for measuring the positions of stars and planets by determining angles between the directions of

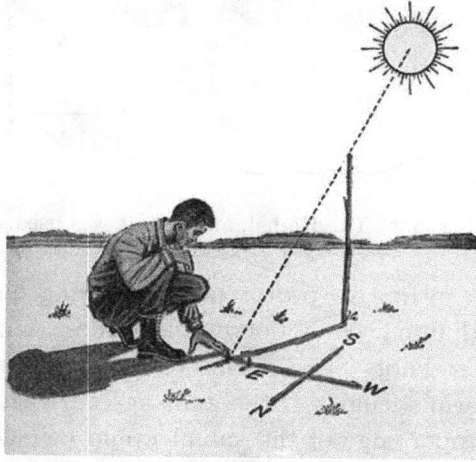

Fig. 5.3 The gnomon.

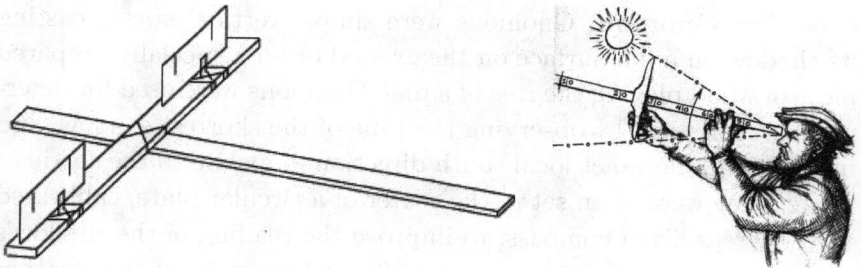

Fig. 5.4 Left: The cross staff, or Jacob's staff; Right: How to use Jacob's staff for evaluation of Sun's altitude.

sight and the horizon, or an angle these directions formed with the meridian or with certain characteristic points in the sky.

The sextants used by sailors were not as precise as the huge mural sextants used in astronomical observatories built on stable foundations and calibrated up to a few arcseconds. A marine sextant is an essential tool for celestial navigation and is used to measure the angle between the horizon and a visible object, or two objects at sea. The working principle of marine sextant is shown in Figure 5.5. One ray of light is reflected by a plane mirror, and the angle of the incident

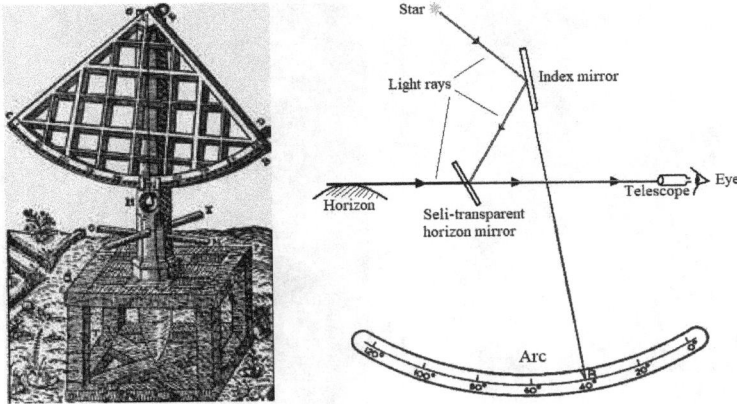

Fig. 5.5 Left: Huge stationary sextant; Right: The construction of marine sextant.

ray is equal to the angle of the reflected ray. When a ray of light undergoes two successive reflections in the same plane by two plane mirrors, the angle between the incident ray and the reflected ray is twice the angle between the mirrors. This is why marine sextants can measure the angles up to 120°, the double of the instrument's arc. The precision of marine sextants was substantially improved when micrometers, based on the principle discovered by Pedro Nunes, were added in order to control the angle between the mirrors.

The observer looking through the small telescope sees two images at once: the horizon directly through the semi-reflecting mirror, and the sun (or the Polaris star) reflected by the inclined upper mirror. By adjusting the angle between the two mirrors, the image of the Sun appears so that le line of the horizon cuts it in halves; the pointer shows then Sun's exact altitude at the time of observation.

• **The Quadrant**. A quadrant was in principle an enlarged version of the sextant, serving a similar goal: to determine with the maximal precision available the angular distances between celestial objects, the only difference being that they covered the entire 90° instead of the 60° of sextants. Mural quadrants were often preferred because they occupied the entire right angle between the floor and the adjacent wall, making their construction easier and the angle

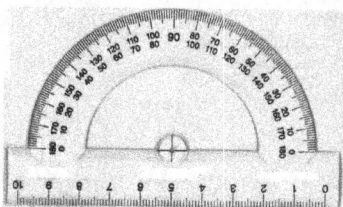

Fig. 5.6   Left: A protractor used in schools during geometry lessons; Right: The great mural quadrant in Tycho Brahe's observatory Uraniborg.

control more precise. Huge mural quadrants erected by Ulugh-Beg and Tycho Brahe in their respective observatories had radii close to 40 meters (Figure 5.6). It is easy to evaluate the gain in precision offered by increasing the size of the instrument. A usual scholar protractor has a radius about 5 cm (two inches), so that the one degree marks on its rim are separated by less than 1 millimeter. The exact value is easily computed by dividing the length of one radius by the number of degrees in one radian. A radian is worth $360°/2\pi = 57.3°$. The radius of the rim of the protractor shown in Figure 5.6 is 4.5 cm, so that one degree on its rim is $4.5/57.3 = 0.0785$ cm $\simeq 0.8$ mm wide.

Taqi al-Din, the builder of the observatory in Istanbul, used a sextant for the determination of the equinoxes.

Tycho Brahe used a sextant for his stellar position measurements.

Johannes Hevelius used a sextant with a particularly ingenious alidade to provide stellar position measurements of great accuracy.

John Flamsteed, the first Astronomer Royal, used a sextant at the Royal Greenwich Observatory.

Christopher Columbus used a number of important tools in his navigation across the Atlantic Ocean: a quadrant, maps, a sandglass, and an astrolabe. And of course the compass, which caused some trouble when the North direction it indicated diverged from the North fixed by the Polaris star. Columbus noted the discrepancy in his log, when he sailed progressively to the West, mentioning the beginnings of panic among the crew. In fact, Columbus was the first one to discover the difference between the astronomical North and the direction towards the northern magnetic pole, although he could not understand the phenomenon.

It seems that neither the quadrant, nor the astrolabe were of much help in determining latitude: according to the measurements of Columbus the islands Hispaniola (Dominica) and Cuba were at latitudes 42° and 34° North while their real locations are much closer to the equator.

- **The Astrolabe**. The word "astrolabe" comes from the Greek *astro* (star) and *labein* (taking) — "a star-taker". It represents a sophisticated inclinometer, more sophisticated than a sextant, serving to evaluate various angles and angular distances between celestial objects. It was also calibrated as a small planetarium, enabling to find the positions of major stars and planets and identify them. It could serve many other purposes, too, like calculating distances, the position of the local meridian, find local directions toward Holy Places, and producing horoscopes. It may be said without exaggeration that the astrolabe was a prototype of analogue calculators. In Figure 5.7 the astrolabe's main parts are shown. These are:

- **1** — the *mater*, "mother" in Latin, serves as a base of the device, making it possible to easily hold the astrolabe vertically when pointing a star or the Sun with the *alidade*. In old astrolabes it was usually decorated and is equipped with a ring or a cord.

- **2** — the *limb* surrounding the base is scaled in hours and in degrees by quadrant. It gives a value when aligning the ruler.

- **3** — the *womb* is the bottom of *mater*, over which an engraved plate can be placed. The engravings contained a coordinate system and the positions of prominent stars suitable for a given latitude. The circles of altitude above the horizon and circles of azimuth,

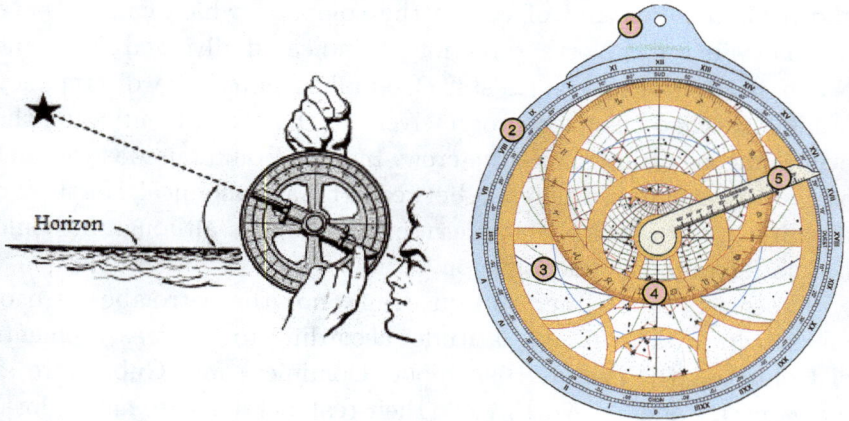

Fig. 5.7   Left: An astrolabe in use; Right: Schematic representation of an astrolabe, with main parts displayed.

arcs of twilight and sometimes arcs of unequal hours were also displayed. An astrolabe was usually provided with 2 or 3 interchangeable plates engraved on both sides for different latitudes.

Each plate came engraved with two kinds of circles. The first were circles of constant altitude called *almucantars*, with the horizon being the most important one. The second were *azimuths*, which met the almucantars at right angles. The most important azimuth was the meridian.

- **4** — the *rete* is a disc with many holes through which the plate can be seen. It can rotate around the axis and has pointers for the main stars. The ecliptic circle is scaled in ecliptic longitude and usually marked with Zodiac signs.

- **5** — the *ruler* makes it possible to point to a value on the limb or a position on the plate.

- **6** — not visible in the figure, the *alidade* is another ruler, on the back side of the bottom, to observe star's positions. In late versions it was endowed with a miniature telescope. On some astrolabes, a clocklike hand called the *rule*, marked with declinations from −30° to +70°, lay on top of everything. A pin passed through the center of the instrument, holding all of the pieces together, but allowing the rule and the rete to rotate over the plate.

• **Armillary Sphere.** The armillary sphere is a three-dimensional version of an astrolabe. It looks like a sphere circled by a number of rings, all set upon a base. Armillary spheres represented the visible sky as seen by terrestrial observer, often with the globe at its center, accordingly to the Ptolemeian View of the world. Much later Copernican armillary spheres were produced, with the Sun in their center.

Hipparchus reports that the armillary sphere was invented by Eratosthenes; certain historians attribute its invention to the Chinese, who used it since ancient times, and most certainly independently of the ancient Greeks. From the Middle Ages on, armillary spheres were widely used as navigation tools by the Portuguese, Spaniards and throughout the Islamic world. It was widely used either as a navigation tool, or for pedagogical purposes.

The devices were made with different numbers of circles arranged at various angles, with the possibility of turning them around the

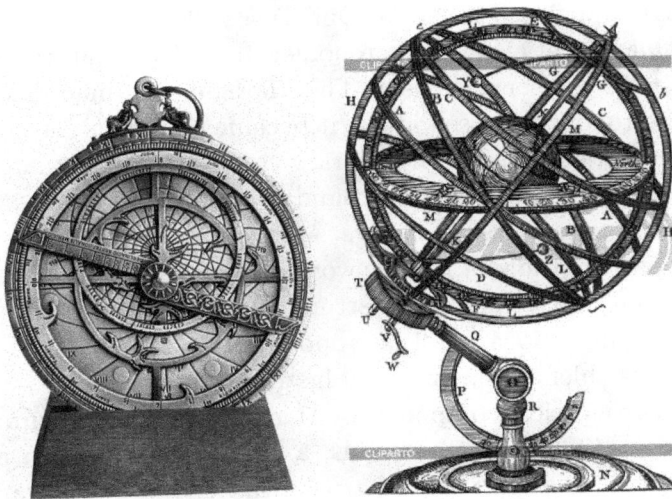

Fig. 5.8   Left: A brass astrolabe, XVIII-th centure; Right: Armillary sphere, with nine circles: the horizon, the equator, the ecliptic, two tropics, the local meridian, the equinox meridian, and two polar circles. Credit: Wikimedia Commons.

axes provided by nuts. Spheres with four up to nine circles have been known to exist, as well as ones with other numbers. These rings would then be adjusted in order to trace the path of heavenly bodies. An armillary sphere in its minimal version would comprise circles representing the celestial equator, the Zodiac (i.e. the ecliptic circle), the local horizon and meridian. More sophisticated versions would display the circles representing the Tropics of Cancer and Capricorn, and the arctic and antarctic parallels. Sometimes the equinoxial and solsticial colures, i.e. the great circles passing through the equinoxes and through the solstices, respectively, were added, too.

5.4 Medieval Islamic observatories

5.4.1 *The Maragha Observatory*

At the time of decline of science in medieval Islam, when Baghdad ceased to be the flourishing cultural center after the Mongols sacked it in 1258 thereby ending the reign of the Abbasides, a new influential institution called the Maragha Observatory, was created in northern Persia under the patronage of Hulagu, Gengis Khan's grandson.

Hulagu appointed Nasir al-Din Tusi (1201–1274), an eminent Persian mathematician and astronomer, to be Observatory's director as well as his personal advisor. The Maragha was much more than just an astronomical observatory: it became a unique center of scientific research, endowed with a library containing more than 100 000 volumes, the main building surrounded by auxiliary ones, including accomodation quarters for visitors. Astronomers and mathematicians from all parts of the vast Islamic world participated in the design and construction of astronomical instruments, many of which were genuinely new inventions. Al Tusi appointed Mu'ayyad al-Din al-Urdi (d. 1266) as chief astronomer and instrument designer.

The astronomical equipment of Maragha included a mural quadrant with a 40-meter large radius, a solstitial armilla, an azimuth ring, a parallactic ruler (called *triquetrum* in Latin), and an armillary sphere with a radius of about 160 cm.

The Maragha Observatory represented a new period of scientific activities in the Islamic world in the mid 13th century. A number of sophisticated pre-Copernican, but non-Ptolemaic systems were elaborated there, explaining the planetary motions with greater accuracy than ancient Greek astronomers were able to produce. Several observatories in Persia, Asia Minor and Central Asia were built with Maragha serving as a model.

5.4.2 *Ulugh Beg*

Ulugh Beg was born in northern Iran in 1394, as the oldest son of Shahrukh, one of Tamerlan's sons. As a young prince, he was designed to rule over the vast province of Transoxania (today in Uzbekistan). He assumed his full responsibilities in 1411, although he continued to be subordinate to his father, who ruled the empire from Herat (today in Afghanistan). After his father's death in 1447, Ulugh Beg succeeded him, but survived only two years as an independent ruler before being overthrown and beheaded by his own son in 1449. He is remembered mostly for his extraordinary achievements in astronomy, and the construction of the greatest and the most modern astronomical observatory and *madrasa*, a high school, a library and a reaserch center combined. The first director of Ulugh Beg's observatory was Qazizadeh Rumi, a native of Anatolia, who was also one of Ulugh Beg teachers. After his master's death he left Samarkand for Istanbul, followed by other astronomers employed in the madrasa.

Ulugh-Beg's observatory was built from 1420 till 1430 on a hill to the north of Samarkand, and was in use until his death in 1447. Its exact location is unknown, because it was almost completely destroyed a few generations after his death. Ulugh Beg constructed the largest mural sextant (or it might have been rather a quadrant) in the 15th century, which had a radius of 40.4 meters. He took inspiration from the first known mural sextant constructed in Ray, Iran, by Abu-Mahmud al-Khujandi in 994, to measure the obliquity of the ecliptic.

Housed in the main building, it had a finely constructed arc with a staircase on either side to provide access for the astronomers who performed the measurements. It would have been used to measure the angle of elevation of major heavenly bodies, especially at the time of the winter and summer solstices. Light from the given body, passing through a controlled opening, would have shone on the curved track, which was marked very precisely with degrees and minutes. Fainter objects were observed directly and the elevation angle reported to the sextant. Travelers and scientists who visited Samarkand and the observatory have described many astronomical instruments which were lost afterwards. There were several armillary spheres of various sizes and huge astrolabes, among others.

Ulugh Beg was certainly the most important observational astronomer of the 15th century. The obliquity of the ecliptic measured by his astronomers was almost perfect, within half a minute discrepancy with what we know at present. His permanently mounted astronomical instruments were first of their kind in the world, and it was their firm position that ensured greater precision of observation. Among his great achievements was the star catalogue containing more than thousand stars (1018 precisely), with exact coordinates attributed to each of them. His masterpiece *"Chronology"* was translated into Latin and edited in Europe in 1650, and the publication of his star tables followed in 1675. The determination of the obliquity of the ecliptic was much better than those of Hipparchus and Ptolemy. His observations of planets remained unsurpassed until Tycho Brahe's more than a century later.

5.5 Europe takes over

For various reasons, scientific and cultural activity in the Islamic world began to decline at the end of the so-called "Golden Age", i.e. in the 13th century, the sack of Baghdad by the Mongols in 1258 marking the turning point. By about the same time (1259) the Portuguese reconquista recovered the Algarve, in the southern part of the Iberian peninsula, while Spain recovered most of the territories under Muslim rule since the 8th century. The Berber dynasties of Almoravids and Almohads which came from the northern part of Africa resisted the reconquista, but were much less interested in scientific endeavours.

In parallel, the first European Universities were founded in Bologna (Italy, in 1088), Salamanca (Spain, in 1134), Paris (France, in 1150), Oxford (England, in 1167), followed by dozens of others in Italy (Padova, Naples), France (Toulouse), Portugal (Coimbra), Spain (Valladolid), Cambridge (England), so that by the end of the 14th century Europe was literally covered by a network of academic schools of the highest quality, where young people studied Medicine, Law, Theology and Sciences (called "Arts", and later on, "Philosophy"). The four types of doctorates corresponding to those four faculties remain still in use. Astronomy in its Ptolemaic version was taught as a part of mathematics; the works of Arab astronomers were translated into Latin and commented upon.

The need for better timekeeping, for religious and economical reasons alike, motivated the rulers to employ court astronomers and mathematicians, and to sponsor constructions of astronomical observatories hosting huge instruments with solid ground supports. The first among these centers of astronomical studies was created by the Landgrave Wilhelm IV of Hesse-Kassel, one of the many German principalities. Wilhelm knew well the major works of Muslim astronomers; but he was aware of large errors in the predicted positions of the Sun, Moon and planets with respect to the stars based on their observations. Wilhelm was the first to tackle this problem by systematic more accurate observations. Such a task required the resources of a prince. Wilhelm built the first observatory in Europe 1560–1561 on the balconies of his castle in Kassel. He personally observed the positions of 58 stars, published in the first *Kassel Catalogue*. The errors were about 10′ up to 12′ as we now know from comparison with modern observations, i.e. a bit smaller errors than Ulugh Beg's observations performed a century earlier.

Tycho Brahe, who visited Kassel and met the Landgrave, called Wilhelm "the most important astronomer in Europe", and decided to build an even bigger observatory in Denmark. He succeeded in convincing the King of the great importance of such an endeavour, and got the entire island of Hven as well as a substantial financial help for setting up an astronomical observatory with excellent and

performant tools produced according to Tycho's plans. The observatory was called "Uraniborg" and became a universally admired
center of astronomical science. Its description is given in Chapter
9, entirely devoted to Tycho Brahe's life and and discoveries, which
were all made before the invention of the telescope.

5.6   The first telescopes

Two basic types of telescopes are used by astronomers: a refrac-
tor telescope that uses a large objective lens to gather light, and a
reflector telescope invented by Isaac Newton in 1661 that uses a large
mirror to gather light. The eypiece, or the ocular, is a smaller lens, a
diverging one in Galileo's telescope, and a converging one in Kepler's
and Newton's telescopes.

An ideal lens transforms a strictly parallel beam into a collection
of converging light rays crossing at one point, called the *focus*. For a
divergent lens, the prolonged rays meet not behind, but in front of the
device, and the distance to the focal point is taken to be negative, as
shown in Figure 5.9. A similar effect can be obtained with a concave
mirror, transforming incident parallel rays into a convergent beam
with focal point in front of the mirror, as shown in Figure 5.9.

A refracting telescope is a tube that has two optical elements at its
ends, an objective and an eyepiece. The objective is a large lens that
collects light from a distant object and creates an image in the focal
plane. The eyepiece is a small lens through which we view this image.
It should be stressed that a telescope by itself is not an image forming
system; it serves only to transform an almost parallel beam of light

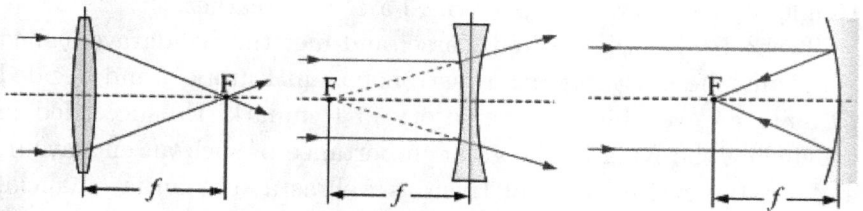

Fig. 5.9   Focal distances of converging and diverging lenses, and of a parabolic
mirror.

coming from a very distant object, which makes a small angle α_0 with respect to the optical axis, into a beam which emerges a larger angle α_e with respect to the axis. The ratio $M = \alpha_e/\alpha_0$ is called the *angular magnification*, or simply *magnification*. The real image is formed on the retina in the observer's eye, or on the photosensitive film or plate in a camera attached to the eyepiece.[1]

The magnification can be determined from the objective's and ocular's focal distances according to the formula:

$$M = \frac{\alpha_e}{\alpha_0} = \frac{F_1}{\mid F_2 \mid},\qquad(5.1)$$

where F_1 and F_2 are the objective's and ocular's focal lengths, respectively.

Another important parameter of a telescope is its *resolution power RP*, or simply *resolution*. It depends on the wavelength λ of the incoming rays and on the diameter of the objective lens D. Expressed in radians, it is equal to $RP = 1.22 \cdot \lambda/D$. The same formula can be applied to the human eye considered as image producing optical device. The diameter of the iris (playing the role of objective lens) is (at a day time) about 0.3 cm, and the average wavelength can be taken as 500 nanometers. This gives eye's resolution power equal to $(1.22 \cdot 500 \cdot 10^{-9}\,\mathrm{m})/(3 \cdot 10^{-3}\,\mathrm{m}) \simeq 2 \cdot 10^{-4}\,\mathrm{rad} = 42''$. In reality it does not go below $1'$, one arcminute. But for an average telescope with diameter of the objective about 10 cm, the resolution power is about 33 times higher, i.e. close to 1.3 arcsecond.

Although optical devices combining two lenses in a tube were proposed by the German born Dutch spectacle maker Johann Lippershey in 1608 in Holland, the invention of the *telescope* as astronomical device destined to observe celestial objects is universally attributed to Galileo Galilei. Galilei's telescope was the combination of two lenses, the convex lens serving as objective and the concave lens as eyepiece. It produced a direct image, which makes it the preferred device for nautical telescopes or theatre or army binoculars widely used today. The telescope with which Galilei made his most

---

[1]Replaced nowadays by digital cameras using the so-called CCD, or charge-coupled devices.

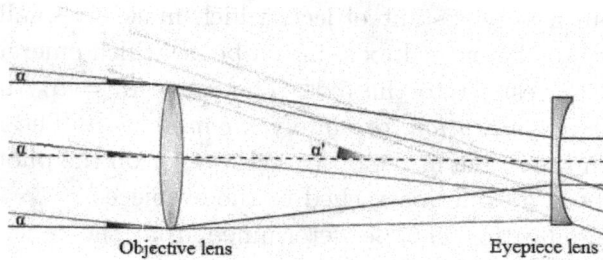

Fig. 5.10   Galilei's telescope.

remarkable astronomical discoveries in 1610 and 1611 had the following characteristics: total length: 127.1 cm; diameter of the biconvex
objective lens: 5.1 cm; its focal length: 133 cm. The eyepiece was a
plano-concave lens, with diameter 2.6 cm and (negative) focal length
−9.4 cm.

Galilei kept the construction of his telescope secret while he published most of his remarkable astronomical discoveries. Johannes
Kepler, with whom he shared most of the sensational news, decided
to build his own optical device, and after a few trials succeeded,
producing another version of the astronomical telescope, using two
converging lenses. Kepler investigated the optical properties of the
human eye and of lenses and telescopes in his books *Astronomiae
Pars Optica* (1604) and *Dioptrice* (1611), in which he explains
the principles of how a telescope works, as well as other optical
phenomena.

The Keplerian telescope was not accepted until Christoph
Scheiner, a German Jesuit mathematician, experimented with telescopes having only convex lenses for his study of sunspots. He found
that when he viewed an object directly through such an instrument
the image was flipped upside down. But it was much brighter and the
field of view much larger than in a Galilean telescope, as Kepler had
predicted. Since for astronomical observations an inverted image is
no problem, the advantages of the Kepler telescope led to its general
acceptance by the astronomers in the middle of the 17th century. The
principle of Kepler's telescope is shown Figure 5.11.

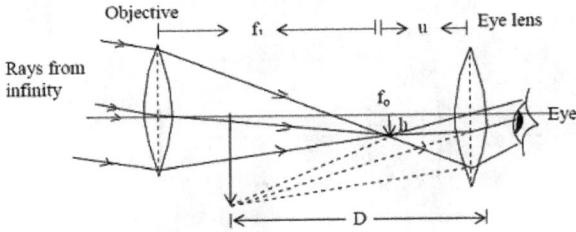

Fig. 5.11   Kepler's telescope.

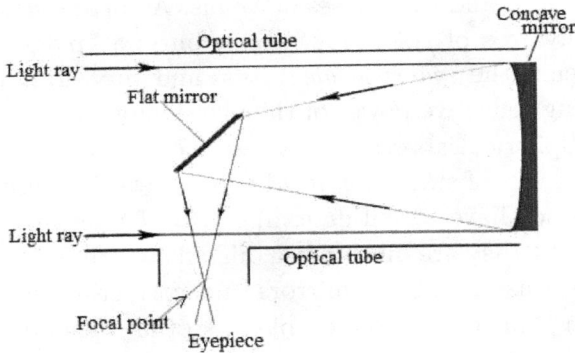

Fig. 5.12   Newton's reflector telescope.

The refraction telescopes, Galilean and Keplerian alike, suffered from *chromatic aberration* which blurred the image with a colored fringe.

A revolutionary telescope construction was proposed by Isaac Newton in 1668, when he was only 26 years old. Instead of a lens, it uses a concave *prime* spherical mirror to gather light. The reflected rays are concentrated towards the focal point, but intercepted before reaching it by a small *secondary* flat mirror placed in the center of the tube, directing them perpendicularly towards the eyepiece lens placed on the side of the tube.

Although the first telescope constructed by Newton was very small — its concave mirror made of copper-tin alloy was only 33 millimeters (1.5 inch) in diameter, it could show the four Galilean

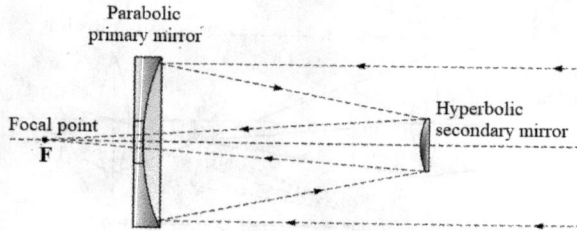

Fig. 5.13   Cassegrain's telescope.

moons of Jupiter and the phases of Venus. As predicted by Newton, it was totally free of color aberration, but was plagued with two other problems: the *spherical aberration* and very dim image due to the weakening reflective power of the metal alloy under the influence of the air. Spherical aberration causes the so-called "coma" which makes the images of stars look like little comets, because a spherical mirror does not have a well-determined focal point: only parabolic mirrors do, but they are much more difficult to grind. Modern reflector telescopes use parabolic mirrors and correcting lenses between the secondary mirror and the eyepiece, so that spherical aberration does not take place anymore.

An improved version of Newton's reflector was introduced by the French astronomer Laurent Cassegrain (1629–1693) who replaced the flat secondary mirror inclined by 45° reflecting the rays laterally by a hyperbolic mirror with the same optical axis as the primary spherical (and later on, parabolic) mirror, reflecting the rays straight back toward the small hole in the primary mirror where the eyepiece was mounted (see Figure 5.13).

Cassegrain's construction permitted to shorten telescope's length by almost a factor two.

5.7   Johann Hevelius

Johann Hevelius (1611–1687) was born in Gdańsk (the Hanseatic Baltic port, known also under its German name *Danzig*) to a rich family of beer brewers. After an excellent education in his city and in Königsberg, Hevelius started University studies in Leyden, during which traveled to London and Paris, where he met famous scientists, among whom John Wallis and Pierre Gassendi. In Holland, he became a friend of the Huygens family. He intended to visit Italy and meet Galileo, but his father urged him to come back in 1634.

An early widower (his first wife Katharina died in 1662), Hevelius married again a young lady of rich Dutch family, Elisabeth Koopmans, who not only gave him four children, but also helped him in his astronomic activities.

Hevelius' was a rich and respected citizen, elected to the membership of the City Council in 1651, where he supervised the hospitals and medical care, and judiciary, but his lifetime passion was the science of astronomy. He started to build his observatory with his own means, but soon got an extra financial support from the Polish King Jan Kazimierz who visited Hevelius in 1650. Hevelius' fame was steadily growing, and he was granted a special annual 1200 francs stipend by the French King Louis XIV, awarded during eight years, from 1664 till 1672. In 1664 Hevelius was elected to the Royal Society of London as its first foreign member. In 1666 he was invited to come to Paris to help with the construction of the astronomical

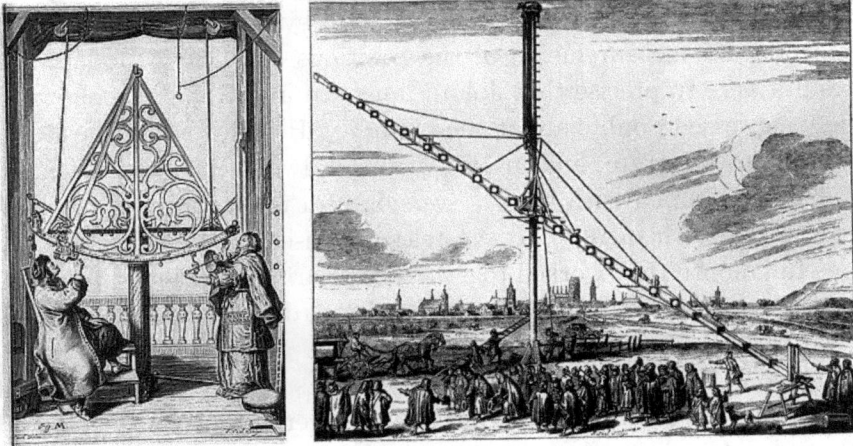

Fig. 5.14 Left: Hevelius and his wife Elisabeth observing the sky with brass sextant; Right: The 140-feet long telescope without tube constructed by Hevelius in Gdańsk.

observatory and to become its first director, but he declined the offer, which was then extended to the young Italian astronomer Giovanni Domenico Cassini.

Jan Sobieski, the head of Polish Parliament, visited Hevelius' observatory in 1668 and ordered a set of optical devices, a microscope, a telescope and a polemoscope (a kind of periscope enabling a lateral vision), and also two globes, celestial and terrestrial. Hevelius manufactured all these object by himself. After being elected to the Polish throne in 1674 as Jan III Sobieski, the King have not lost interest in Hevelius' work, and granted him with annual income of 1000 florins. In gratitude, Hevelius dedicated his celestial atlas to the King, naming it *Firmamentum Sobescianum*. In 1679 a great part of Hevelius' house and observatory, including the great sextant, were destroyed by a fire. Both Polish King Jan III and French King Louis XIV donated several thousands of florins for the reconstruction of Hevelius' observatory, which lasted many years. Hevelius died in 1687 leaving a rich legacy of astronomical observations and discoveries.

While a century earlier Tycho Brahe was the last great astronomer before the invention of the telescope, Hevelius was using at first mostly ancient observational tools like his great sextant, but very soon started to use telescopes, whose quality was constantly improving. Moreover, he had exceptionally good eyes, which permitted to sharpen the measurements of the positions of celestial bodies. His results were so precise that John Flamsteed and Robert Hooke publicly expressed doubts about their validity. Hevelius asked the Royal Society to arbitrate, and in 1679 young Edward Halley was sent to Gdansk equipped with the last version of a telescope with micrometer, superior in quality to the telesopes used by Hevelius. During one full month they worked together, comparing their measurements, which came out to be of the same precision, despite the more sophisticated device used by Halley.

Among the most important achievements are the observations of seven comets, four of which Hevelius discovered first; he also proved that comet trajectories are not straight lines as believed erroneously by Kepler. Another important contribution was the atlas of the Moon, observation of solar spots and flares, and the monumental stellar atlas with 1564 stars catalogued and their positions determined with utmost accuracy; it cites some of the positions determined by Ulugh Beg. Hevelius introduced seven new constellations, among which the *Scutum Sobescianum* honoring Jan III Sobieski's triumph over the Turkish army at Vienna in 1683, and the *Sextant*, in memory of his instrument lost during the fire. Perhaps the most original of Hevelius' endeavours was the construction of a giant tubeless telescope, with a focal length of 150 meters. It had to be handled by many assistants, but was too unstable to produce any valuable observations.

5.8 John Flamsteed and the Greenwich Observatory

John Flamsteed (1646–1719) was the first Astronomer Royal, appointed by the King of England. Born in Dunby as the only son of a malting business owner, he received a good education at the free school of Derby, ran by Puritans. After a few years pause due to health problems Flamsteed was admitted to the University of Cambridge, where he completed his education, developing interest in mathematics and astronomy in parallel and reading a number of books, from ancient to modern ones. He was particularly impressed by the works of Jeremiah Horrocks, a talented English astronomer who had died in 1641 at the age of 22, but was the first to observe the transit of Venus (in 1639) and to check the validity of Kepler's laws by determining the ellipticity of the Moon's orbit.

In 1674 the King Charles II of England appointed a Royal Commission to examine the proposal of building an astronomical observatory destined to improve the measurements of longitudes using new methods based on the analysis of the motions of the Moon. In August 1675 Flamsteed laid the foundation stone, and in July 1676 the main building of the Royal Greenwich Observatory was completed, and Flamsteed was appointed its first Director and Astronomer Royal, with an allowance of 100 pounds a year. He lived in the Observatory until his death in 1719.

There was an open rivalry between England and France also in the scientific field. The construction of the Royal Observatory in

Paris was ordered a few years earlier, with Cassini its first Director,[2] but it was the Greenwich one that became operational first. The great meridian line on the floor of the central room of the Parisian observatory was intended to become the universal zeroth meridian, but it was Greenwich that established its precise meridian circle first.

In 1681 Flamsteed concluded that the two great comets observed in November and December 1680 were in fact one and the same object, travelling on an elliptic trajectory, first towards the Sun, then away from it, thus confirming Kepler's laws.

5.9   William Herschel

A notable observatory built and operated by an individual was that of Sir William Herschel (1738–1822), assisted by his sister, Caroline Herschel, in Slough, England. Known as Observatory House. Its largest instrument had a mirror made of speculum metal, with a diameter of 122 cm (48 inches) and a focal length of 17 metres (40 feet). Completed in 1789, it became one of the technical wonders of the 18th century.

Born as Friedrich Wilhelm Herschel in the German electorate of Hannover to a musician father, Herschel first got a musical education and followed his father in the local military band, but after the French invasion he emigrated to England as a refugee, and started a carreer as music teacher and composer. His sister and brothers followed him to England, too. His career as a musician was quite promising, and nothing seemed to predict an extraordinary turn of fate which made him one of the most important astronomers of his time, and immortalized his name more than all his musical creations.

After studying Robert Smith's *Harmonics* and *A Compleat System of Opticks*, William Herschel soon developed an interest in astronomy and the techniques of telescope construction. He built his own telescope and eyepieces that had a magnifying power close to 6000. Herschel undertook two preliminary telescopic surveys of

---

[2]The Paris Observatory is descibed in Chapter 14.

the heavens, and in 1781, he discovered an extraordinary object, the planet Uranus, and its two moons, Titania and Oberon.

The discovery earned him the prestigious Copley Medal in 1781 and a fellowship at the Royal Society of London.

Herschel later studied the nature of nebulae and discovered that all known nebulae were formed of stars, hence rejecting the long-held belief that nebulae were composed of a luminous fluid. He also discovered two moons of Saturn, namely Mimas and Enceladus, and introduced the term "asteroid". By comparing apparent changes of positions of the stars closest to the Sun Herschel discovered that the solar system is moving through space and determined the direction of that movement pointing out from a point in the sky called *solar apex*. This point is situated in between the Hercules and Lyra constellations, between the stars Vega and Albireo (β Cygni).

Herschel's observations of stellar positions and comparisons with former ones showed the apparent centrifugal motions from the point situated in the Hercules constellation, as shown in Figure 5.15. Later on, Prévost proposed his version of the apex position, but Herschel's estimate was closer to what we know now.

With a great 45 inch telescope Herschel discovered two new moons of Saturn, Mimas and Enceladus. One of his most remarkable

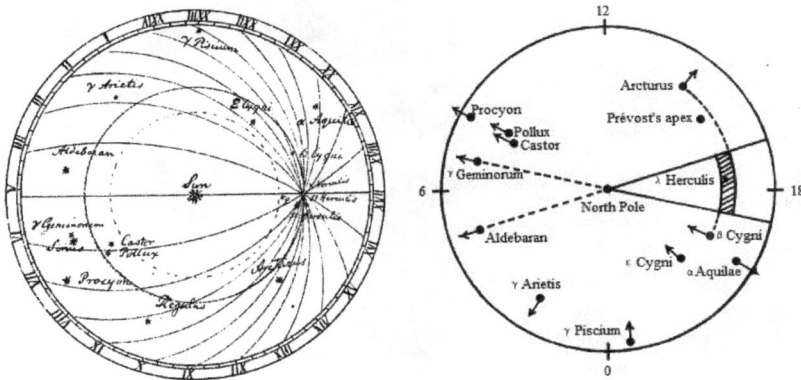

Fig. 5.15 Left: Herschel's drawing exposing stellar displacements, Right: the apex position circumscribed. Vectors indicate just the direction, not the absolute value of displacements.

dicoveries was that of infrared light. Herschel used a prism to split the light coming from the Sun, then placed a small thermometer in the region under the red color, and saw that his thermometer reacted as if it were heated.

He also suggested that the Milky Way was shaped like a disk. Therefore Herschel can be granted the title of the first cosmologist.

Chapter 6

Measuring time

Determining time of the day - Sundials and water clocks - First mechanical clocks - Babylonians and lunar calendar - The Metonic cycle - Days of the week - Lunar and solar months - Sidereal and tropical years - Ancient and modern calendars.

6.1 Days and nights, hours and minutes

The regular rythm of sunrise and sunset, mornings and evenings, days and nights, is what a newborn baby is confronted with from the very first hours of existence, imprinting a mark of time second in importance after one's own heartbeat. Days and nights have equal length only twice a year, in autumn (September 22/23) and in springtime (March 21/22). These dates are called *equinoxes*, and the equality occurs everywhere on the Earth, independently of the latitude, on both sides of the equator, in Northern and Southern Hemispheres. But we should remember that when we speak of days and nights, we refer to the relative position of the Earth with respect to the Sun — if the Earth was turning on its axis on its own, among the distant stars only, the very phenomenon of nights and days would be unknown.

It is easy therefore to understand why one full rotation of Earth with respect to the distant stars should be shorter than the time between two subsequent noons: exactly like in the case of lunar months, the rotation of the Earth around the Sun results in a discrepancy between sidereal and synodic days. Only here the difference

179

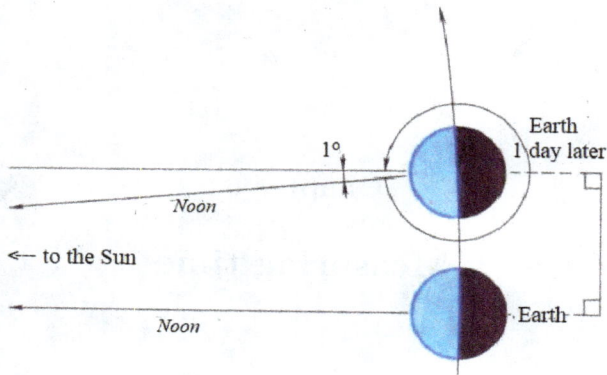

Fig. 6.1 Sidereal and solar days are slightly different.

is smaller, because there is only one extra revolution per year to be
added, as shown in Figure 6.1.

More than three thousand years ago the Babylonians were the
first to observe that the Sun's position with respect to the constellations of the Zodiac moves slowly, with the speed of roughly two solar
discs a day. This most probably led to the division of full circle into
360 degrees, this number being the closest to 365.2422 days in a tropical year, and the division of the day and night in 24 hours, an hour
in 60 minutes and the minute in 60 seconds. The same system was
adopted for measuring angles, 360 degrees in full circle, 60 arcminutes in one degree, and 60 arcseconds in one angular minute — this
system of measuring time and angles is still in use after more than
3000 years. It does not mean, of course, that the Babylonians had
instruments able to measure time, like our watches, up to a second,
or even a minute — the division in 60 parts was simply their way of
counting numbers, including fractional ones.

By a happy coincidence, the Sun's diurnal motion is also easily
expressed in terms of degrees and minutes. If we divide the Sun's
daily cycle into 24 hours, it will move by 15° per hour. Dividing by
60, we see that the Sun moves by a quarter of a degree per minute,
and as its disc is about half a degree large, this means that every two
minutes the Sun covers a path the size of its disc.

The ancient Egyptians used the shadows of obelisks to track the passing of time during the day. They also used vertical rods called *gnomons*, casting a shadow that was long in the morning, shortest at noon, and getting longer towards evening. Gnomons were known in Babylon as early as 3500 B.C.E.; later sundials were used, allowing a more precise distinction between different parts of a day. The Egyptians divided the daily cycle into twelve day hours and twelve night hours; they were not exactly of the same length. The day started at dawn, shortly before sunrise, and ended after twilight at night. No mention of times shorter than half an hour is known. Sundials were also in use very early in ancient China, and universally in ancient Greece and the Roman Empire. They can be seen until now in central squares of European cities, or on the walls of churches, city halls or other important buildings.

Later on, water clocks (also known under their ancient Greek name *clepsydra*, meaning "water thieve") were widely used, measuring time by checking the level of water in a special conical vessel from which water was allowed to escape through a hole in the bottom (Figure 6.2). The time was read by checking the level of the remaining water using a grading on the vessel's interior surface. These clocks were not very precise, because the pressure of the remaining water decreased with time, so the flow through the hole would slow down.

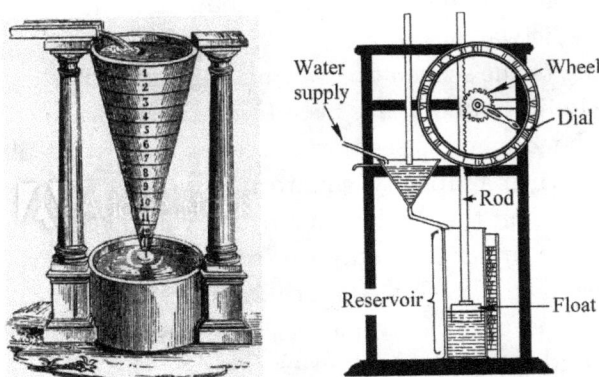

Fig. 6.2 Left: The simplest water clock used in Ancient Egypt; Right: The water clock invented by Ktesibios.

The best water clocks were constructed by Ctesibius or Ktesibios (Greek: Κτησίβιος, 285–222 B.C.E.), a Greek inventor and mathematician living in Alexandria during the Ptolemaic rule in Egypt. Similar water clocks were invented independently in China, but about 1000 years later. In these improved versions, the level of water in the upper vessel was controlled by a valve floating in the lower vessel so as to keep the level of water constant. This ensured that water dripped stadily, filling slowly the third vessel. A floater with vertical pole, rising while following the level of water, and a horizontal pointer at the pole's top indicated time on a vertical scale.

Sand clocks, or *hourglasses* were invented in early medieval Europe. The invention is traditonally attributed to Liutprand, a monk living in France in the 8th century C.E. They are in common use until today, but can measure only short periods of time, usually no more than half an hour. However, to regulate everyday life, especially in a community, people needed devices that would indicate hours of the day more precisely, and working during longer periods of time. Sundials responded to this need during the day, under the condition of cloudless skies prevailing most of the time.

Sundials are of three types: *equatorial, horizontal* or *vertical*. The equatorial sundial is the simplest, in a way: its surface should be parallel to the plane of celestial equator. The hours on such a dial can be marked with regular angular span, 15° per one hour — because the Sun completes one full turn in 24 hours, and not in twelve hours like on our wristwatches.

According to the schematic drawing in Figure 6.3, the photograph shot of the Beijing sundial has been taken in wintertime, because the shadow of the gnomon is seen on the lower face of the plate.

In spite of their simplicity, equatorial sundials are not widely used due to the fact that the shadow cast by the central pole (the gnomon) appears on the upper side only during the summer, when the Sun is high enough in the sky — more precisely, between the celestial equator and the celestial tropic of cancer, which happens between the vernal and autumnal equinoxes. During the remaining part of the year, from September to March, the shadow will be cast on the bottom face of the sundial, which makes the reading uncomfortable.

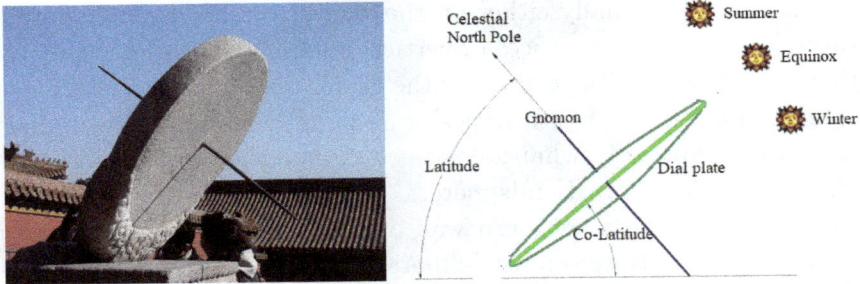

Fig. 6.3   Left: Equatorial sundial in Beijing, China; Right: Schematic view of an equatorial sundial with the Sun's positions in different seasons.

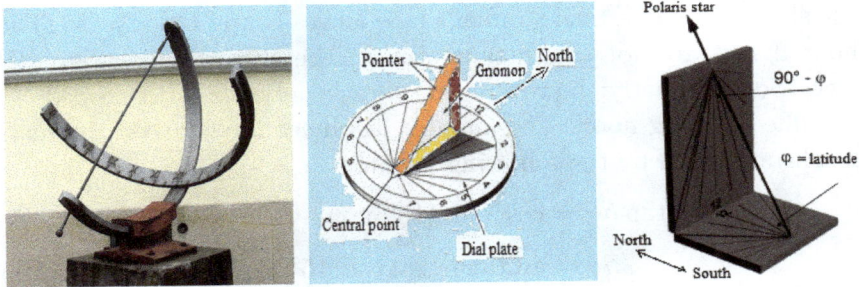

Fig. 6.4   Left: Equatorial bow sundial. The shade cast by the central gnomon is visible on the bow independently of season; Middle: Horizontal sundial; Right: Vertical and horizontal sundials combined.

This shortcoming can be overcome by replacing the inclined dial plate by a concave semi-circular dial, with the gnomon passing through its center. Such an *equatorial bow sundial* is shown in Figure 6.4.

Two other kinds of sundials, the *horizontal* and *vertical* ones, both have their advantages and shortcomings. Horizontal sundials are endowed with a triangle combining the vertical gnomon with a pointer making an angle φ with the horizontal plane which should be aligned with the direction towards the celestial North Pole, as shown in Figure 6.4. The hour marks on horizontal sundials must take into account the latitude at which a given sundial is supposed to serve.

When a sundial is supposed to be seen from a distance by many people, it should be put as high as possible, preferably in the central part of town — usually either on the wall of the local city hall or a cathedral. The performances of vertical sundials depend strongly on their positioning with respect to the South (in the Northern Hemisphere; in Australia, South Africa or South America vertical sundials should face North. In what follows, we discuss the properties of sundials in the Northern Hemisphere.). The best choice is of course to construct them on a southern wall of the building, but not all city halls or cathedrals were built deliberately parallelly to principal geographic directions.

In a vertical sundial the gnomon should be aligned along the axis of the Earth's diurnal rotation. It is easy to see in Figure 6.4 that while the gnomon's shadow in horizontal sundials moves clockwise, the shadow on a vertical sundial's face moves counter-clockwise. The hour marks are not equidistant even in the case when the sundial faces South exactly; however, they are symmetric with respect to the line marking noon. The angular distances h_V between the hour marks are given by the following formula:

$$\tan h_V = \cos \lambda \tan(15° \times t), \quad \text{whence}$$

$$h_V = \arctan\left(\cos \lambda \tan(15° \times t)\right), \tag{6.1}$$

where λ is the local latitude and t solar time. If a vertical sundial is not oriented towards the South (usually it would be directed to South-West, because people are more active in the afternoon and not in the early morning), the hour marks are neither equidistant, nor symmetric. Sometimes vertical sundials are oriented to the West. In such a case the lines marking the hours are parallel, and of course not equiduistant. They become parallel to the gnomon. The three types of vertical sundials are shown in Figure 6.5.

But there is an extra problem which makes the synchronization of sundials with astronomical time quite complicated. As we know from the previous chapter, when the Sun reaches its highest point it does not occupy the same point in the sky every day of the year. This would be the case only if Earth's axis of rotation was strictly perpendicular to the plane of its orbit, and if the orbit was perfectly circular with the Sun at its center. But neither of these assumptions is true: the axis of rotation is inclined by 23°26′, and the Earth's orbit

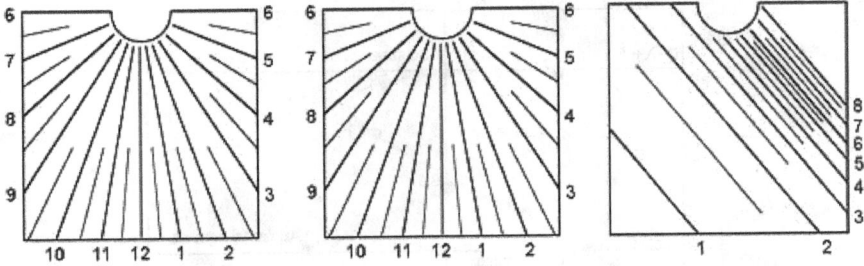

Fig. 6.5 Three vertical sundials: facing the South (left), South-West (middle) and West (right).

is a low eccentricity ellipse with $e = 0.0167$. Were the orbit perfectly circular, but the tilt of Earth's axis like it is now, the Sun would reach its maximal elevation at different angles, closer to the horizon during the winter and higher during the summer, but noon would always coincide with maximal elevation i.e. when the Sun crosses the local meridian. The set of all peak positions during the year would be a vertical segment in the sky, contained between the upper limit $\lambda + 23°26'$ on June 21 and the lower limit $\lambda - 23°25'$ on December 21, where λ is the latitude of the observer expressed in degrees. Plotted as a function of time from December 21 till next December 21, the Sun's elevation in the sky would be represented by a sinusoidal curve, with minimum at winter solstice (December 21) and maximum at summer solstice on June 21.

The *mean Sun* is what would be observed from a circular orbit and with zero tilt. Due to the tilt, between December and March the real Sun is slightly behind the mean Sun, it coincides with the mean Sun at vernal equinox, between March and late June it is slightly ahead of the mean Sun, then again slightly behind from summer solstice till autumnal equinox, and slightly ahead from late September till the winter solstice.

Suppose now for a while that Earth's axis is strictly perpendicular to the plane of its orbit (the ecliptic), but its orbit is slightly elliptic. According to Kepler's second law, the velocity of the Earth on the orbit is more rapid at perihelion, and slower at aphelion. This will cause a horizontal displacement of the Sun's highest point in the

186 *Our Celestial Clockwork*

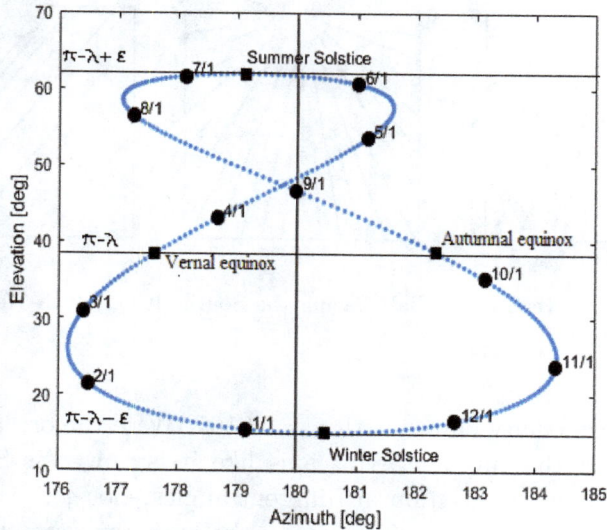

Fig. 6.6 Analemma taken at Greenwich Observatory. Note the maximal (June) and minimal (December) elevations and their relation with local latitude 51°. Note also four coincidences with the local meridian.

sky towards the East between January and July, and towards the West between July and January. The combined effect of the two displacements of the peak positions, which coincide with mean solar noon only four times a year, is shown in Figure 6.6.

The curve in the form of a figure 8 is called an *analemma*, and is a result of the composition of two sinusoidal curves. Its vertical component is due to the tilt of Earth's axis, and its horizonal component is due to variations in angular speed of apparent solar motion with respect to the fixed stars according to Kepler's second law of planetary motion. The eccentricity of the Earth's orbit is quite small, $e = 0.0167$, so these variations are not very big, but they can cause a difference between the "mean noon" and the actual passage of the Sun through the local meridian up to 16 minutes.

The difference between the mean Sun and the observed highest Sun's positions can be plotted against sidereal time through the period of one year. The resulting curve is called the "equation of time", and is represented in Figure 6.7.

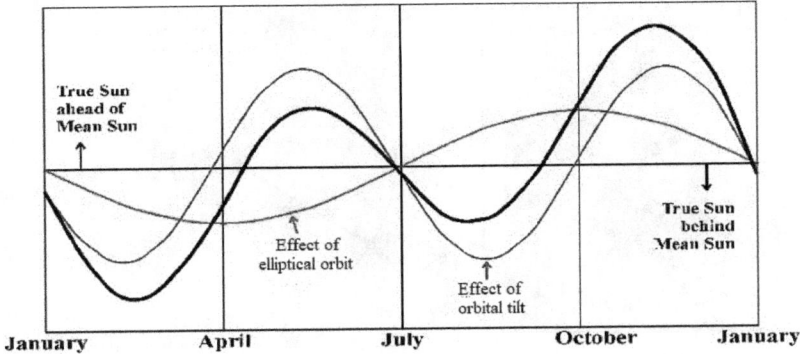

Fig. 6.7 The equation of time, i.e. the difference between the mean Sun and the observed one, as a sum of two effects: the ellipticity of the Earth's orbit and its orbital tilt.

Sundials were of no use after sunset, of course; but another device, called *merkhet*, was in use in ancient Egypt. It was a plumb line attached to a piece of wood, enabling the observation of the crossing of the meridian by a chosen star — usually the northern one, fixed with another "merkhet" directed so that it showed the Pole Star (which at that time was not the Polaris belonging to the Little Bear constellation; it was another circumpolar star called *Thuban*, in the tail of the constellation of Draco). This slow displacement of the North Pole is due to the phenomenon of *precession*, the slow rotation of the Earth's axis which completes a full circle in 26 000 years.

The circumpolar stars form by themselves a universal clock, provided that night is cloudless and the Great Bear and Polaris star are perfectly visible. Quite obviously, they will indicate the sidereal time, from which solar hour can be deduced if the longitude of the observer's position is known. Here is how to use the stellar clock (Figure 6.8):

1. Look up at the Polar star and the Great Bear constellation. Imagine a great dial, centered on the Polar star. The vertical line on the celestial dial should coincide with the South-North direction, and marks the local observer's meridian. The 0 hour is at the lowest point of the dial, its highest corresponds to 12 hours, the 6-th hour

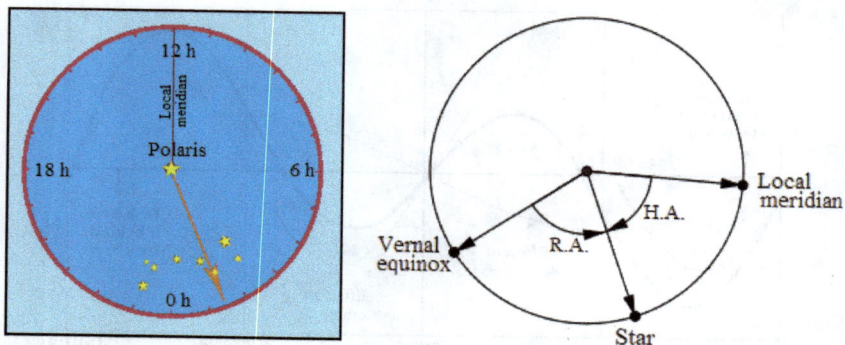

Fig. 6.8   Left: Great Bear and Polaris star as celestial clock. Note that the arrow following the constellation turns counter-clockwise; the dial is divided into 24 hours (not 12 as usual). Right: Stellar time = Right Ascension + Hour Angle.

mark is on the right, and the 18-th hour mark on the left. Note that stars and constellations turn counter-clockwise.

The Reader can ask at this point, "How come that we see the **meridian** that is usually marking the South, while looking in the North direction?" — the answer is that the local meridian is a part of the great circle joining the North Pole with the southern point of the horizon. In the figure above, only a segment of the local meridian is marked by thin vertical line, starting from the Polaris star and ending at the digit 12 on the dial. It covers approximately 35° of arc; the remaining part of the meridian goes towards the South point of the horizon. The line (not represented in the figure) going down, joining the Polaris star with the 0 digit on the dial, is a part of the great circle opposite the local meridian.

2. The imaginary hand of stellar clock should pass between the stars known under their Arabic names *Megrez* ($\delta\,Ursae\,Majoris$ and *Phecda* ($\gamma\,Ursae\,Majoris$, or "bear's thigh"). Its arrow points then in the direction of autumnal equinox — in the example above, it marks 1:30. The vernal equinox is in the opposite direction, in the figure above it would correspond to 13:30 on our imaginary celestial dial.

3. According to the definitions, the local sidereal time S is equal to the right ascension minus the hour angle of the stellar object we are looking at. The right ascension of the imaginary hand of the clock pointing at the autumnal equinox is 12 hours (counted from the vernal equinox). As the hand turns counter-clockwise, we see that it will take 10 hours and 30 minutes until the point between the stars *Megrez* (δ *Ursae Maj.*) and *Phecda* (γ *Ursae Maj.*) will reach the highest position passing through the meridian. This means that the hour angle HA is at the moment negative, $HA = -10.5$, and the local sidereal time is indeed $S = RA + HA = 12\,\mathrm{h} - 10\,\mathrm{h}\,30\,\mathrm{min} = 1\,\mathrm{h}\,30\,\mathrm{min}$, as shown in Figure 6.8.

4. Now we have to translate the local sidereal time into local solar time. To do this, we need to know local geographical longitude and the right ascension (RA) of the Sun at the moment of observation. Suppose we know our local longitude (easy to find out on any geographical map); what is Sun's right ascension? We can evaluate it quite easily keeping two things in mind: first, that the Sun's RA is 0 h on March 21, 6 h on June 22, 12 h on September 22 and 18 h on December 22; secondly, that after every 24 hours the Sun's right ascension increases by about 4 minutes (remembering that Sun completes a full turn along the ecliptic during 365.24 days, the hour angle it covers per day is

$$24\,\mathrm{h}/365.24 = 0.0657\,\mathrm{h} = 3.94\,\mathrm{min} \simeq 4\,\mathrm{min}.$$

5. Let us suppose for example that the snapshot of the circumpolar region was taken on October 12 in Torino, Italy (whose geographic coordinates are 45° latitude North and 14° longitude East). The time zone is +1 (+2 in summer). Now, the Sun's right ascension was 12 h 00 on September 22, since then, 20 days have passed, so that solar R.A. grew by 20 × 4 minutes = 80 minutes, or 1 hour and 20 minutes, so at the date of October 12 solar right ascension is 12 h 00 min + 1 h 20 min = 13 h 20 min. From the equation $HA = S - RA$ we obtain

$$HA = 1\,\mathrm{h}\,30\,\mathrm{min} - 13\,\mathrm{h}\,20\,\mathrm{min} = -11\,\mathrm{h}\,50\,\mathrm{min}. \qquad (6.2)$$

This means that the Sun will reach the highest point by passing across the local meridian in 11 hours and 50 minutes from now.

But this result is valid for the Universal time, as defined by the meridian for Greenwich (longitude 0). If noon is expected in 11 hours and 50 minutes, it means that actual solar time is 0 h 10 minutes A.M. ("ante meridiem") in Greenwich. Torino lies on the meridian 14° East of Greenwich, which corresponds to 56 minutes of time, which should be added to the Universal Greenwich time. Therefore local time in Torino is 1 h and 06 minutes A.M. However the official time is just one hour later than the Universal time, everywhere in the time zone +1 to which Torino belongs. In summer time an extra hour should be added due to the introduction of the "daylight saving time" intended to save electric energy.

In principle we could take into account the equation of time, but the precision of our stellar clock is not good enough to distinguish the small difference the use of time equation would imply (no more than 16 minutes).

In late medieval Europe mechanical clocks were invented, based on a new approach to measuring time. Instead of relying on some sort of repetitive process, like the rising and setting of the sun for sundials, and water dripping from a hole at the bottom of a stone vessel in water clocks, the clock makers had shifted from the continuous flow of water or sand, to the measurement of small increments of time. One of the first mechanical clocks was already working in Milano, Italy, by 1335. The oldest mechanical clock still in use (though after serious restauration works) can be visited in Salisbury, England. It works via a system of weights and pulleys. Two long ropes are wound upon a couple of spools, or barrels, and each is then threaded through a pulley about 7 meters above the clock. At the bottom end of each rope was a weight, made from lead. Each rope was fully wound back around the barrels every morning, pulling the weights to the top. The crucial element of this new kind of clock, as shown in Figure 6.9, was the mechanism called *"verge and foliot"*. It was critical in the development of time keeping. It was the first device that used periodic repetition of the same elementary motion to keep the time, and paved the way for pendulum clocks. This clever mechanism right inside the clock regulated the weights' slow descent and time keeping. It is composed of two small weights on either side of a small, horizontal

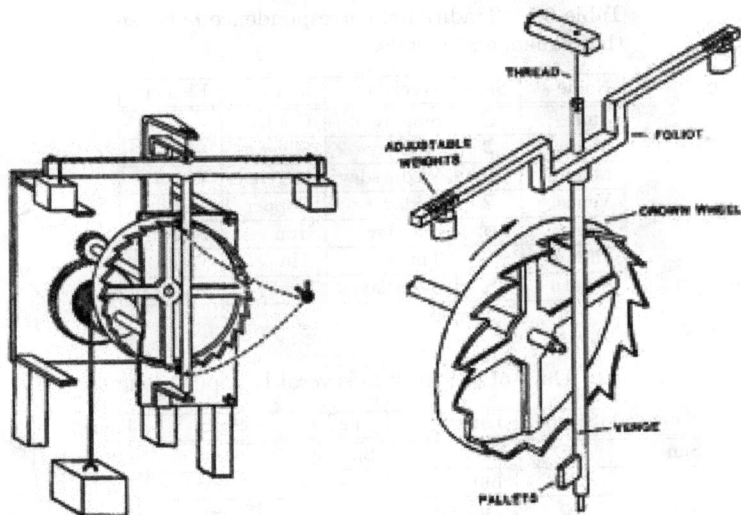

Fig. 6.9 Left: Verge and foliot mechanism, general view. Right: The detailed view of verge with two pallets and the crown wheel.

beam that is suspended by a thread, called the *foliot*. The weighted beam is pushed back and forth while a vertical rod in the middle of the beam, the *verge*, then is caught by the teeth of a rotating wheel every four seconds.

6.2 Weeks and months

The origin of the division of months into four shorter, seven days long units called weeks goes back to Babylon. The four phases of the Moon offer a unique and universal tool for such a division of time. If we start counting days from the first crescent after the new moon, the first quarter will occur after 7 days, the full moon after 14 days, the third quarter after 21 days, and on the 28-th day we will see the last, thin disappearing crescent. Although the synodic month with its 29.53 days is slightly longer than 28 days, it is close enough to make the division into four weeks universally accepted.

This choice was reinforced by an extra coincidence: there are exactly seven celestial bodies that travel along the ecliptic among the fixed stars: the Sun, the Moon, and the five planets visible by

Table 6.1   Traditional correspondence between
the planets and metals.

Name	Sign	Weekday	Metal	Element
Sun	☉	Sunday	Gold	—
Moon	☽	Monday	Silver	—
Mercury	☿	Wednesday	Mercury	Water
Venus	♀	Friday	Copper	Metal
Mars	♂	Tuesday	Iron	Fire
Jupiter	♃	Thursday	Tin	Wood
Saturn	♄	Saturday	Lead	Earth

Table 6.2   Days of the week in several European languages.

Name	Symbol	Latin	English	Spanish	French
Sun	☉	Solis	Sunday	domingo	dimanche
Moon	☽	Lunae	Monday	lunes	lundi
Mars	♂	Martis	Tuesday	martes	mardi
Mercury	☿	Mercurii	Wednesday	miercoles	mercredi
Jupiter	♃	Jovis	Thursday	jueves	jeudi
Venus	♀	Veneris	Friday	viernes	vendredi
Saturn	♄	Saturni	Saturday	sabato	samedi

the naked eye. They were identified with divinities and worshipped as such, and were believed to influence the fate of humans. They were also put into correspondence with days of the week, metals and in the Chinese tradition, the five planets were connected to five "elements": water, fire, metal, wood and earth (Table 6.1).

The "Planetary Week" was imported into Hellenistic Egypt from Babylon, where the "magic" number 7 was already in use. The days of the week were put into correspondence with planets, Sun and Moon, in the following order: Saturn, Sun, Moon, Mars, Mercury, Jupiter, Venus. In 321 C.E. Constantine the Great grafted this system on the Roman calendar which at that time was in use since Julius Caesar imposed it in 46 B.C.E.

The Latin names of the days correspond to the seven celestial bodies, and they gave rise to similar names in many European languages, see Table 6.2.[1]

---

[1] In Spanish and French Sunday is named "The day of the Lord", and Saturday after the Jewish Sabbath.

6.3 Tropical and sidereal years: Precession

The sidereal year is of little practical use, because it is the Sun that rules the times of day and night, as well as the seasons. A good calendar should take this into account and define the *tropical year* as the period after which the Sun occupies exactly the same position with respect to the fixed stars as one year before. Sidereal day is the period of time after which a chosen star passes through the same meridian (but not at the same altitude). The sidereal year is defined as the period of time after which a chosen star passes at midnight through the local meridian at the same altitude.

The most practical way to define the tropical year is to count the time between two consecutive equinoxes: it takes 366.242 sidereal days.

It turns out that the tropical year is shorter than the sidereal year by 1224 seconds, or roughly 20.5 minutes.

This small difference is the result of *precession* of the Earth's axis of rotation. At this point we shall not discuss the physical origin of this phenomenon, which was discovered by the ancient Greek astronomer Hipparchus of Nicaea who lived from 190 till 120 B.C.E. in Greece, which at that time was part of Roman Republic. What is important here is the acknowledgement of the fact that the Sun's apparent motion with respect to the fixed stars contains a small extra component, increasing its velocity. When Hipparchus made his observations in order to produce a new catalogue of stars, he compared them with the positions of constellations made by his predecessors more than 150 years before, and noticed that the positions of all stars have shifted with respect to the point of spring equinox, which at that time was in the constellation of Aries. The equinox point moved by more than one angular degree towards the Pisces constellation, which hosts the Sun earlier. The phenomenon is called *precession*, which is derived from Latin verb "*precedere*" (to precede). Hipparchus estimated the shift as slightly more than one angular degree per century.

By now the spring equinox point is located in the middle of Pisces. This is about 32° from the center of Aries, where it was in ancient times. Considering that the positions of the equinoxes, first described

Table 6.3   Ancient and contemporary annual positions of the Sun in the Zodiac.

Constellation	Symbol	Ancient dates	Contemporary dates
Aquarius	♒	20. I - 18. II	21. II - 18. III
Pisces	♓	19. II - 20. III	19. III - 21. IV
Aries	♈	21. III - 19. IV	22. IV - 21. V
Taurus	♉	20. IV - 20. V	22. V - 21. VI
Gemini	♊	21. V - 21. VI	22. VI - 22/. VII
Cancer	♋	22. VI - 22. VII	23. VII - 23. VIII
Leo	♌	23. VII - 22. VIII	24. VIII - 22. IX
Virgo	♍	23. VIII - 22. IX	24. IX - 24. X
Libra	♎	23. IX - 22. X	25. X - 23. XI
Scorpio	♏	23. X - 21. XI	24. XI - 21. XII
Sagittarius	♐	22. XI - 21. XII	24. XII - 22. I
Capricorn	♑	22. XII - 19. I	23. I - 20. II

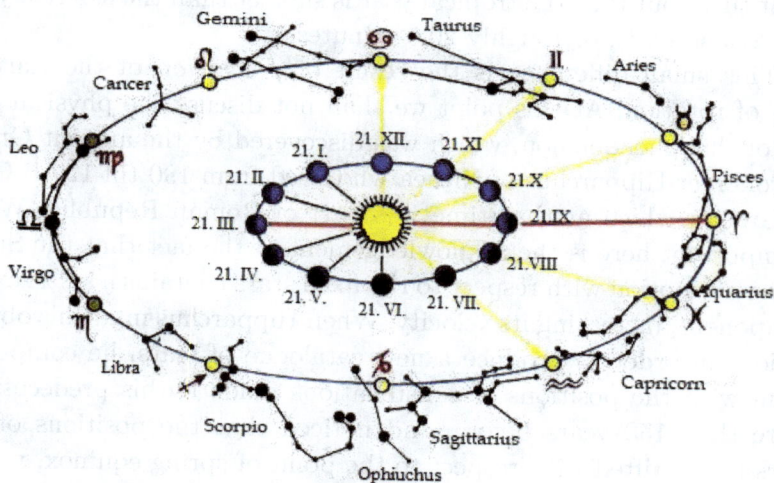

Fig. 6.10   Positions of the Sun during the year, relative to the constellations of the Zodiac.

by the Babylonians, became canonically fixed by the ancient Greeks shortly after Alexander the Great conquered Babylon around 331 B.C.E., we can evaluate the time it takes the equinoxes (and the solstices as well) to complete a full revolution of 360° (see Table 6.3 and Figure 6.10).

Twenty three centuries separate us from Alexander the Great. The effect of precession after 23 centuries amounts to a 32° shift. Therefore, to complete a full circle, we must wait for

$$2\,300 \times \frac{360}{32} = 25\,875 \text{ years.} \qquad (6.3)$$

This, of course, is a rough estimate; in what follows, we shall see that it turns out to be quite fair. Figure 6.11 shows the Earth rotating around its own axis with an angular velocity of 2π per day, and the Earth's axis rotating slowly around the axis of precession pointing towards the center of the constellation Draco, with angular velocity 2π per 26 000 years. The angular radius of the circular path of the perecession axis in the sky is $23°26'$, the obliquity of the Earth's axis with respect to the ecliptic, or the angle between the celestial equatorial plane and the ecliptic.

Let us recall again the precise lengths of sidereal year, sidereal day, tropical year and tropical day (in seconds):

$$\text{Sidereal year}: \ T_{Earth} = 31\,558\,150 \text{ seconds};$$

$$\text{Sidereal day}: \ P_{sid} = 86\,164.09 \text{ seconds};$$

$$\text{Tropical year}: \ T_{trop} = 31\,556\,926 \text{ seconds}.$$

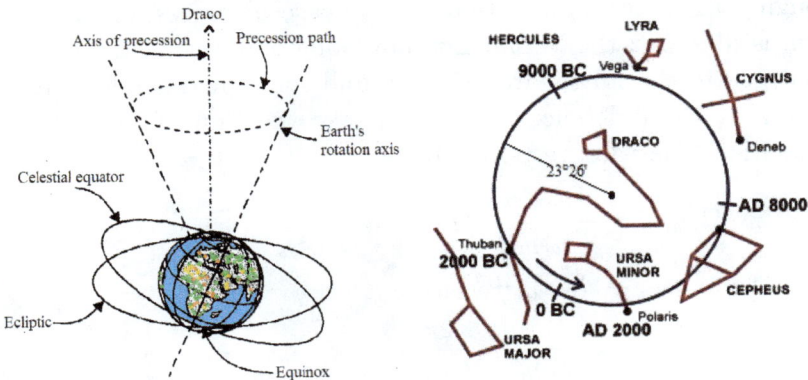

Fig. 6.11   Left: Earth's axis precession; Right: The path of the North Pole in the sky during the precession cycle.

The difference between the sidereal and solar ("tropical") days can be easily explained: the rotation of the Earth with respect to the fixed stars is the sum of two contributions: the proper rotation of Earth around its axis, and one extra rotation around the Sun, which takes one tropical year to complete (with respect to the fixed stars, too). As the two rotations with respect to the Sun are in the same direction, the absolute rotation with respect to the fixed stars is the sum of these two terms, so that

$$\frac{1}{P_{sid}} = \frac{1}{P_{Sun}} + \frac{1}{T_{trop}} \rightarrow \frac{1}{P_{Sun}} = \frac{1}{P_{sid}} - \frac{1}{T_{trop}},$$

$$P_{Sun} = \frac{P_{sid}T_{trop}}{T_{trop} - P_{sid}}, \tag{6.4}$$

and the exact answer is: 1 tropical day $= P_{Sun} = 86\,400$ seconds, as it should be.

We remind that a tropical year (the exact time between two consecutive vernal equinoxes) is by 1224 seconds, or about 20.5 minutes shorter than the sidereal year.

These data enable us to find out the precise duration of the *precession* cycle T_x. As usual, we should take into account two absolute (i.e. defined with respect to the stars) angular velocities in order to find the relative one. In the case of planets or the Moon we took the difference, because the rotations of the planets and the Earth were roughly in the same direction. In the case of precession the situation is different: the Vernal Equinox point x moves in the direction contrary to the Earth's rotation around the Sun. Therefore instead of taking the difference, we should take the sum of these angular velocities. They are, respectively:

$$\Omega_{Earth} = \frac{2\pi}{T_{Earth}}, \quad \Omega_{Aries} = \frac{2\pi}{T_{Aries}}$$

$$\rightarrow \Omega_{trop} = \Omega_{Earth} + \Omega_{Aries}, \tag{6.5}$$

therefore $\dfrac{2\pi}{T_{trop}} = \dfrac{2\pi}{T_{Earth}} + \dfrac{2\pi}{T_{Aries}}$, or $\dfrac{1}{T_{Aries}} = \dfrac{1}{T_{trop}} - \dfrac{1}{T_{Earth}}$,

so that $T_{Aries} = \dfrac{T_L \cdot T_{trop}}{T_{Earth} - T_{trop}} = 8.13636 \cdot 10^{11}$ sec. $\tag{6.6}$

To get the period of precession expressed in years, we divide its value in seconds (6.6) by the length of sidereal year in seconds:

$$\frac{8.13636 \cdot 10^{11}}{31558150} = 25\,782 \text{ years}, \tag{6.7}$$

in a very good agreement with our rough evaluation made on the basis of historical data (6.3).

Now we can find the exact number X of solar days in tropical year — the days that a good calendar should count, in order to ensure that the days of solstices and equinoxes fall always on the same date — e.g. the vernal equinox should occur invariably on March 21. The exact number of days in tropical year is obtained by dividing the length of tropical year by the length of solar day:

$$X = \frac{31\,556\,926 \sec}{86\,400 \sec} = 365.242199\ldots \tag{6.8}$$

One recognizes easily the first three digits, the number of days in a year. The first two decimals after are taken care of by adding a leap year every four years: this gives 365.25 days per year on the average. However in the long run the small difference between the exact value and this approximation will accumulate resulting in a growing discrepancy between real time and the date shown by the calendar. In the following section we recall how different civilizations coped with the challenge posed by the harmonization of stellar, lunar and solar timekeeping.

6.4 The first calendars

All calendars began with people trying to do their best to record time by counting days, months, and years (solar cycles). The year is 365 days, 5 hours, 48 minutes and 46 seconds, or 365.242199 days long. The time between consecutive full moons is 29.53 days. By "days" we mean here the tropical solar days 84 600 seconds long.

These numbers being incommensurate, the major problem for astronomers was to find a system combining months lasting an integer number of days and years lasting an integer number of months, approaching in a best way lunar months and solar days and years.

A year lasting 12 lunations lasting 354.36 days is too short, and a year with 13 lunar months is too long (383.89 days), and the prime number 13 is very impractical. Various stratagems were adopted by ancient astronomers to get around the incommensurability of natural time units given by the celestial clocks. Here are the most remarkable solutions to the calendar problem by ancient astronomers:

• **The Egyptian calendar** — The Ancient Egyptians are credited with the first calendar of 12 months, each consisting of 30 days, comprising a year. They added 5 days at the end of the year to synchronize somewhat with the solar year. By making all their months an even 30 days, they abandoned trying to synchronize them with lunar cycles and concentrated instead on alignment with the solar (tropical) year.

The year was divided into three seasons, four months long, corresponding to important agricultural periods: the Inundation, the Emergence and the Harvest. The date was then defined as e.g. "Day 12 of the Month 2 of the Harvest season", and the years were usually counted as e.g. "6th year of the reign of Tutmosis XII".

The Egyptians recognized that this calendar didn't quite align with the actual year. They called the 1461 Egyptian years it took to re-align with the 1460 solar years a Sothis Period.

Eventually, the Greek rulers of Egypt under Ptolemy added the concept of a leap year, adding a day every 4 years. The Romans kept this concept when they later ruled Egypt.

• **The Babylonian calendar** — Among ancient calendars the most sophisticated was the one set forth by the Babylonians, who improved the Sumerian calendar originating about 21 centuries B.C.E. It was based on the solar annual cycle and on lunar phases. It contained twelve lunar months, which started after the sunset when the first thin lunar crescent after the new Moon could be seen. The years started at spring equinoxes, which could be detected by observing the motion of the Sun and the comparative length of day and night, which become equal twice during each year. Similarly the summer and winter solstices were identified long time before the regular calendars appeared.

Twelve synodic months last $12 \times 29.53 = 354.36$ days, while the full year, from one equinox to another, lasts 365.24 days. Substracting one from another we get 10.88 days missing to complete the full year cycle. To choose 13 lunar months instead of 12 would have been even worse, because the excess would be greater: $13 \times 29.53 = 383.89$ days, which makes it 18.65 days more than the solar year. Besides, the ancient Babylonians used the sexagesimal arithmetical system based on multiples of 12, divisible by 2, 3, 4, and 6, which is extremely practical, while 13 is a prime number, thus ill-suited for creating divisible units of anything.

In order to keep in pace with the solar cycle Babylonian astronomers used to add an extra lunar month every three years. This improved the calendar quite substantially, but was still unsatisfactory in the long run. Let us check how closely we can approximate the length of solar year by adding one synodic month every three years. Without an additional month, three years would be equivalent with 36 synodic months, which gives $36 \times 29.53 = 1063.08$ days. This is more than 30 days short as compared to three solar years, $3 \times 365.24 = 1095.72$; more precisely, one has to compensate $1095.72 - 1063.08 = 31.92$ days. Adding one extra synodic month would reduce the discrepancy to $31.92 - 29.53 = 2.39$ days. This is still too much: after ten years, the discrepancy will be again almost 10 days.

The ancient astronomers of Babylon found, probably by trial and error, two integer numbers that would minimize the difference between the number of days contained in N years, and the same number contained in $12N + L$ synodic months, which leads to the following equation defining the discrepancy after adding to the calendar L extra synodic months in N years:

$$\Delta T = 365.24N - 29.53(12N + L)$$

$$= (365.24 - 12 \times 29.53)N - 29.53L. \qquad (6.9)$$

The discrepancy ΔT can be made small at will if one chooses N and L great enough, but for practical reasons N and L should be as low as possible. The solution found by the Babylonians was $N = 19$ and $L = 7$. This means that in order to keep a lunar calendar containing twelve synodic months with the solar one, based on the years containing 365.24 days each, one has to add 7 extra synodic months during the time span of 19 years. Let us check the error ΔT resulting from this choice.

Twelve synodic months defining the "lunar year" last $12 \times 29.53 = 354.36$ days; therefore 19 such "lunar years" would last $19 \times 354.36 = 6732.84$ days. This is to be compared with the number of days contained in 19 solar years, which is $19 \times 365.24 = 6939.56$. According to the equation (6.9), substituting $N = 19$ and $L = 7$, we get

$$\Delta T = 19 \times (365.24 - 12 \times 29.53) - 7 \times 29.53$$

$$= (6939.56 - 6732.84) - 206.71$$

$$= 206.72 - 206.71 = 0.01, \tag{6.10}$$

one hundredth part of one day after 19 years! This is an amazing precision, meaning that a discrepancy between the Babylonian calendar and the solar one is attained after *one hundred* 19-year cycles, i.e. after 1900 years, or nineteen centuries. No wonder that Babylonians were considered as best astrologists and astronomers in the ancient world, and their calendar was adopted by the ancient Hebrews and Greeks, with minor modifications.

What is added in practice to the calendar must be counted in days, and not just in months. The synodic month does not coincide with an integer number of days, so the additional months should be of different length — usually either 29 or 30 days, in order to arrive at 29.53 days on the average.

First primitive schemes of regular insertion were used about 535 B.C.E., using the *octaeteris* (with three intercalary months in every eight years). The 19-year cycle (with seven additional months in each cycle) was introduced after about 500 B.C.E. Counting from the first year of Nabonassar (747 B.C.E.), the years 3, 6, 8, 11, 14 and 19 were augmented by adding an extra Addaru month at the end of the year and an additional Ululu month in the middle of year 17.

• **The Hebrew calendar** — The ancient Hebrews adopted the calendar after their Babylonian exile. Its slightly modified version is still in use in the Jewish religious calendar.

The Hebrew calendar, or modern Jewish religious calendar which determines Jewish Holy Days, reproduces the 19-year long cycle and 12 synodic lunar months, with an extra month, called *Adar II* added once every two or three years. The years that count 12 synodic months last only 354 days, which is 11 days less than the solar year which is 365.24 days long. The twelve months, whose names in Hebrew calendar are taken from the Babylonian calendar, last 30 or 29 days, adding up to 354. Leap years contain 13 months and last 384 days. The additional month, *Adar II*, can be 29 or 30 days long, and 7 leap years occur in each 19-year long period, so that the Jewish calendar catches up with the Gregorian calendar everytime such a period is completed.

The first day of the first year of the Jewish calendar is interpreted as the first day of creation. According to Maimonides (1138 − 1204 C.E.), the famous Jewish scholar of Cordoba, Spain, the creation of the Universe occurred 4938 years back from the day he wrote down his calculation, which was on the 3-rd of Nissan, corresponding to March 22 of the common calendar year 1178. Consequently, the year 2020 of the Gregorian calendar universally in use in modern times, coincides with the Jewish year 5780. (To check this result, just add up 4938 + 2020 and substract 1178: 4938 + 2020 − 1178 = 5780.)

• **The Greek calendar** — Among many calendars used in ancient Greece the *Attic*, or *Athenian calendar* is best known due to the immense cultural influence and heritage of Athens during the golden period of Greek history, form 500 till 300 B.C.E. According to the scarce sources left, it was a mix of three different systems, a lunar calendar used to fix the dates of festivities, containing twelve months based on the phases of the Moon, a democratic calendar with ten arbitrary months, and an agricultural stellar calendar based on the observation of stars rising at chosen points of the horizon. The year started on the first new Moon after summer solstice.

Various types of calendars were in use locally, in different city states of ancient Greece.

In 331 B.C.E. the Macedonian king Alexander the Great conquered the city of Babylon, which had already been in the Persian zone of influence. Young Alexander took lessons from Aristotle, and was very respectful towards science and philosophy. In his military expedition to Asia he took along many Greek scholars who were eager to learn about the scientific achievements of more ancient civilizations. In particular, the Greeks adopted the Babylonian calendar based on the metonic cycle.

• **The Roman calendar** — The early Romans followed the Greeks (and therefore the Babylonians) by synchronizing the months with the first crescent Moon following a new Moon, and introducing some months of 29 days and some longer ones.

Every other year, February was shortened and a leap month (Intercalaris) was added in an attempt to realign the lunar cycles with the solar calendar. The lengths of the years in a four year cycle of this lunisolar calendar were 355, 377, 355, and 378 days. This added up to 4 days too many to stay in line with the solar year.

When Julius Cæsar came to power, he asked an Egyptian astronomer, Sosigenes of Alexandria, to invent a better calendar. What resulted is called the Julian calendar. He abandoned aligning the months with lunar cycles, and adopted months of 30 or 31 days length, keeping the last month, February, at 28 days. He introduced an extra day in February in leap years.

Julius Cæsar re-named the fifth month (Julius) after himself. His successor, Augustus Cæsar, re-named the sixth month (Augustus) after himself. Both months became 31 days long; as a result, the time duration between spring and autumn equinoxes (March 21/22– September 22/23) became longer than the time lapse between the autumn and spring equinoxes. In the Northern Hemisphere winter is shorter and summer is longer; in the Southern Hemisphere the opposite is true.

The first day of each month was called Kalendae, or calends. Debts were due on this day, so books to track payments were called calendarium (account book) from which we get our modern day calendar. The years were counted "ab Urbe condita" (abbreviated as AUC), from the legendary date of foundation of the City of Rome by Romulus and Remus in 753 B.C.E., so that the foundation of the Empire by Augustus which occurred in 27 B.C.E. was at that time in the year 727 AUC.

Table 6.4 The names and lengths of ancient Roman months.

Old Roman names	Length	Julian calendar names	Length
Martius	31	Martius	31
Aprilis	29	Aprilis	30
Maius	31	Maius	31
Iunius	29	Iunius	30
Quintilis	31	Julius	31
Sextlis	29	Augustus	31
September	29	September	30
October	31	October	31
November	29	November	30
December	29	December	31
Februarius	28	Februarius	28
Februarius*	23	Februarius*	29

After the Christianity was adopted as state religion by Constantine the Great in 323 C.E., the beginning of the year in the Julian calendar has undergone some modifications in order to be more in harmony with the Christian narrative and beliefs. So, the birth of Christ coincided with winter solstice, which was celebrated by many Europeans as the most important astronomical event marking the return of longer days and shorter nights. Christ's circumcision 8 days after the birth was taken as the first day of the New Year. As a consequence, January became the first month of the calendar, March the third, and December the twelfth. But we continue to use the old names of months coming from Latin numerals, shifted by two: the ninth month of our calendar is called September (from Latin "septem", seven), the tenth month is called October (after Latin "octo", eight), etc.

The Julian calendar was the most precise one at the time it was introduced. The extra day added to February every four years (the leap years) increased the 365 days duration up to $365 + 1/4 = 365.25$ days on the avarage, which is pretty close to the exact duration of tropical year: 365.2422 solar days. Nevertheless, with the difference between the two values being equal to 0.0078 solar days per year, it is easy to see that the discrepancy between Julian date and the astronomical count will slowly accumulate, reaching a one-day difference after $(0.0078)^{-1} = 128.2$ years. This is why it was replaced by a more precise calendar in 1582, named after the Pope Gregory XIII who decided to skip the accumulated 10 days difference and to introduce a better system, which bears his name and is used until now worldwide. However the Julian calendar is still in use in the Orthodox Church.

 • **The Chinese calendar** — The ancient Chinese used a quite complicated lunar calendar, which needed systematic corrections in order to catch up with the solar calendar used in parallel. The lunar calendar contained twelve synodic months, of lengths varying from 29 to 30, which resulted in a backlog of about ten days per solar tropical year (12×29.56 days $= 354.72$ days instead of 365.24 days). In order to fill in the missing days and suppress the discrepancy with respect to the Sun, a leap month was added every 32 or 33 months; usually it was placed as second or third month after the Chinese New Year, usually in the month corresponding to our February.

A year containing a leap month lasted 384 or 385 days, thus catching up the delay accumulated by shorter years of 12 lunar months. Besides, the year was divided in 24 solar terms, 15 or 16 days long, centered on equinoxes and solstices. Because $24 \times 15 = 360$, adding 5 terms of 16 days is enough to come to 365 days per year.

Since ancient times the Chinese were aware of the Moon's daily progress in the sky by about 13° eastwards. The path followed by the Moon was divided into 27 (in some versions 28) "stations" corresponding to every day in sidereal month. The "stations" were thus more than twice as numerous as the European twelve Zodiac constellations.

The Chinese ware also aware of the Sun's changing position with respect to the motionless stars, but the 12-year long cycle of changing animal signs attributed to each year does not have any direct relationship to the annual change of the Sun's position; nevertheless it is often referred to as the "Chinese Zodiac". It was established more than 2000 years ago during the reign of Emperor Qin, and was based on Jupiter's 11.86 years long cyclic motion with respect to the fixed stars. The years bear the names of animals, which are: *Rat, Ox,*

Tiger, Rabbit, Dragon, Snake, Horse, Goat, Monkey, Rooster, Dog, Pig. The year 2021 is the Year of the Ox.

Due to the difference between 11.86 and 12 full solar years, after about 7 Jupiter cycles its position with respect to the constellations representing animals will undergo a shift by one, which was taken into account by Chinese astrologists.

The ancient Chinese divided the day into twelve unequal hours. Daylight was divided into six equal parts, two hours in each, no matter how long it was. As a result, the summer hours were almost twice as long as the winter ones. The same division was applied to darkness at night. This system was adopted also in Japan in the 7th century C.E. Both day and night hours were given names of various divinities or animals.

• **The Hindu calendar** — Most calendars used in ancient India were lunisolar, with twelve synodic lunar months of 30 days each. The months were divided into two fortnights corresponding to the waxing Moon (called "the bright fortnight") and to the waning Moon (called "the dark fortnight").

Solar months named after the Zodiac constellations in which the Sun dwelt during the year were also taken into account; they had names different from those used in Europe following the tradition inherited from ancient Greece.

In order to synchronize the lunar and solar calendars, a leap year was added once in three years. However, as the 12 lunar months last 354.367 days, and 13 lunar months last 383.923 days, it is easy to find out that one leap year in three years still does not set the count even. The difference between the tropical solar year (e.g. between two consecutive vernal equinoxes, or between two consecutive winter solstices) and the 12 lunar months is $365.242 - 354.367 = 10.875$ days. After three years the retardation amounts to $3 \times 10.875 = 31.1$ days, so that after adding 29.56 days of an extra lunar leap month 1.54 days are still missing. This is why extra leap years were added from time to time to keep calendars aligned with solar years.

• **The Mayan calendar** — The so called *Mayan calendar* was in use in most pre-Colombian cultures of Central America, and was probably inherited by the Mayan civilization of the Yukatan peninsula from previous cultures. Of all ancient calendars, this is the strangest and perhaps the most complicated. It is actually a superposition of three different ways of counting time: the somewhat mysterious *Long Count*, the divine calendar called *Tzolkin*, and the civil calendar called *Haab*. The last one contained 365 days divided into 18 months 20 days long each, plus an extra "month" with 5 days. As a result, the *Haab* was behind the solar time by 0.24219 days per year, being much less accurate than the Julian calendar with its leap years in one out of four, adding the missing 0.25 days per average year.

The *divine calendar* represented a 260-day cycle with 20 periods of 13 days each, corresponding to waxing and waning Moon. It was used exclusively for determining the dates of religious events.

The *Long Count* has little parallels in other cultures. This particular time count was based on the idea that our Universe is cyclically destroyed and then re-created, with periodicity of 1 870 625 days, i.e. about 5125 solar years. The last creation was supposed to have taken place on August 11, 3114 B.C.E., the destruction according to Mayan beliefs should have occurred on December 21, 2012. This prediction created some anxiety among superstitious people, but nothing particular happened on that date. Perhaps the next cycle will be more conclusive?

• **The Arabic calendar** — The Muslim religious calendar is a strictly lunar one, with twelve synodic months per year. A month starts with the first thinnest visible crescent after new Moon. The months are divided in four weeks, labeled by numbers. The first day (*al 'Ahad*, "the first" in Arabic) coincides with the Christian Sunday; the second (*al 'Ithnaim*, "the second" in Arabic) falls on Monday, and so forth, with the exception of Friday, called *al Jummah* ("the gathering"), which is the day of prayer, and Saturday, called *as Sabt* ("the rest"), inherited from the Jews as the day off. In many Muslim countries the official weekend is fixed on Thursday and Friday (e.g. in Saudi Arabia, Egypt, Jordan...), while in some others (Iran, Afghanistan) only Friday is free of work. Many Muslim countries (e.g. Morocco, Turkey, Pakistan) follow the Western trend, with weekends on Saturday and Sunday, with a long pause for prayer on Fridays.

The years are counted from the date of *Hijra*, 622 C.E. The years are thus shorter than years in Roman or Gregorian calendar by as much as almost eleven days (this is because $29.53 \times 12 = 354.36$ instead of 365.2425 of the Gregorian calendar). One of the consequences is that Muslim holidays change periodically, from one year to another, by this shift. The date of the Holy Month of Ramadan changes every year, completing the full cycle in about 35 years. The difference between the Islamic religious calendar and the Gregorian calendar in use now is quite impressive. In the year 2020 of the Gregorian calendar the Muslim year is 1441, whereas it is easy to check that there are only $2020 - 622 = 1398$ years since the starting point of the Muslim calendar. The difference accumulated since the beginning in 622 C.E. equals by now 43 years.

6.5 The Gregorian and improved Julian calendars

• **The Modern (Gregorian) calendar** — In October of the year 1582 A.D. according to the Julian calendar Pope Gregory XIII introduced a new calendar which would better take into account the duration of real tropical year, including the precise knowledge of the precession of the Earth. The contemporary astronomers noticed that the spring equinox had changed its position with respect to the stars since it was set in the constellation of Aries about sixteen centuries earlier.

At the end of 16th century the astronomical spring equinox occurred about ten days earlier, when the Sun was still in the constellation of Pisces (nowadays it is in the middle of Pisces at spring equinox). For religious reasons it was necessary to adopt the calendar in a way that would ensure the Nativity to coincide with the winter solstice, and to make the spring equinox occur at March 21 and not to drift towards April like it happened already according to Julian calendar.

Clearly, the discrepancy was due to the leap years occurring too frequently — one leap year in four years produced a standard year containing 365.25 days, just a little bit too much, but it was enough to add 11 minutes 14 seconds per year, i.e. an extra day every 128 years. Since Constantine officially imposed the Julian calendar on the entire Roman Empire in 320 C.E. The deviation between the Julian year of 365.25 days and the tropical year of 365.2422 days gradually produced significant errors; some learned people in Byzance and Western Europe alike became aware of it since the Middle Ages. The discrepancy mounted at a rate of 11 minutes 14 seconds per year until it attained 10 days in 1545, when the Council of Trent authorized Pope Paul III to take corrective action. No satisfactory solution was found for many years. In 1572 Pope Gregory III agreed to issue a papal bull introducing the new calendar drawn up by the Jesuit astronomer Christopher Clavius. Ten years later, when the edict was finally proclaimed, October 5 of the year 1582 A.D. became October 15, 1582 A.D., to bring the calendar back in line.

The new calendar was much more precise than its predecessor: the average length of Gregorian day became now

$$365 + \frac{1}{4} - \frac{3}{400} = 365.25 - 0.0075 = 365.2425. \tag{6.11}$$

The difference between this value and the observed average length of solar day is $365.24250/365.24219 = 0.00031$ day per year, as compared with 0.00781 day per year difference of Julian calendar. The ratio of two differences is $0.00781/0.00031 = 25.19$, which means that the precision of Gregorian calendar is 25 times better than that of Julian calendar.

In spite of its obvious superiority, the new calendar was not immediately accepted outside the Catholic countries. At the time of its creation England had already separated itself from Rome after King Henry VIII declared himself head of the Church of England. Protestantism was also spreading in Northern Europe, and Eastern Orthodox Churches were in traditional opposition to Rome since the Great Schism of 1054 C.E.

Nevertheless after almost two centuries, in 1750, the *Calendar New Style Act* was proclaimed in England, which entered in practice in 1752, when the date September 2 was changed into September 14, 1752. The difference between the Gregorian and Julian dates had grown up by two days since 1582, but as 1752 was a leap year, only 11 days were really omitted (with the date skip imposed, the year counted 365 instead of 366 days).

In Russia the orthodox Julian calendar was in use until January 1918, when the newly established Soviet power abolished the "old style" Julian calendar in favour of the Gregorian one. The date of January 31, 1918 was changed into February 13, 1918, which added up 13 days. This is why the Bolshevik "October Revolution" that occurred on October 25 according to Julian calendar, was celebrated on November 7 every year, until the end of Soviet Union in 1992.

But the Orthodox Church celebrates all its religious holidays according to their Julian calendar dates. By now the difference between the two calendars grew up to 14 days, so that Orthodox Christmas in 2020 is celebrated on January 7, and the New Year on January 14.

• The improved Julian calendar

In 1932 Serbian scientist Milutin Milanković proposed what he called "revised Julian calendar", much more precisely staying aligned with the tropical year than the Gregorian calendar widely in use. The difference between Milanković's version and the Gregorian calendar resides in the exceptions to leap years, occurring also every fourth year. In the Gregorian calendar the leap years are divisible by four, and the exceptions are those divisible by 100, but not by 400. This removes from leap years 3 out of 400, and not just 1 out of 100. But the result is still a bit too high as compared with the real length of tropical year: 365.2425 solar days instead of 365.24219 days in tropical solar year. The error is very small: only 0.00031 days per year. A one day difference will be accumulated only after 3226 years.

The improvement proposed by Milanković was as follows: the leap years are still one out of four, with the exception of years divisible by 100, of which however are excluded years whose numbers not divisible by 900 give the remainder 200 or 600; such years remain leap ones. For example, 1800 and 1900 are NOT leap years, but 2000 is leap (by the way, in this particular case it was a leap year in Gregorian calendar, too, because is is divisible by 400. In fact, the first difference between the two calendars, Gregorian and "improved Julian" would occur only in the year 2800: being divisible by 400, it is a leap year according to Gregorian calendar, but IS NOT a leap one according to the improved Julian calendar, because when divided by 900 the remainder is 100.

It is easy to determine the average year length according to Milanković's improved Julian calendar. The exceptions from the four-year leap rule occur now twice every 900 years, so only the remaining 7 out of 900 years are not to be counted as leap. This results in the following formula for the average length of solar year:

$$365 + \frac{1}{4} - \frac{7}{900} = 365.25 - 0.007778 = 365.242222. \qquad (6.12)$$

The difference between this approximation and the actual value is $365.242222 - 365.24219 = 0.00003$ day per year, which is TEN TIMES less than the similar difference of Gregorian calendar. It is so tiny that the question arises whether it makes sense at all to measure time with a precision so great that it may interfere with the natural variations of the Earth's own rotation.

Chapter 7

Measuring space

Ancient units of length - Two great theorems of geometry: Thales and Pythagoras - Applications in everyday life - Euclid and the foundations of geometry - Evaluation of distances - Eratosthenes and the Earth's radius - How Aristarchus measured distances to the Moon and Sun - Dimensions of Ptolemy's world - Explorers and cartographers

7.1 Preamble

Measuring time does not require much space: in fact, everybody carries a biological clock given by one's own heartbeat. It is not very reliable, because it can go faster or slow down depending on the circumstances, but it is always with us and stops definitely only when we do not need it anymore. There were no better movable clocks able to measure time up to second-like intervals until the 16th century. Galilei used his pulse along with the water clocks in his experiments on the free fall and accelerated motion.

In contrast, measuring space , especially at large distances, cannot be performed instantly. Small lengths, as long as they are comparable with our own dimensions, can be evaluated with calibrated objects of known size: hence the length units evoking parts of human body, like foot, cubit (from Latin *cubitus* meaning "elbow", the length of an arm from elbow to the tip of the middle finger), palm (the width of a hand), and thumb (English inch, etymologically derived from Latin *uncia* meaning "one twelfth" because it was defined as a twelfth part of one foot in ancient Rome, but in many European languages similar units of length are designed by the word meaning "the thumb", like the French "pouce", Dutch "duim" or Danish "tomme"). Following the biblical description of Noah's ark (*Genesis* 6 : 14–16), it was 300 cubits long, 50 cubits wide and 30 cubits high, which means 160 meters long, 26.5 meters wide and 16 meters high.

In ancient Greece distances were expressed in *stadia*; unfortunately, different Greek cities used different definitions at different historical times, so we cannot be sure what was its exact length — usually between 150 and 160 meters.

At present, the decimal metric system prevails in science, with 1 meter defined as one part in $4 \cdot 10^7$ of the Earth's circumference. A more precise definition was that of a kilometer, or 1000 meters, as one ten-thousandth (10^{-4}) part of the quarter of the meridian, i.e. the distance from North Pole to the equator. For smaller lengths, centimeters and millimeters are used; for bigger sizes, the decimeters and hectometers, and for distances on Earth or in nearby space, the kilometers. This system was introduced in France during the Revolution, and in less that half of a century became standard all over continental Europe. The choice of rational units of length imposed also the most natural definition of units of weight, based on water as the most common substance available. One metric ton was defined as the mass of one cubic meter of water, one kilogram is the mass of one cubic decimeter, and one gram is the mass of one cubic centimeter.

The English speaking nations continue to use traditional length standards in everyday life. A *foot* is 30.48 cm long, it is divided into 12 *inches* 2.54 cm long each. A *yard* equals three feet, or 0.914 meter. In 1592 Queen Elizabeth I proposed to adopt the distance of *statute mile* as official unit of distance. Approved by the Parliament, the statute mile has the length of 1760 yards, or 5280 feet, which corresponds to 1609 meters. It is still in use in Great Britain and in the USA.

There are no landmarks on sea, so only the stars in the sky can provide a reliable spatial reference. The *nautical mile* is therefore the most natural way of measuring distances at sea. Its length corresponds to one arcminute of a meridian. Remembering that Earth's circumference is 40 000 km, the distance corresponding to one degree is $40\,000/360 = 111.11$ km, and one arcminute is $111.11/60 =$

1.852 km = 1 nautical mile. The speed of ships is measured in *knots*, with 1 knot = one nautical mile per hour.

When ancient astronomers started to evaluate distances to the Moon, Sun and planets, they could base their evaluations exclusively on angular sizes and distances, taking as the basic unit Earth's radius (in what follows, we shall often denote it by "e.r."), which was seriously underestimated until the 17th century. The distance to the Moon was determined quite well by Aristarchus as about 60 e.r., but he seriously underestimated (by the factor of 20) the distance to the Sun. Nevertheless, it reached the order of thousands of e.r., and for the distances to the planets a new unit became necessary. Quite naturally, the best measuring rod for this purpose was the *astronomical unit*, the average radius of Earth's orbit, denoted by A.U. since then.

7.2 Thales and Pythagoras

Notwithstanding the richness and sophistication of ancient Greek geometry developed by great mathematicians and exposed in the magnificent treaty by Euclid, almost all astronomical measurements and their subsequent interpretation are based on two great geometrical theorems attributed traditionally to Thales and Pythagoras.

Thales of Miletus (Θαλης, 625 ∼ 547 B.C.E.) became famous for his prediction of the solar eclipse of 585 B.C.E. and for his ability to evaluate dimensions of objects at a distance, by comparing their shadows with the shadow of a stick of known dimension.

The relationship of proportionality used by Thales to determine the height of the Great Pyramid is also an introduction of *linear dependence*, the essence of linear algebra.

In fact, Thales performed an important physical experiment relating different definitions of geometry; to put it more precisely, the notions of straight lines and right angles. It turns out that the phenomena involved in this experiment belong to quite different domains of physics: gravitation, which served to define the right angle between vertical and horizontal lines; quantum mechanics that explains how solid states can be formed out of atoms, and electromagnetism explaining propagation of light rays along the straight lines.

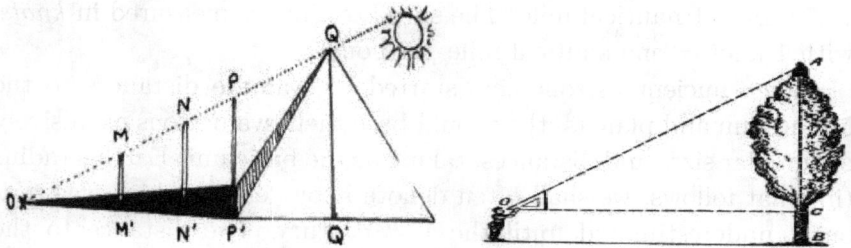

Fig. 7.1   Left: A schematic representation of Thales' experiment: the ratios between the lengths of vertical objects and their shadows are constant, MM' : $OM = QQ' : OQ$; Right: The simplest practical application of Thales' theorem.

The fact that they lead to three different, but compatible definitions of geometry, suggests that these distinct aspects of physical reality are deeply related. This fact gives rise to one of the most important and fundamental questions concerning physics and our perception of physical world, still open after more than twenty-five centuries.

In its pure mathematical formulation Thales' theorem states the following: when two intersecting straight lines are cut by two parallel lines, the resulting triangles are conformally similar, so that their corresponding sides are all related by the same proportionality coefficient. In Figure 7.2 two straight lines intersect at the point O. Three parallel lines p_1, p_2 and p_3 are drawn at an arbitrary angle, two of them (the red one and the green one) on one side from the intersection, the third one (blue), on the other. The resulting triangles have all their similar angles equal, and the triangles are conformally similar, i.e. they differ only in *size*, but have the same *shape*.

In Figure 7.2 we can compare three triangles having common summit at 0: ΔAOB, ΔCOD and ΔEOF. The Thales theorem tells us that the following series of identities are satisfied: Comparing the triangles ΔAOB and ΔCOD gives

$$\frac{AB}{CD} = \frac{OB}{OD} = \frac{OA}{OC}. \tag{7.1}$$

Comparing the triangles ΔAOB and ΔEOF gives

$$\frac{AB}{EF} = \frac{OB}{OE} = \frac{OA}{OF}. \tag{7.2}$$

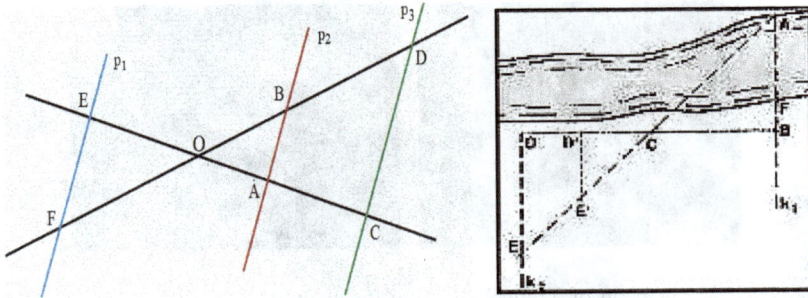

Fig. 7.2 Left: The triangles created by the straight lines and three parallel lines are similar; Right: Application of Thales' theorem for evaluation of distances. The width of a river AB is determined without crossing it, just by the construction of two similar triangles with parallel sides on the bank: here $AB = OE$.

It is worthwhile to note that while the triangles $\triangle AOB$ and $\triangle COD$ are conformally similar by a simple change of scale, i.e. the triangle $\triangle COD$ can be obtained by an appropriate blow-up of the triangle $\triangle AOB$, the triangles $\triangle AOB$ and $\triangle EOF$ are similar after a reflection with respect to the point O is performed. This situation is often referred to as "conformally symmetric".

From the identities (7.1) and (7.2) one can deduce another set of identities by multiplying and dividing both sides by appropriately chosen items, e.g.

$$\frac{AB}{OB} = \frac{CD}{OD}, \quad \frac{OB}{OA} = \frac{OC}{OD}. \tag{7.3}$$

The famous application of Thales' theorem for determining the height of the pyramid corresponds to the first identity in (7.3).

Thales' theorem provides a simple and efficient method to evaluate the sizes of the Sun and Moon, as well as the corresponding distances — but only in terms of their diameters, which cannot be evaluated by Thales' theorem alone, because it concerns only dimensionless ratios, or angles, not the real dimensions, unless we can measure at least one of the segments. On a sunny day, walking in the shadow under the trees, one can often see a collection of luminous spots on the ground. A closer look shows that many have more or less elliptical shape, with almost the same size. Some people observed similar phenomenon during partial solar eclipse, and noticed that the

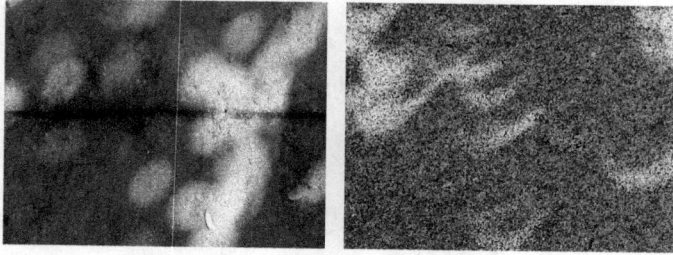

Fig. 7.3   Application of Thales' theorem for evaluation of distance to the Sun.

spots have the form of crescents — exactly like the solar disc par-
tially covered by the Moon (see Figure 7.3). What we see are the
images of the Sun produced by the effect known as *camera obscura*
and used by painters since the Middle Ages. If a pinhole is made in
one of the walls of a room without windows, or in a box with the
side opposite to the pinhole made of semi-transparent paper, the rays
coming from outside will form an inverted image, whose dimensions
are in the same proportion to the sizes of real objects as the ratio of
distance from the pinhole to the wall to the distance to the object —
just another application of Thales' theorem.

According to the same principle, the luminous spots under the
trees are produced by natural "cameras obscuras" with randomly
positioned holes in the otherwise dense canopy of leaves (Figure 7.4).
Applying Thales' theorem, we can find out how high over the side-
walk these holes are: the distance from the image to the "pinhole"
in the leaves in terms of spots' radii is the same as the distance to
the real object (the Sun) in terms of solar diameters.

The average size of the sunspots on the ground is $\simeq 10$ cm, the
average distance to the slot between the leaves serving as a hole in
"camera obscura" is about 10 meters; therefore the distance to the
Sun, by similarity between triangles, is about 100 solar diameters.
The exact value is 108.

[* Another theorem of Thales is a special case of the inscribed
angle theorem, and is mentioned and proved as part of the 31-st
proposition, in the third book of Euclid's Elements. It is generally
attributed to Thales of Miletus, who is said to have offered an ox

Fig. 7.4 Application of Thales' theorem: the *camera obscura* device.

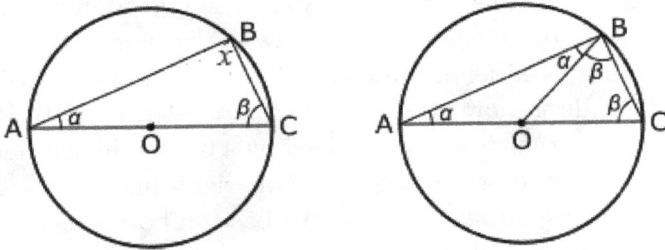

Fig. 7.5 Proof of Thales' theorem stating that a triangle inscribed in a circle built on one of circle's diameters is rectangular.

(probably to the god Apollo) as a sacrifice of thanksgiving for the discovery. However some historians of science attribute it to Pythagoras.

In geometry, Thales's theorem states that if A, B, and C are distinct points on a circle where the line AC is a diameter, then the angle $\angle ABC$ is a right angle (see Figure 7.5).

The proof needs a *lemma*, or an auxiliary theorem, which is also one of the bases of Euclidean geometry: the sum of angles in a triangle is equal to two right angles. One of the simplest versions of proof makes use of similarity and symmetry, typical ingredients of ancient Greek geometry. It goes in two steps: first, one proves that the sum of angles of a *right-angled* triangle is equal to two right angles; second, the theorem is extended to an arbitrary triangle. As shown in Figure 7.6, the first step consists in constructing a rectangle by doubling the original right-angled triangle, i.e. by adding an identical

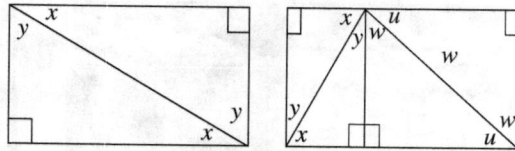

Fig. 7.6　Proof of second Thales' theorem by pure geometric construction.

triangle inverted and joining them by the common hypothenuse. The resulting quadrilateral has obviously rectangular shape, because it contains two opposite right angles belonging to the triangles. By symmetry, the sum of the angles of the rectangle is equal to four right angles, which is divided equally between the two triangles, each one having the sum of its angles equal to two right angles.

The proof of the lemma for an arbitrary (not necessarily right-angled) triangle is now straightforward. As shown on the right of Figure 7.6, any triangle can be divided by a perpendicular line drawn from the summit to the base, into two right-angled triangles. By drawing two lines perpendicular to the base on both sides, and a line parallel to it passing through the upper summit, we reproduce two situations similar to that on the left. In the two right-angled triangles thus obtained, we have $x + y =$ right angle, and $u + w =$ right angle, too. Then the sum of our original triangle's angles is $x + u + (y + w)$ = two right angles, and this completes the proof.

Now we can come back to the main theorem. We want to prove that the angle x of the triangle ABC is a right angle, i.e. $\frac{\pi}{2}$. Let us draw a radius OC dividing our triangle into two triangles AOC and BOC. These two triangles are isoceles, the first one with two identical angles α and the third angle p, the second one with two identical angles β and the third angle q. Obviously, $p + q = \pi$. The sum of the angles of the first triangle is $2\alpha + p = \pi$; the sum of the angles of the second triangle is $2\beta + q = \pi$. Therefore adding up the last two identities we get $2\alpha + 2\beta + p + q = 2\pi$. But $p + q = \pi$, therefore $2\alpha + 2\beta = \pi$, and we get the result $x = \alpha + \beta = \frac{\pi}{2}$, which completes the proof.

The fact that the sum of angles of a triangle is equal to two right angles is equivalent to the fifth postulate of Euclidean geometry.

It does not hold for triangles drawn on a sphere, which always have the sum of their angles greater than π. *]

7.3 Pythagoras and trigonometry

Pythagoras (570–495 B.C.E.) and his followers believed that "the world is ruled by numbers", and explored physical and astronomical phenomena seeking numerical rules and coincidences.

As of today, there exist more than 500 different proofs of his well-known theorem; one of the most elegant ones was given by Leonardo da Vinci.

The fact that triangles with one right angle have special proportions was known to the Babylonians, who gave several examples of right triangles satisfying the equation $a^2 + b^2 = c^2$, with integer values for a, b, c, for example $(3, 4, 5)$, $(5, 12, 13)$, $(20, 21, 29)$. The Chinese were able to prove this theorem in special cases, for example by simple counting how many small squares are needed to cover the square with side equal to 5, as shown in Figure 7.7.

Pythagoras' theorem opens the door to *trigonometry*. Let us divide the identity $a^2 + b^2 = c^2$ by c^2, and we get a new identity involving dimensionless ratios:

$$\frac{a^2}{c^2} + \frac{b^2}{c^2} = 1 \;\to\; \sin^2 A + \cos^2 A = 1. \tag{7.4}$$

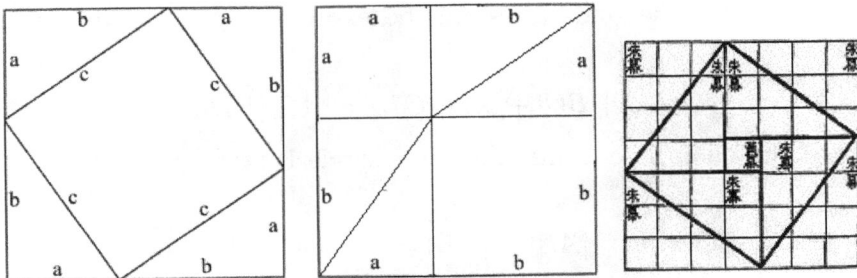

Fig. 7.7   The most convincing elementary proof of Pythagoras' theorem.

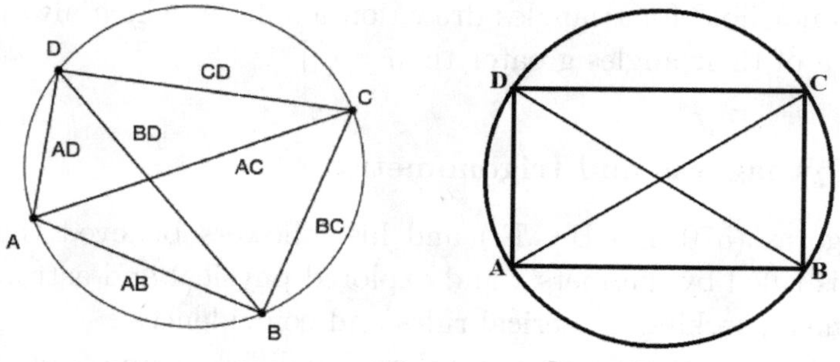

Fig. 7.8   Left: Ptolemy's theorem, general case; Right: Ptolemy's theorem for a rectangle inscribed in a circle, from which Pythagoras' theorem follows directly.

Introducing another trigonometric function, $\tan \theta = \frac{\sin \theta}{\cos \theta}$, one can express both $\sin \theta$ and $\cos \theta$ with $\tan \theta$ only:

$$\sin \theta = \frac{\tan \theta}{\sqrt{1 + \tan^2 \theta}}, \quad \cos \theta = \frac{1}{\sqrt{1 + \tan^2 \theta}}. \qquad (7.5)$$

[* What is less widely known is the theorem proved by Ptolemy, and bearing his name, which is stronger than Pythagoras' theorem. The corresponding geometrical construction is shown in Figure 7.8.

The theorem states that if a quadrilateral can be inscribed in a circle, then the product of the lengths of its diagonals is equal to the sum of the products of the lengths of the pairs of opposite sides. The converse is also true: if the sum of the products of the lengths of two pairs of opposite sides of a quadrilateral is equal to the product of the lengths of its diagonals, then the quadrilateral can be inscribed in a circle. Such quadrilaterals are called *cyclic*.

What the theorem states can be expressed with the following formula:

$$|AD|\,|BC| + |AB|\,|CD| = |AC|\,|BD|. \qquad (7.6)$$

If the quadrilateral inscribed in a circle is rectangular, then we have

$$|AD| = |BC| = a, \quad |AB| = |CD| = b, \quad \text{and} \quad |AC| = |BD| = c, \qquad (7.7)$$

therefore $a^2 + b^2 = c^2$.

This theorem was used by Ptolemy as an aid to creating the table of chords that he applied to astronomy. It played the role of trigonometric tables used in modern times. Also Copernicus used this theorem for the same trigonometric purposes. *]

7.4 Euclidean geometry

Euclid's (325–270 B.C.E.) *magnum opus* entitled *"The Elements"* became the basis of geometry teaching for the next millenia, translated into all major languages and studied by all mathematicians in Ancient, Christian and Islamic civilizations. The legend says that when Ptolemy the king asked Euclid whether he could learn geometry rapidly, without going through the "Elements", Euclid answered "there is no royal road to geometry".

The exposition of geometry written by Euclid summarized mathematical knowledge accumulated since many centuries by Egyptian, Babylonian and Greek scientists. Perhaps the greatest of Euclid's achievements was a unique and clear systematization of geometry, separating the most basic definitions (as many as 23 in total, beginning with a point being "that which has no parts", a line defined as "a length without breadth", and so on) from the rest of the text containing geometrical constructions and theorems. The definitions are followed by five unproved assumptions that Euclid called *postulates* (now known as *axioms*), which we present below:

I. Given two points, there is a straight line that joins them.
II. A straight line segment can be prolonged indefinitely.
III. A circle can be constructed when a point for its center and a distance for its radius are given.
IV. All right angles are equal.
V. If a straight line falling on two straight lines makes the interior angles on the same side less than two right angles, the two straight lines, if produced indefinitely, will meet on that side on which the angles are less than the two right angles.

The axioms (I) and (V) were refined by German mathematician David Hilbert as follows:

I. For any two different points, (a) there exists a line containing these two points, and (b) this line is unique.

V. For any line L and point p not on L, (a) there exists a line through p not meeting L, and (b) this line is unique.

The last (fifth) postulate seemed somehow less obvious than the first four. It is easy to see that it is equivalent to the statement that the three angles of a triangle sum up to two right angles.

Spherical geometry describes relationships between lines and points drawn on the surface of a two-dimensional sphere. There are no straight lines on a sphere, but this notion can be replaced by the *geodesics*, i.e. the shortest distance curves relying two arbitrary points on a sphere. They are synonimous with *great circles* passing through two given points. Triangles and polygons can be defined, and it turned out that such two-dimensional geometry follows the first four Euclidean postulates, but not the fifth one: the sum of angles of a triangle is always greater than two right angles, and the excess grows with size: for example, an equilateral triangle with each of its sizes equal to the quarter of total circumference has all his angles right (the sum being therefore three right angles, or 270°, and not 180° like in the Euclidean plane).

[* The *Elements* written by Euclides contained not only geometric theorems and constructions, but also basic rules of arithmetic and the wide collection of known facts concerning the theory of numbers. One of the most famous results is the proof that prime numbers form an infinite set, in other word, there is no such thing as the last (greatest) prime number. The proof is by construction and *reductio ad absurdum* (reduction to absurdity): let us suppose that N_f is the last and greatest prime number, divisible only by 1 and itself. Assume there are a finite number, n, of primes, the largest being p_f. Consider the number that is the product of these, plus one: $N = p_1 \ldots p_n + 1$. By construction, N is not divisible by any of the p_k, $k = 1, 2, \ldots, n$. Therefore, it is either prime itself, or divisible by another prime greater than p_n, contradicting the assumption.

Another famous proof by contradiction contained in Euclid's treaty is the theorem stating that square root of 2 cannot be represented by any finite fraction, which means that $\sqrt{2}$ is an *irrational number*. Ancient Greek mathematicians made a clear distinction between "rational" numbers, including integers and fractions, and what they called "magnitudes": lengths, areas, volumes, which could be rational or irrational. *]

The mathematical knowledge developed in Ancient Greece represented a tremendous step forward not only due to its applications in everyday life, but also as a necessary tool for next generations of astronomers and geographers. Armed with this invaluable tool, they were able to meet the challenge of evaluating great distances on land and seas, then Earth's dimensions and shape, and finally finding out the distances to the Moon and Sun. Although the last distance, found by Aristarchus and improved by Hipparchus, was seriously underestimated (about 20 times less than the real astronomical unit, A.U., established in a fully reliable way only by the end of 18th century), it represented an enormous leap, unparalleled until the Copernican revolution happened more than fifteen centuries later.

7.5 From a flat to a spherical Earth

The fact that the horizon becomes more distant with observer's growing elevation was known long ago, especially to sailors looking for a port to land. Greek geographer Strabo (ca. 64 B.C.E.–24 C.E.), who worked mostly in Rome, mentions the fact that at greater distances, but still visible, ships partly disappear under the horizon so that only their upper part can be seen above water, and concludes that the Earth has a spherical shape.

Let the radius of the Earth be $R = 6370\,\text{km}$, let the observer's position be h above the ground, and let the distance to the horizon, where the tangent line meets the Earth's surface, be x (all three lengths should be expressed with the same units, e.g. kilometers). From the rectangular triangle in Figure 7.9 with adjacent sides' lengths R and x and hypotenuse $R + h$ we get, using Pythagoras' theorem,

$$(R + h)^2 = R^2 + x^2 \;\rightarrow\; x^2 = 2Rh + h^2 \;\rightarrow\; x \simeq \sqrt{2Rh}. \qquad (7.8)$$

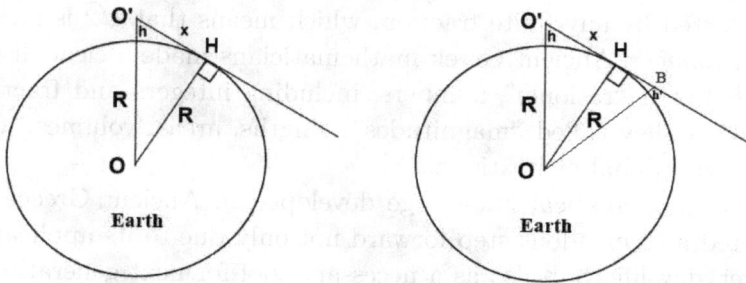

Fig. 7.9 Distance to the horizon depends on the observers' altitude (left). Higher objects can be seen from greater distance (right).

Table 7.1 Distance to the horizon as a function of the observer's altitude.

Height h in m	Horizon in km	Height in km	Horizon in km
1.5 m	4.5 km	0.3 km	62 km
5.0 m	6.1 km	0.5 km	80 km
10 m	11.3 km	1.0 km	113 km
50 m	25.2 km	3.0 km	195 km
200 m	50.5 km	10 km	357 km

The last approximation is valid for altitudes h negligible as compared to Earth's radius, which is true even for jets traveling at altitudes between 10 and 12 kilometers. With $h = 10$ km and $R = 6\,370$ km the expression $2Rh$ amounts to $127\,400$ km^2, while $h^2 = 100$ km^2, so that for the ratio $h^2/(2Rh) = h/2R$ we get $0.000785 \simeq 0.08\%$, a fairly acceptable margin of error. Table 7.1 shows the distances to the horizon for an observer whose point of view is at the height h above the ground.

The fact that the visibility range increases with the observer's altitude above the sea was known by the sailors since ancient times. The Vikings used to take a small flock of ravens on their boats when heading to unknown waters. Periodically, they would free the birds and observe them. As long as there was no land in sight, the ravens turned around above the boat; but as soon as they saw the land, they would fly in that direction, and the boat followed. Combined with the birds' excellent vision and the altitude of flight of 30 to 40 meters, new land could be discovered at distances up to 30 kilometers. Polynesian sailors crossing great distances between Pacific islands used birds (other than ravens, of course!) for the same purpose.

When sea traffic across the Mediterrenean Sea flourished, many lighthouses were erected in order to help sailors to find their way. The most famous was the *Pharos* built by the Ptolemaic kings in Alexandria, about 100 meters high and considered as one of the seven wonders of Ancient World, and remained there during almost a millenium until it was definitely destroyed by several earthquakes. Its light could be seen at open sea from a distance at least 40 km, and sailors with good eyesight could observe how its lower parts seemed to disappear under water when seen from a great distance — another proof of Earth's spherical shape.

7.6 How Eratosthenes measured the Earth

The first reliable estimate of Earth's dimensions were given by Eratosthenes of Cyrene (276–194 B.C.E.), who lived and worked in Alexandria. The Egyptian civilization flourished around the mighty Nile river which by its regular inundations leaving fertile mud deposit ensured abundant crop harvests and the wealth of the country and its inhabitants. The Nile valley extends thousands of kilometers in the South-North direction, almost along the same meridian. This particular configuration gave Eratosthenes the opportunity not only to prove that the Earth has a spherical shape, but also to give a fair estimate of its dimension.

People living in the southern city of Syena (called Assouan today) have since a long time noticed an interesting phenomenon: one day in the year, on June 21 (we are using today's name of the month and day, another calendar was in use in Hellenic Egypt at the time of Eratosthenes, but we know now that it happened on the day of summer solstice) the Sun was so high in the sky that its rays penetrated deep wells, illuminating them down to the bottom. No such phenomenon was ever observed in Alexandria, located more than 800 kilometers north of Syena. With the help of a gnomon Eratosthenes

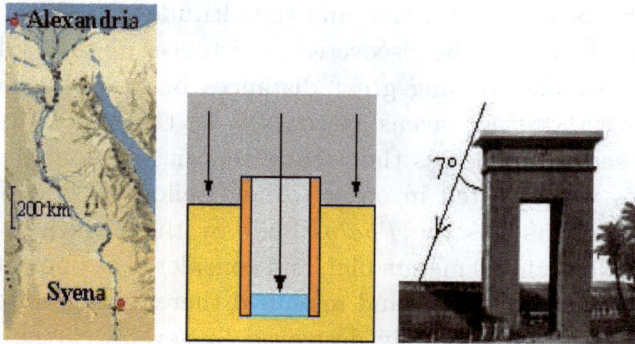

Fig. 7.10   Left: Map of ancient Egypt, with Alexandria and Syena; Right: Different shadows observed at summer solstice in June, in Syena and and Alexandria.

found the angle between the gnomon's top and the extremity of its short shadow equal to 1/50-th of a full circle (expressed in degrees, this gives $360/50 = 7.2°$.

The fact that the shadows at the same time of the year are longer in the North than in the South was already known to Anaxagoras, who supposed that the distance to the Sun was commensurable with the terrestrial scale, and the shadows were shorter in places under the Sun or close by, and longer far from the luminary. When the Sun moves from East to West, shadows change direction and length, too. Eratosthenes understood that the difference of shadows' lengths between Syena and Alexandria were not only a proof of Earth's spherical shape, but gave also the oppoturnity to measure the Earth's circumference, as shown in Figure 7.11.

Supposing that the Sun was so far away that its rays arriving at Earth's surface can be considered as being parallel, and using the similarity of triangles with the same angles, Eratosthenes concluded that the distance from Syena to Alexandria represented the segment of the meridian subtended by the angle between the vertical gnomon in Alexandria and solar rays at noon, i.e. one fiftieth part of the total circumference. What remained to be defined was the exact distance separating Syena from Alexandria. There were two other assumptions which were tacitly made, namely, that Syena was exactly on the Tropic of Cancer (almost true, today's Assouan

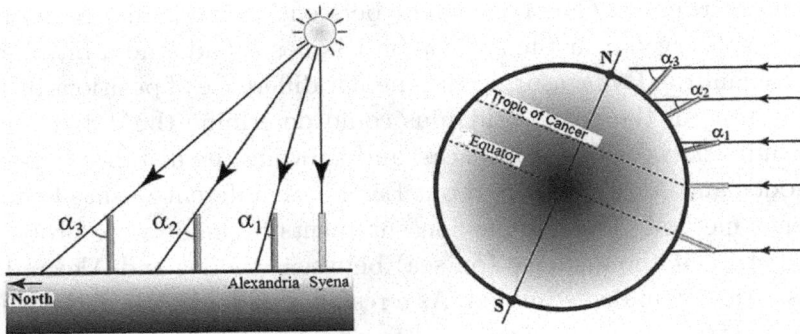

Fig. 7.11   Explanation of different shadows in Alexandria and Syena, by Anaxagoras (left) and Eratosthenes (right).

is on the 23.7° parallel), and that Alexandria and Syena are on the same meridian (less exact, Alexandria's longitude is 29.9° E, and Assouan's longitude is 32.72° E, more than 2.5° to the East). But the most difficult piece of the puzzle was the distance between the two cities. It was known to be five days of camel caravan, but such an estimate is not very reliable.

In order to make a reliable measurement Eratosthenes hired *bematists*, specially trained walkers with equal length steps. They found that the distance was about 5000 stadia, which enabled Eratosthenes to determine Earth's circumference to be 50 times greater, i.e. to be about 250 000 stadia long. No final agreement has been reached among the historians concerning the length of a stadium used by Eratosthenes. Different Greek cities at different epochs named so the specific measure of length then used. Values between 150 and 180 meters have been suggested, corresponding to Earth's circumference as measured by Eratosthenes to be between 37 500 km and 45 000 km, as compared with the actual value 40 000 km. So until now we do not know for sure whether Eratosthenes got it perfectly right, but his result was remarkably close in any case.

Sometimes "better" is the enemy of "good enough". Another Greek geographer and astronomer, Posidonius (135–51 B.C.E.) wanted to improve Eratosthenes' measurement using the bright star Canopus' position in the sky as seen from Alexandria and from the island of Rhodos. Canopus cannot be seen in mainland Greece, but can be observed near the horizon on Rhodos island, and a bit higher in Alexandria. By measuring the angular difference of positions at the same time of the year Posidonius could determine the difference of latitudes between the two places, and knowing the distance between Rhodos and Alexandria, find out Earth circumference using Eratosthenes method. The angular measurements he made were probably correct,[1] but the distance (on sea) between Rhodos and Alexandria was seriously underestimated. As a result, Posidonius' gave the value of terrestrial circumference equal to about 28 800 km, which is 0.72 times smaller than previous result established by Eratostenes.

This value was included by Ptolemy in his treatise on geography, and has become universally accepted. Perhaps it is because Columbus took this smaller size of our globe for granted that he dared to sail around it to reach India, and arrived in the New World. We should keep in mind that Columbus planned to sail westwards at latitudes between 40° and 30°, where the circumference (along a parallel) is reduced by a factor $\cos \lambda \simeq 0.8$. Ignoring the existence of the American Continent and above all, of the Pacific Ocean, and being aware of the distance eastwards from Spain to India and China, Columbus could reasonably hope to arrive at the eastern shores of Asia crossing the great ocean in western direction.

Precision measurements of the Earth's size were resumed in modern times using the method of *triangulation*. In order to determine the distance between two points O_1 and O_2 belonging to the same meridian, one has to start by determining as precisely as possible, the distance between the first extremity O_1 of the meridian and some point A, chosen not too far away from O_1 (e.g. the point A in Figure 7.12). The distance is measured using special measuring tapes. Triangulation towers are built in other places on both sides of the meridian, forming a triangular network with average distance of several kilometers, so that from each one of them at least two or

---

[1]Although they did not take into account the atmospheric refraction close to the horizon, which for Canopus resulted in a non-negligible effect (up to half a degree).

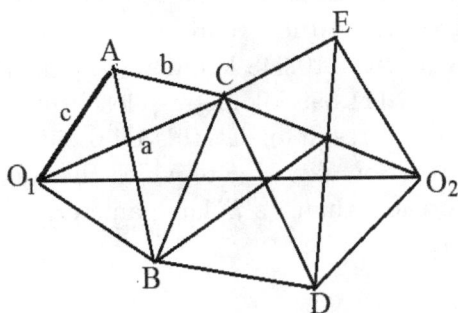

Fig. 7.12 Determining the length of meridian's segment O_1O_2 using the triangulation method. On the right, a typical triangulation tower.

three others are visible. Then the angles are measured using special optical devices called teodolites. With one side and two adjacent angles known, one can solve for the triangle and get the lengths of the other two sides, then proceed in the same way farther, until all distances in the network are determined. Finally, projecting them on the meridian O_1O_2, we get its length with as little error as possible. In Figure 7.12 the basis is chosen to be the line O_1A. Let its length be c, and the angles and sides of triangle O_1AC be $\hat{O}_1$, $\hat{A}$ and $\hat{C}$, and a, b and c, respectively. The law of sines gives then the following equalities:

$$\frac{\sin \hat{C}}{c} = \frac{\sin \hat{A}}{a} = \frac{\sin \hat{O}_1}{b}. \tag{7.9}$$

The first measurements of a meridian by triangulation were performed in 1615 in Holland by Snellius, and since then in many European countries, and finally everywhere over the world. Triangulation was particularly simple in Holland because of the ideally flat landscape. In countries where hills and montains prevail, great care must be taken to project the distances between triangulation towers onto the imaginary sphere reproducing sea level everywhere.

Of particular interest was establishing the exact length of one angular grade along a meridian. After many comparative measures it turned out that although 1° should correspond to 111.11 kilometers if the Earth was a perfect sphere, the reality is somewhat more complicated. In fact, one degree in latitude represent the distance of 110.567 kilometers at the equator, 110.948 kilometers at each of the tropics (latitudes $\pm 23.5°$), and 111.699 kilometers close to the poles. This can be understood as the result of Earth's ellipsoidal shape, slightly squeezed along its rotation axis, due to which the equatorial diameter of our planet is greater than its polar diameter, although by a relatively small amount.

7.7   How Aristarchus found distances to the Sun and the Moon

Aristarchus of Samos (310–230 B.C.E.) was not aware of Eratosthenes' estimate of the diameter of the Earth, made almost one century after his lifetime. But he understood the nature of solar and lunar eclipses as respective occultations of the Sun by the Moon, or the occultation of the Sun by the Earth as seen from the Moon. Aristarchus' own reports on his astronomical findings survived in the unique work *On the Sizes and Distances of the Sun and Moon*. We know about his discoveries also from Archimedes (288–212 B.C.E.). He decided to use these simple facts in order to find out the size of the Moon and its distance from the Earth in terms of Earth radii. After several observations of the shadow of the Earth during lunar eclipses Aristarchus came to the conclusion that:

1) the Moon receives its light from the Sun.
2) the Earth is the center of the sphere of the Moon (i.e. Moon's circular orbit).
3) when the Moon appears to us to be exactly halfway the great circle (Moon's meridian) dividing the light and dark portions of the moon is in line with the observer's eye.
4) when the moon appears to us to be halfway, its distance from the Sun is less than a quadrant by 1/30 part of a quadrant ($90° - \frac{1}{30} \times 90° = 90° - 3° = 87°$).

5) the width of the Earth's shadow (during eclipses) is that of two Moons.
6) the Moon subtends 1/15 part of a sign (of the Zodiac, i.e. 2°).

The last point seems strange, because already the Babylonians found that the Sun's angular diameter was close to $\frac{1}{720}$ part of the full circle, i.e. about 0.5°. Perhaps the error was committed by a scribe who copied the treatise, or it might be that the Greek word used by Aristarchus meant "one quarter of a Zodiac sign", which would correspond to $\frac{1}{15} \cdot \frac{1}{4} \cdot 30° = 0.5°$. Perhaps we shall never know.

Taking these assumptions as the starting point, Aristarchus deduced the following estimates of the Moon's size and distance from the Earth, using the geometrical construction displayed in Figure 7.13.

- The ratio of Moon's radius to Earth's radius is more than $\frac{19}{60}$ and less than $\frac{43}{108}$, i.e. close to one third. (Actually, this ratio is $1/3.65 = 0.274$.)
- The distance to the Moon is between 22.5 and 30 lunar diameters, or about 20 Earth's radii. (Actually, it is about 60 e.r.)

The distance to the Moon being known, Aristarchus used the assumption 4) to evaluate the distance to the Sun. which he found to be between 18 and 20 times greater than the distance to the Moon. Expressed in terms of Earth's radii it gives 380 e.r. (Actually, the

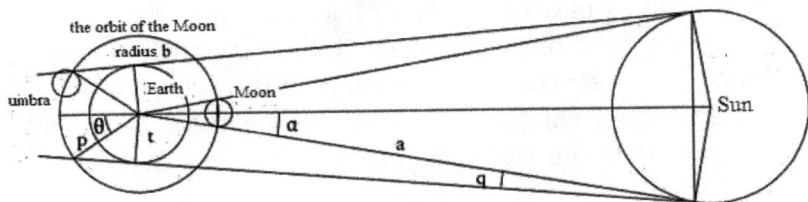

Fig. 7.13 Aristarchus' method of measuring the distance and the size of the Moon using Earth's shadow during lunar eclipses.

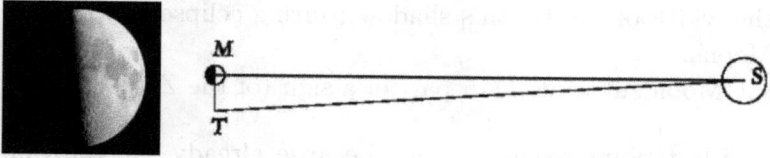

Fig. 7.14   Aristarchus' method of evaluating the distance to the Sun using the Moon's dichotomy.

mean distance to the Sun is equal to 23 500 Earth's radii — quite beyond imagination not only for the ancient Greeks, but even for Copernicus more than sixteen centuries later.)

7.8   Modern parallax measurements

Aristarchus' method based on the estimate of the small deviation from the straight angle between the direction towards the Moon and the Sun at the precise time when exactly half of the Moon's disc is illuminated by the morning Sun was not precise enough, and led to the strongly underestimated values of the distance separating us from the Sun. This in turn created the same error in determining the sizes of the orbits of all planets: the accepted dimensions of the Solar System were 20 times smaller than their real size. Even so, the Sun appeared to be more than five times larger than Earth, which served as a strong argument for the heliocentric model of the Universe proposed by Aristarchus. In spite of its simplicity and elegance, it was rejected by his contemporaries The stellar parallaxes are so small that they could be measured only with a new generation of powerful telescopes that were produced in the 19th century. As a matter of fact, even the nearest stars in the vicinity of the Solar System are so far away that their angular parallaxes are below one arcsecond; no wonder that neither in the 17th, nor in the 18th century telescopes had a sufficient resolution to discover it.

Let us recall the basic trigonometric relations between the distance to an object and its angular size. Dividing the circumference of a circle into 360 degrees, let us establish the ratio between the circle's radius and the length of the segment of the same circle corresponding to 1 degree. If the circle's radius is R, then the length of its circumference is equal to $L = 2\pi R$. The length of the small segment subtended by 1 degree is therefore

$$l = \frac{2\pi R}{360} = \frac{2 \cdot 3.14159}{360} R = 0.017453\,R, \qquad (7.10)$$

a very small fraction of the radius. The inverse of this fraction defines an alternative unit of angle, called the *radian*. When expressed in degrees, its value is $57.297°$, or $57°17'48''$; usually the value 57.3 gives a sufficient precision.

When we look at an object of given size d from a distance D, we see it under certain angle x, as shown in Figure 7.15. When the angle x is not very small, a substantial difference appears between the length of the arc subtended by x on the circle subscribed on the object, centered in the observer's eye, and the object itself, whose length is that of the corresponding chord. As follows from the figure,

$$\frac{d}{2} : D = \tan\frac{x}{2}, \quad d = 2D\tan\frac{x}{2}, \quad \text{length}_{(ab)} = xR\,\text{radians}. \quad (7.11)$$

Also, $D = R\cos\frac{x}{2}$, and because $\cos\alpha = \frac{1}{\sqrt{1+\tan^2\alpha}}$, we have $R = D\sqrt{1 + \tan^2(\frac{x}{2})}$.

The length of the arc (ab) is always greater than the size of the object $d = 2D\tan\frac{x}{2} < xR$; however, for small angles one can approximate $\tan\frac{x}{2}$ by $\frac{x}{2}$, and the two lengths will become equal, and we can

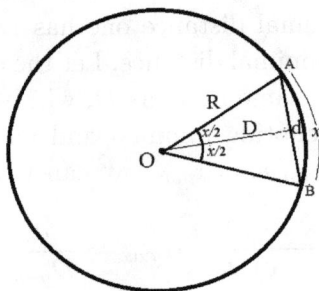

Fig. 7.15 Geometric relation between the size of AB an object and the angle x under which it is seen from a distance AO.

neglect terms of the order of x^2 and higher, setting $R \simeq D$. Then the simplified relation between the object's real size d, the angle θ under which it is perceived and the distance D separating it from the observer is

$$\theta = \frac{d}{D} \text{ radians}, \quad \text{or} \quad \theta = 57.3 \times \frac{d}{D} \text{ degrees.} \qquad (7.12)$$

Unfortunately, when we observe celestial bodies, only their angular sizes are accessible to our instruments, so that only *ratios* between sizes and distances can be evaluated, not the real dimensions in kilometers. Nevertheless, a lot of interesting information can be drawn from angular sizes, too, if we have a reliable geometrical model of what we actually see.

Let us evaluate the eccentricities of the orbits of the Earth and Moon comparing the maximal and minimal angular sizes of the Sun and Moon. Precise measurements give the following angular sizes:

- The Sun: minimal angular size is $31'27''$, maximal angular size is $32'32''$.
- The Moon: minimal angular size is $29'20''$, maximal angular size is $34'06''$.

The variations of angular sizes are due to the changing distance to a given celestial object, its size supposed to remain constant. Without precluding the shape of the orbit, which can be circular or elliptic, the eccentricity e measures the departure from the average distance D as follows: for maximal distance one has $D_{max} = (1 + e)D$, and $D_{min} = (1-e)D$ for minimal distance. Let the diameter of the object be d (measured in the same units as D, which may be Earth's radii e.r. or kilometers). Denoting minimal and maximal angular size of the same object by α_{min} and α_{max}, we can write

$$\alpha_{min} = \frac{d}{D_{max}} = \frac{d}{(1+e)D}, \quad \alpha_{max} = \frac{d}{D_{min}} = \frac{d}{(1-e)D}. \qquad (7.13)$$

Dividing the first equality by the second one, we eliminate both unknowns d and D, and get the equation relating the eccentricity

e and the two angular sizes:

$$\frac{(1+e)}{(1-e)} = \frac{\alpha_{max}}{\alpha_{min}},$$

$$(1+e)\alpha_{min} = (1-e)\alpha_{max},$$

$$e = \frac{\alpha_{max} - \alpha_{min}}{\alpha_{max} + \alpha_{min}}. \qquad (7.14)$$

Inserting the values of α_{min} and α_{max} for the Sun, we get

$$e_S = \frac{32'32'' - 31'27''}{32'32'' + 31'27''} = \frac{1952'' - 1887''}{1952'' + 1887''} = \frac{65''}{3839''} = 0.01693. \qquad (7.15)$$

A similar calculation gives the eccentricity of Moon's orbit $e_M = 0.076$.

7.9 Discovering the Globe

For a very long time the ancient Egyptians traveled exclusively along their great river Nile, so that their view of the world was limited by the Mediterrenean Sea in the North and Red Sea in the East, and seemingly limitless deserts on both sides of the Nile. Their world appeared so unique that when the first Egyptian explorers reached Mesopotamy, they were stunned by what they considered an unnatural phenomenon: a river can flow backwards — from North towards South, contrary to what the Nile does. Under later dynasties Egyptians progressively discovered other lands in Africa and Asia, conquering new territories, but were also invaded by others.

The world of the ancient Mesopotamian civilizations was larger, and in several conserved clay tablets one can recognise parts of Africa, the Middle East and Southern Europe.

A radical enlargement of horizons occurred with the development of the maritime nations, mostly Phoenicians and Greeks. With swift and reliable ships endowed with improved sails, Phoenicians colonized the North Africa and Spain, establishing colonies along the southern shores of the Mediterrenean Sea and as far as the Iberic peninsula's Atlantic coast (today's Portugal).

The ancient Greeks colonized most of the Eastern Mediterrenean and Black Sea. After the rise of the Roman Empire, it extended the control over the totality of the Mediterrenean shores, from North to South, so that it was called by the Romans *Mare Nostrum* — "Our Sea". The enlargement of horizons was acknowledged by written descriptions of distant lands and countries as well as of cities, people, trees and animals. When the Islamic world took the lead, the Arab and Persian navigators sailed regularly to India and Indonesia, and their travelers and polymaths contributed mightly to our knowledge of faraway lands and people.

The most prominent authors of geographical literature were at the same time great astronomers whose works we discussed in Chapter 4.

7.9.1   *The route to India*

Until the Renaissance, European sailors did not travel outside the Mediterrenean in the South, and the North Atlantic and Baltic Sea in the North. The essential knowledge about Asia was due to the Italian adventurer, traveler and diplomatic envoy Marco Polo who visited the Mongol Empire and China and in his famous report "*The Description of the World*" related the existence of a great island Cipangu (Japan) in the far East.

Since the conquest of the Middle East and North Africa all trade routes to India and China, were controlled by the Muslims, and sailing in the Southern Mediterrenean had become hazardous for the Europeans. The alternative ways to reach the fabulous world of India and China with highly appreciated spices, silk and precious stones would bring wealth and power to those who could discover them. This became the major challenge for the dominant European sea powers of 15th century, Spain and Portugal, During the reign of Henry the Navigator, the Portuguese conquered the north African cities Tanger and Ceuta, discovered Madeira and the Açores archipelago, and explored the Atlantic African coast down to the Cape of Good Hope (in 1487, by Bartolomeu Dias). The name was given later because the new way to reach India by sea was supposed to be at hand.

The route to India surrounding the African continent was explored by Portuguese sailor Vasco da Gama (1480–1524), The exclusively oceanic trajectory allowed the Portuguese to avoid the highly disputed Mediterranean waters and traversing on land the dangerous Arabian Peninsula. The sum of the distances covered in the outward and return voyages made this expedition the longest ocean voyage ever made until then, far longer than a hypothetical full voyage around the world along the equator.

Portugal's Iberian rival Spain followed the steps of its neighbor conquering the Canary Islands and some parts of the African coast, which remained dominated by the Portuguese.

7.9.2 *Columbus' discoveries*

Christopher Columbus (Cristoforo Colombo), was born in 1451 in the Republic of Genoa. As a young man he moved to Portugal, and later to Spain, where he became a renowned sailor. Having explored the Atlantic African coast and the Canary Islands, Columbus gathered a solid knowledge of currents and winds, arriving to the conviction that it was possible to reach India and the Far East by navigating westwards and circumnavigating the Earth. It remains a matter of discussion whether Columbus underestimated the size of our globe using the data of Ptolemy, who did not follow former results of Eratosthenes and Posidonius, his own estimate of Earth's circumference being 29 000 km, less than 75% of the real value, or deliberately did not share his knowledge with his sponsors-to-be. He proposed to open the new route to India to the Portuguese, but King Manuel was not interested, being certain that his Empire would control the route around Africa soon (and he was right: Vasco da Gama arrived by circumnavigating Africa to Calicut in India in 1498); but the total length of travel was greater than the equator, i.e. more than 40 000 km. Columbus made the same proposal to Spanish rulers, Isabel of Castille and Fernando II of Aragon, who accepted to finance the endeavour.

On August 3, 1492 Columbus and his crew set sail from Spain in three ships: the Niña, the Pinta and the Santa Maria, with the total crew of 86. On October 12, the ships made landfall not in the East Indies, as Columbus assumed, but on one of the Bahamian islands. He and his men looked for gold, silver, exotic fruit and spices, and tried to trade with the native people who were peaceful and had no arms at all. He also explored the northeast coast of Cuba, landing on

Fig. 7.16   Christopher Columbus, Vasco da Gama and Ferdinand Magellan.

October 28, 1492, and the north-western coast of Hispaniola, present day Haiti, by December 5, 1492. After the Santa Maria sank, 39 men were left to establish a fort, *La Navidad* (the Santa Maria sank on Christmas eve).

Columbus used the compass, the wonderful Chinese invention known to the Chinese sailors since the 11th century, and adopted by the Europeans hundred years later. The Portuguese King Henry the Navigator (1434–1460) recommended his sailors to use the magnetic compass and to draw accurate nautical maps. The constant use of the compass enabled navigating westwards with only three short intermissions, one caused by a contrary wind, two others due to false signs of land southwest. Columbus used also a quadrant, but apparently with very poor results: his estimates of latitude were often more than 10° off the real value. Nevertheless Columbus made an important astronomical discovery! Sailing westwards, he noted that the magnetic North did not coincide with the direction indicated by the Polaris star, which was about 3.5° off the North pole. Columbus crossed the Atlantic three more times, in 1493–1496, 1498–1500 and in 1502–1504, exploring the Carribean and the northern coast of today's Venezuela. Until the end of his life he was convinced that those islands were close to East India; this delusion remained in the name "Indians" he gave to the natives, and used until today in many languages. It was the Italian merchant, explorer and cartographer Amerigo Vespucci (1454–1512) who recognized that the lands found by Columbus behind the first Carribean islands were part of the huge continent which he named the *New World*; soon after the German cartographer Martin Waldseemüller (1470–1520) paid tribute to Vespucci putting the name "America" on the great unexplored continent beyond the Atlantic Ocean.

7.9.3 *Magellan's ultimate proof*

The ultimate proof that the Earth is round and can be circumnavigated belongs to the great Portuguese sailor and discoverer Fernand Magellan (Fernão de Magalhaes), who proposed to explore the route to East Asia and India going westwards and finding a way around the newly discovered lands.

Magellan was born to a noble Portuguese family; since the age of 15 he started sailing, first to India where he took part in battles with Muslim ships controlling the Indian Ocean; later he sailed to Malacca (today's Malaysia) in search for spices which were sold by their weight in gold at that times. In 1513 he took part in battles in Morocco, but accused by King Manuel of selling a part of the booty to the adversary, he was dismissed, and saught the protection of Spain.

The Portuguese were not interested having monopolized the route along the African coast, so it was only in the service of the Spanish King Carlos I that Magellan was able to realize the great expedition. On September 2, 1519 the fleet of five ships with 270 men aboard, mostly Spaniards, and about 40 Portuguese, set sail for the west. The names of the ships were: *Trinidad*, Magellan's flagship; *San Antonio, Concepciòn, Victoria* and *Santiago*. The fleet reached the bay of Rio de Janeiro in December, then continuing southwards discovered the La Plata estuary, interpreted erroneously as the strait they were looking for. Having understood that it was a great river, they navigated southwards until the place called now Port San Julian and stayed there for 5 months awaiting the summer in the Southern Hemisphere. A mutiny of Spanish captains was quelled by Magellan, but one of the ships was lost and its crew left behind.

Sailing farther to the South, on October 21, 1520 Magellan found the strait that opened the way to the west, and the great Pacific Ocean. Another ship defected, her captain and crew deciding to go back to Spain, and the flagship *Trinidad* had to be abandoned being badly damaged. The two remaining ships, *Victoria* and *Santiago*, continued in dire conditions — hunger, thirst, scurvy — until after 99 days they reached the first islands of Polynesia, with fresh water and food available at last.

Magellan decided to head for the Philippines, and finally the Moluccas, which he already visited before, sailing from the west. Unfortunately, in a fight with one of the tribes of the Mactan island Magellan was killed by a poisoned arrow. Magellan's companion Juan Sebastian Elcano arrived back in Spain with one ship (*Victoria*) loaded with spices and with only 19 men after three years of navigation. According to their records, the arrival occurred on Friday, September 5, 1522, but in Spain it was already Saturday, September 6. The Church imposed penance on Elcano and his crew, because they ate meat on Fridays and celebrated Easter on Monday. This was the first time when the round shape of the Earth was directly proven. However, the loss of one day due the rotation of the Earth was not interpreted properly: in the Ptolemaic model, with the Sun orbiting around the Earth one time a day, the result would be the same.

Magellan's name was given to the strait between the Atlantic and Pacific oceans, and to the dwarf galaxies seen only in the southern sky and called the "Magellan's clouds" today.

7.10 Maps and cartographers

The first representations of the inhabited world can be attributed to the Babylonians. A clay tablet with a circular map endowed with cuneiform inscriptions is conserved in the British Muesum; it is more than 2600 years old. The world of Babylonians was not vast by our standards: it was centered on the Euphrates river, flowing from the north to the south. The city of Babylon is shown on the Euphrates, in the northern half of the map. Susa, the capital of Elam, is shown to the south, Urartu to the northeast. Mesopotamia is surrounded by a circular ocean, and eight regions, depicted as triangular sections, are shown as lying beyond the ocean.

The world of the ancient Greeks was centered on the Meditterrenean Sea. For the great ancient geographer and historian Herodotus (active between 440 and 425 B.C.E.) the *Ecumene*, or the inhabited world, was divided in three parts, Europe, Asia and Africa (which he called Libya), surrounded by great ocean. He heard about India, but apparently ignored the existence of the Far East and China.

With time, navigators needed better maps guiding them in the open sea, and showing their whereabouts as precisely as possible. The same can be said of merchants traveling on land, sometimes for months and even years, often through yet unexplored countries. Although maps representing the known parts of the world were drawn since very ancient times, they were primitive and gave a very distorted image of lands and seas.

600 year after Herodotus Ptolemy's described the lands and seas known to the Hellenistic culture at the peak of the Roman expansion in the second century C.E. in his eight-volume work *Geography.* Although the original maps did not survive, their content was fairly well related by later Arabic translations and Byzantine editions of Ptolemy's work; the Latin translation appeared in 1406. The inhabited world known to Ptolemy comprised not only India and Ceylan, but also China and other Far Eastern lands, however without Japan, of which the existence was still ignored. It seems that Ptolemy used something like the conic projection, to take into account the Earth's spherical shape. The most important innovation consisted in adding the lines of constant longitude (meridians) and parallels, endowing maps with an elementary coordinate system.

The beginning of professional cartography can be traced to the middle ages in Europe, where the first navigational charts were produced by Petrus Vesconte at Genoa in 1311. These charts, called also *portolan charts* or *rhumb charts* were characterized by rhumb lines radiating from a given port, indicating to the pilots the directions towards different harbours. Most were made in Italy, in Catalonia, and a few in Portugal. They frequently had descriptive legends, which usually consisted of drawings integrated into the actual map. One of such maps was the heavily decorated Catalan Atlas, which was produced in 1375. It was presented as a gift to the King of France from the King of the Crown of Aragon.

Throughout the 15th and 16th centuries the two Iberian naval powers, Portugal and Spain, dominated the seas; quite naturally, they became also the centers of map-making. The most prestigious one was the "School of Reinel" in Lisbon, founded by Pedro Reinel (1462–1542) and continued by his son Jorge Reinel, who worked also in Sevilla, where he collaborated in the preparation for the circumnavigation of Ferdinand Magellan in the service of Castille. Many extraordinary 16th century maritime charts are conserved at the University libraries in Coimbra (Portugal) and Salamanca (Spain).

After the discovery of America and the Pacific Ocean, the full image of our globe could be reproduced using spheres made of metal or wood. But for practical use the two-dimensional maps printed on paper were much better adapted. This required geometrical methods of projecting a sphere onto a plane. In the case when a territory described by a map does not extend more than a few degrees in latitude and longitude, the distortion due to the difference between planar (the map) and spherical (the Earth's surface) geometry is not very important; but as the area of represented regions of the globe grows, the distortion grows as well, so that distances and directions do not correspond to the real ones. One must adapt the way of mapping a sphere on the plane so as to make the distortion as small as possible, having to choose between keeping the angles but deforming the distances, or keeping distances but with angles slightly deformed.

Johannes Ruysch (1460–1533) was probably the first cartographer to apply what can be recognized as a true conic projection (Figure 7.17) in 1507.

A new and original representation of terrestrial globe on a plane, with special properties making it particularly useful for navigation, was created by Mercator in early 17th century.

Geert de Kremer, universally known as Gerardus Mercator, was born in 1588 in the Country of Flanders. His interests covered geography, cartography and cosmography. He became a successful producer of instruments for navigation and globes; but above all, he owes his fame to the invention of cylindrical projection which permits representing the spherical surface of the Earth on a plane. The Mercator projection consists in enveloping the sphere by a cylinder of the same radius, then to project the surface of the sphere parallelly onto the

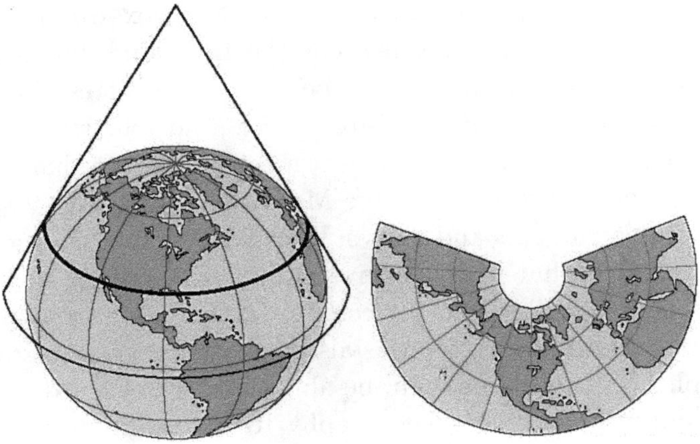

Fig. 7.17 The principle of the conic projection. Credit: Wikimedia commons.

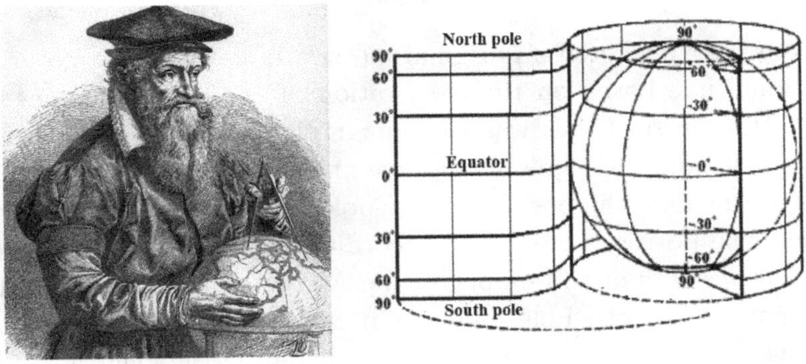

Fig. 7.18 The principle of Mercator's projection.

cylindric surface by drawing rays from the cylinder's axis, as shown in Figure 7.18.

The Mercator projection has a unique property of conserving angles and directions with respect to meridians and parallels. A straight line joining two arbitrary points on such a map crosses meridians and parallels under a constant angle. This angle corresponds to the angle sailors navigating in the open sea have to keep with respect to the fixed direction, e.g. the Polar Star in the North. The resulting trajectory is called a *loxodrome*, or a *rhumb line*;

it crosses the meridians under the same angle. A loxodrome is not the shortest way joining two given points on the globe, but it is much easier to follow when one is at open sea. The shortest line on the surface of the Earth is always a great circle, but following it requires constant changes of direction with respect to the meridians.

During the first part of his life Mercator was successfully selling navigation instruments and globes; they were so popular and in such a great number, that many survived until today in the universities, libraries or city halls. Also his celestial globes were considered the finest in the world. Mercator was also an accomplished engraver and calligrapher. Curiously enough, he almost never travelled, owing his extraordinary knowledge of geography to the information gathered from his visitors and from vast correspondence in six languages not only with other scholars, but also with merchants and seamen. He owned a library of more than a thousand books and maps of various origins.

Mercator authored an original method of determining geographical longitude based on the observation of the terrestrial magnetic field. His central thesis was that magnetic compasses are attracted to a single pole along great circles passing through it. He then showed how to calculate the position of the pole if the deviation is known at two known positions on Earth. According to the data he gathered at Leuven in Belgium and on one of the Açores islands, he found that the magnetic North Pole must be at latitude 73°02′ and longitude 169°34′. This led to the possibility of calculating the longitude difference between the pole and an arbitrary position, which would solve the longitude problem. Unfortunately, this method was not precise enough to be used by sailors at that time.

Chapter 8

The Copernican revolution

Copernicus' origins - Studies and early career - Scientific prede-
cessors - "Commentariolus", the first exposition of the heliocentric
ideas - "De Revolitionibus" and its revolutionary content - Critics
and enthousiasts - Rheticus and Giordano

8.1 Copernicus' life

Nicolaus Copernicus (Polish name: Mikołaj Kopernik) was born in
1473 in Toruń (Thorn in German), a city that belonged to the
Teutonic Order until 1466, when the Warmia region became a dependence of the Kingdom of Poland. Therefore Copernicus was a Polish
subject from his birth, although he was most probably bilingual,
speaking also German, his mother's name being Barbara Watzelrode.
Copernicus father's family went to Toruń from the Krakow region,
according to some historians.

It should be stressed here that in 15th century Europe people
considered themselves first and above all as Christians, and next as
subjects of the Monarch who ruled their country. They were more
attached to their region than to a kingdom or principality. They
spoke the local vernacular language, but the educated people used
Latin. All official documents were written in Latin, and learned clercs
and priests spoke the so-called medieval Latin quite fluently. Also at
the universities Latin was used by teachers and students alike. Young
people willing to study at the best European universities could follow

245

Fig. 8.1   Left: Copernicus; Right: A romanticized portrayal by Polish painter Jan Matejko (1838–1893).

lectures delivered in Oxford, Paris, Salamanca, Cracow without being forced to learn any extra tongue — Latin was common everywhere.

After the young Copernicus had lost his father, his uncle Lucas Watzelrode took care of him. Having completed his education in St. John's school in his native town Copernicus went to study at the Jagellonian University of Cracow, then the Polish capital city. He learned mathematics and astronomy under Albert Brudzewski. His uncle arranged for him the position of canon (specialist of canon ecclesiastic law), but Copernicus asked for permission to continue studies at the University of Bologna. After three years spent there, Copernicus went to Rome, where he taught mathematics during one year, after which he returned to Poland in 1500, to start his duties as canon of Frombork (Frauenburg), the position ensured by his uncle and protector. He obtained another permission to continue studies in Italy, and came to Padova in 1501, then to Ferrara in 1503, where he was awarded the title of Doctor of Canon Law.

Two years passed before he returned home and stayed there until his death in 1543, working relentlessly on his astronomical theory, but without neglecting official duties. He made an important contribution to economical science of that time, formulating what is also known as Gresham's law: *"Bad money drives out the good one"*. Copernicus was the author of an important economic treaty. The first draft of

"De estimatione monetae", written in 1517 contained an observation that *"bad coins push the good ones out of the market"*.

Copernicus hesitated a lot before he decided to publish the treatise of his life, *"De Revolutionibus Orbium Coelestium"*, perhaps aware of its content being theologically controversial. Nevertheless, since its short version, the *"Commentariolus"* continued to be copied and circulated among learned people, his reputation was steadily growing across the Europe. In 1539 a young professor of the University of Wittenberg, Georg Joachim Rheticus (1514–1574) visited Copernicus in his town Frombork with intention to learn more about the new ideas on the structure of the Universe. He was so impressed that he stayed with his new Master for the next two years, studying diligently, and became a true enthousiast of heliocentrism. Upon his return to Germany Rheticus published the first account of the new views in his *De libris revolutionum Nic. Copernici narratio prima* ("The First Account of the Book on the Revolutions by Nicolaus Copernicus"). This "Narratio Prima" has played an important role in propagating Copernican ideas.

Rheticus encouraged Copernicus to complete his great work, and after another year, took it to Nürnberg for publication. In 1543 he brought the first copies of the great treatise to Frombork when Copernicus was on his deathbed.

8.2 Sources of inspiration

It is hard to tell when exactly the idea of the heliocentricity started to germ in Copernicus' mind. It is highly probable that he started thinking about possibilities offered by the heliocentric model of the universe still during his scholarly years in Cracow Academy, where he got his fundamental astronomical education, being also acquainted with works criticizing the geocentric Ptolemaic system, especially those of Arab and Muslim astronomers, like al-Bitruji (Apetragius) whose works he abundantly cites.

Some historians suggest that inspiration could come from reading a text by Cicero (106–43 B.C.E.) entitled *Somnium Scipionis* ("Scipio's dream"), in which Scipio learns in his dream the true structure of the Universe, with *Sun being almost in the center, accompanied by the orbits of Venus and Mercury.* Another possible influence could be the popular book by Marsilio Ficino (1433–1499) entitled *De sole et lumini* ("On the Sun and light"), in which the Sun is presented as symbol of God, created first and placed in the center of the Universe. The argument is repeated almost literally by Copernicus in the foreword to his great work *De Revolutionibus Orbium Coelestium* as follows:

> "*Not without reason, some people call the Sun the lantern of the world, other call it the world's mind, and yet others its Lord.*"

Copernicus diligently studied ancient astronomers, both Greek and Islamic ones. However critical the last ones could be, they all were tributary to the Pythagorean postulate that no other motion can exist in the heavens beside the uniform circular one — this was like a revealed truth, also for Copernicus. All geometrical constructions tending to explain the planetary motions could be made using exclusively various combinations of circles.

Although Copernicus' work was mostly theoretical, he did not neglect astronomical observations, which rare as they were, enabled him to confirm most of his theoretical constructs and even discover something new, namely the slow motion of the planetary apsides.

It is important to assess the content of Copernicus' treaty in the context of the 16th century and not from the point of view of present astronomical knowledge. His geometrical construction was different from the Ptolemaic model, although it was also based on the use of the epicycles; yet the size of his Universe remained the same, based on Aristarchus' evaluations, adopted by Ptolemy in his *Syntaxis*, and later on by all Arab and European astronomers. This is extremely important, because it sheds light on how Copernicus imagined planetary sizes as compared to the size of the Earth, equally underestimated by many in his time.

In Ptolemy's, as well as in Copernicus' Universe the Earth's diameter was about 9500 km (75% of its real size); the Moon was about three times smaller, with a diameter of about 2650 km. The distance to the Moon as derived from its angular size was quite well estimated in relative terms, as equal to 110 times its diameter, or 60 Earth's radii. The distance to the Sun was still based on Aristarchus' angle of 87°, which gives the result 7 500 000 km, 20 times smaller than the real distance. As a consequence, the diameter of the Sun was believed to be 5.5 times greater than that of the Earth.

Whatever could be said on sizes of the five planets, one thing was certain: their angular sizes, without exception, were under the resolution limit of the naked eye, which means less than one arcminute. Knowing the relative distances from planets to the Sun, it is easy to find out the closest distances to planets from the Earth: for two inferior planets the shortest distance is when the planet is in lower conjunction, and for the superior planets the closest approach occurs in the opposition. The distances to the planets in the heliocentric system could be easily computed in terms of the Earth-Sun distance.

Now, if we observe a celestial body of diameter d from a large distance L, its angular diameter will be $\alpha = d/L$ (in radians). Therefore if $\alpha < 1' = 1/3438$ radian, then $d < L/3438$. The resulting estimates of the diameters of the planets are displayed in Table 8.1 (at distances closest to the Earth).

The picture of the solar system as proposed by Copernican model was quite coherent. The Sun was by far the biggest body occupying the central place; the Earth was next in size, then the Moon. Inferior planets and Mars were by far smaller than the Moon; only the two slowest and largest planets could be compared in size with the Earth.

In his later years Copernicus improved Aristarchus' and Ptolemy's data. In 1535 he gave a new estimate of the Earth-Sun distance,

Table 8.1   Distances, diameters and angular sizes of the Sun and planets according to Copernicus

Planet	Sign	Angle	Distance	Radius	Diameter
Sun	☉	31'	1.0 A.U.	5.5 e.r.	52370 km
Earth	⊕	–	–	1.0 e.r.	9500 km
Moon	☽	32'	0.05 A.U.	0.3 e.r.	2850 km
Mercury	☿	<1'	0.65 A.U.	< 0.3 e.r.	< 1400 km
Venus	♀	<1'	0.3 A.U.	< 0.14 e.r.	< 900 km
Mars	♂	<1'	0.52 A.U.	< 0.24 e.r.	< 1150 km
Jupiter	♃	<1'	4.2 A.U.	< 1.93 e.r.	< 9100 km
Saturn	♄	<1'	8.5 A.U.	< 4 e.r.	< 18400 km

corresponding to solar parallax of 2.3'. This was still 15.6 times less than the actual size, but represented a step forward in the right direction.

8.3   The first attempts

Copernicus' first treaty on astronomy was written as early as in 1512, upon his return from Italy. The text is known under its latin title *"De hypothesibus motuum coelestium a se constitutis commentariolus"*, meaning "Little commentary on celestial motions", or simply as *"Commentariolus"* given to it by Tycho Brahe, who possessed one of its written copies. Perhaps the original title, lost since, was just "Theorica", but we cannot be sure about that. Only three copies of this early astronomical essay by Copernicus survived until today in libraries, in Vienna, in Stockholm and in Aberdeen. During his studies in Ferrara, Copernicus was under a strong influence of the Pythagorean worldview recommending keeping new findings secret for the unprepared minds. An echo of this attitude can be found later in his masterpiece *De Revolutionibus* where Copernicus cites on the title page the mythical inscription at the entrance of Plato's Academy, "Let no one untrained in geometry enter here". So only a few handwritten copies circulated among Copernicus' friends.

In his *"Commentariolus"* Copernicus already presented seven axioms of his new vision of the Universe:

1: No common center exists for all celestial spheres.
2: The center of Earth is not the center of the Universe; it is only the center of the Moon's orbit.
3: All planets revolve around the Sun which is close to the center of the Universe.
4: The ratio between the distance from the Earth to the Sun and the distance to the firmament of stars is much smaller than the ratio between Earth's radius and its distance to the Sun. The stars are much farther away from us than it was believed in Ptolemy's system.

5: The common motion of all stars of the firmament is not real, but results from the Earth's own rotation around its axis, defined by the poles.

6: The Sun's visible annual displacement with respect to the stars is not the result of its real motion, but just an illusion due to the Earth's orbital revolution around the Sun (or around a point not far from it). The motion of the Earth is the combination of its annual and diurnal rotations.

7: The retrograde motion of planets is not their own motion, but an illusion due to the motion of Earth.

These principles found their ultimate realization in Copernicus' *magnum opus* "De Revolutionibus Orbium Coelestium".

Copernicus continued to work on the model announced in his *Commentariolus* for the next twenty-odd years. During all that time he was not entirely convinced about the superiority of his system over the Ptolemaic one. He knew well that he could not expect recognition from other learned men of his time without being able to improve the model in a way that it would give predictions as good or better than the geocentric system in use. The search for improvements took a lot of Copernicus' time and efforts. However, his mind was still under the influence of Aristotle's doctrine about the necessity and naturalness of uniform circular motions in the celestial realm.

8.4 The Copernican heliocentric system

De Revolutionibus Orbium Caelestium, Copernicus' *magnum opus*, is composed of six books, of which the best known is the first one, containing the full and well-argumented exposition of the heliocentric system. Here are the first eleven chapters of Book I:

Chapter 1. The Universe is spherical.

Chapter 2. The Earth too is spherical.

Chapter 3. How the Earth forms a single sphere with water.

Chapter 4. The motion of the heavenly bodies is uniform, eternal, and circular or compounded of circular motions.

Chapter 5. Does circular motion suit the Earth? What is its position?

Chapter 6. The immensity of Heavens compared to the size of the Earth.

Chapter 7. Why the Ancients believed that the Earth remained at rest in the middle of the Universe as its center.

Chapter 8. The inadequacy of the previous arguments and a refutation of them.

Chapter 9. Can several motions be attributed to the Earth? The center of the Universe.

Chapter 10. The order of the heavenly spheres.

Chapter 11. Proof of the Earth's triple motion.

The chapters' titles of speak for themselves. The first arguments for a round Earth are the same as were put forward by Aristarchus, Hipparchus and Ptolemy. Copernicus concludes by saying that *"Therefore the Earth is not flat, as Empedocles and Anaximenes thought; nor drum-shaped, as Leucippus; nor bowl-shaped, as Heraclitus; nor hollow in another way, as Democritus; nor again cylindrical, as Anaximander; nor does its lower side extend infinitely downward, the thickness diminishing toward the bottom, as Xenophanes taught; but it is perfectly round, as the philosophers hold"*.

Next, Copernicus refutes the arguments that were currently used for defense of the idea of Earth's central position and immobility. Some of his arguments announce the ideas put forward by Galileo more than half a century later.

One of the arguments against the Earth's orbital motion which was expressed many times since Hipparchus and Ptolemy is the following: if such was the case, a terrible wind will tear off everything on the Earth's surface. Copernicus replies that, contrary to what was taken for granted, the air does not fill the entire Universe, at least up to the orbit of the Moon, but extends only for a small fraction of the Earth's radius, and is trained along. In fact, Copernicus defines the Earth's atmosphere, comparing it with water, which also follows the Earth's motion.

In Chapter 4 Copernicus follows Aristotle affirming that all motions of heavenly bodies must follow circular paths, and the motion along those circles must be uniform. However he does not exclude the possibility of composition of several such motions.

Chapters 5, 6 and 7 contain the discussion on Earth's position in the Universe, the arguments of ancient astronomers for its central position and immobility, and comparison of distances in the Universe. After the well-documented survey of dominant views, citing all ancient sources — with a very strange omission of Aristarchus of Samos, who proposed a heliocentric system long before, but whose model was rejected by contemporaries — Copernicus proceeds in Chapter 8 to a critical refutation of ancient and contemporary views, citing abundantly Hipparchus and Ptolemy, as well as the Arab astronomers al Bittani and al Bitruji (under their latinized names Albatanius and Alpetragius). Figure 8.2 is taken from the first edition of Copernicus work.

The distances between the planets and the Sun are not to scale, of course. But Copernicus was able to determine the radii of planetary orbits quite well — in terms of the Earth's orbit's radius, i.e. the astronomical unit (A.U.), which according to the evaluation made by

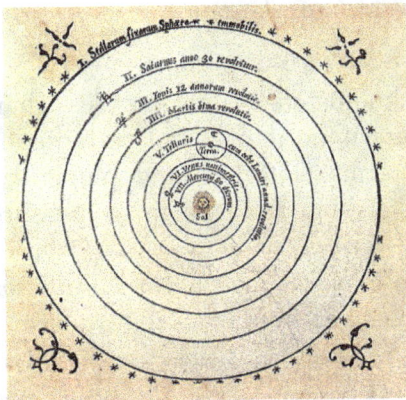

Fig. 8.2   Left: The heliocentric system proposed by Copernicus as it appeared in the first edition of *"De Revolutionibus..."*, printed in 1543 in Nuremberg; Right: The geocentric Ptolemaic system.

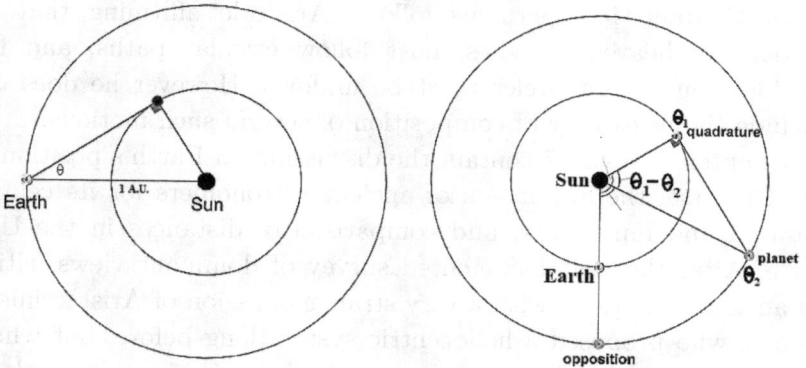

Fig. 8.3   How Copernicus determined the sizes of Venus' and Mercury's orbits in terms of Earth's orbit radius.

Table 8.2   Relative radii of planetary orbits as compared with the actual average distances of planets from the Sun.

Planet	Symbol	Copernicus	Modern
Mercury	☿	0.38	0.39
Venus	♀	0.72	0.72
Earth	⊕	1.0	1.0
Mars	♂	1.52	1.52
Jupiter	♃	5.2	5.2
Saturn	♄	9.2	9.54

Aristarchus, was about 7.5 million kilometers — 20 times less than the actual size we know now. To determine the radii of two inferior planets, Mercury and Venus (the terms "inferior" and "superior" planets was introduced by Copernicus), it was enough to measure their maximal elongations.

As shown in Figure 8.3, the ratio between the radii of an inferior planet's orbit and the Earth's orbit is given by $\sin \theta$, if the two orbits are circular. In the case of Venus it is practically the case; for Mercury, which has a highly eccentric orbit (ca. 0.206), the result was the average of many measurements.

The elaborated model presented in *"De Revolutionibus"* is more sophisticated than the one described in the *Commentariolus*. It was

Table 8.3   Periods of planetary orbital motions, with epicycles: the comparison between the Ptolemaic and Copernican systems.

Planet	Ptolemaic	Geocentric	Copernican	Heliocentric
	Epicycle	Deferent	Epicycle	Deferent
Sun	0	1 year	0	0
Mercury	80 days	1 year	0	80 days
Venus	9 months	1 year	0	9 months
Earth	0	0	0	1 year
Mars	1 year	2 years	0	2 years
Jupiter	1 year	12 years	0	12 years
Saturn	1 year	30 years	0	30 years

clear that circular orbits along which planets were supposed to move with constant velocities could not describe the variation of angular velocities of real planets.

Nevertheless the retrograde motion of planets was taken care of, being the effect of combined motions of planets and the Earth orbiting around the Sun with different periods. Thus the Copernican model made superfluous the "great epicycles" of the Ptolemaic system, as can be seen from Table 8.3

Note that the planets of the Solar System are ordered according to the heliocentric model, and not along the lines of the Ptolemaic system where the Sun is orbiting between the deferents of Venus and Mars. Another advantage of the Copernican system is that it does not anymore need the so-called "great epicycles", as the annual periodicity in planets' motion became the result of Earth's annual revolution around the Sun — in other words, a consequence of observing the planets from a moving "platform".

Copernicus proudly announced at the end of *De Revolutionibus* that his heliocentric system needs only 34 circular motions to give a complete description of celestial bodies' motions: *"Mercury's orbit needs 7 circles, Venus' orbit needs 5, Moon's orbit needs four circles, that of Mars, Jupiter and Saturn — 5 circles each. Therefore 34 circular motions suffice to explain the construction of the Universe and all planetary motions within"*.

8.5 The next five books

The treaty written by Copernicus was a result of a life-long endeavour consisting in meticulous analysis of astronomical knowledge accumulated in ancient and contemporary books, including the improvements made in Islam. The impact of the heliocentric system proposed in the Book I was so great among the generations that followed Copernicus, that the content of the next five Books is less known and commented upon. Yet, the work of Copernicus represents the summary of knowledge available in his time, exposed in a very clear and organized manner, unparalleled since Ptolemy's "Syntaxis", and earning him — though posthumously — the title of the greatest astronomer of his time.

Here is a brief description of the content of the five remaining Books.

- Book I: The last chapters of this Book not mentioned in the previous section, from Chapter 12 till 14, contain an introduction to the geometry of circles and straight lines, theorems on polygons circumscribed by a circle, the proof of Ptolemy's theorem, tables of arcs, chords and half-chords (6 pages, the difference between items being 10′ arcminutes). The last Chapter 14 is devoted to spherical geometry and stereometry.

- Book II: The first five chapters contain the definitions and data concerning the ecliptic, the constellations of the Zodiac, explanations about how to determine the obliquity of the ecliptic and the distance between the tropics, and tables of right ascensions and declinations, meridian angles and differences in noon shadows at different latitudes. Chapter 7 contains the times of risings and settings of planets, celestial coordinates of zodiacal constellations, and a descriptive catalogue (33 pages) of Zodiac signs and stars composing them.

- Book III: Describes precession of the equinoxes, based on the observations of Spica, the brightest star in Virgo constellation. Copernicus compares its position with respect to the equinox as observed by al-Battani 1202 years after Alexander's the Great death, and his own measurements performed 674 years later. The conclusion is that the precession is not uniform: slightly more rapid during the period of time from Aristarchus till Ptolemy, and slower from Ptolemy till al-Battani. Chapter 4 contains the model of the oscillating motion of libration which is obtained by the same method as the Tusi couple, creating rectilinear motion by rolling a smaller circle inside the circle twice as big. It remains unclear whether Copernicus knew Tusi's writings, or re-invented this geometric procedure independently, the last version being the most probable one. The obliquity of the Earth's orbit is discussed, too; Copernicus' attributes the value of $23°28'$ to it.

Book IV: Here Copernicus discusses the Moon's motions, with tables of lunar positions during the past 60 years. The full theory of lunar motion is presented, with epicycles and libration. Solar and lunar parallaxes are discussed. Copernicus also relates his observation of the occultation of Aldebaran by the lunar disc on March 9, 1497. Oppositions and conjunctions of planets are calculated and presented in tables, and eclipses and their durations are predicted.

Book V: The planets and their apparent motions are presented in tables 10 full pages long. The retrograde motion is explained and its duration calculated. This book contains the mathematical tables for *prosthaphaeresis*, which was the method of multiplication and division of great numbers using trigonometric formulas. Later on the prosthaphaeresis was replaced by tables of logarithms, more efficient, but based on the same principle: replacing multiplication by addition, which is much easier to perform, even on great numbers. Here is how it worked: Suppose that we want to multiply two big numbers A and B, both smaller than e.g. 10 000. Divide A and B by 10 000, defining two numbers smaller than 1, identifying them as $\sin a$ and $\sin b$, then use the formula

$$\sin a \cdot \sin b = \frac{1}{2}\left[\cos(a-b) - \cos(a+b)\right]$$

and perform the addition, which is an operation much less complicated than multiplication. The three other formulae can be used, for the products $\cos a \cdot \cos b = \frac{1}{2}[\cos(a-b) + \cos(a+b)]$, etc.

Book VI: In Chapter 1, a general explanation of latitude deviation of the five planets is given. In Chapter 2 the scheme of circles by which these planets move is presented. In Chapter 3 the inclinations of the superior planets are discussed, and in Chapter 4 a general explanation of other latitudes of Saturn, Jupiter and Mars is presented. Chapters 5 and 6 contain similar discussion and data for the two inferior planets, Mercury and Venus, inclinations of their closest and farthest positions with respect to the Earth, and second latitudinal digressions. In Chapter 7 the obliquation angles of both planets are given in a special table.

Both the amount and quality of astronomical knowledge contained in Copernicus' *magnum opus* earned him the acknowledgement of the greatest astronomer since Hipparchus and Ptolemy. The heliocentric Copernican system, although too revolutionary to be accepted by most of contemporaries, prevailed and became universally admitted as physical reality in the next generations.

8.6 Critics and enthusiasts

As we already mentioned, Georg Joachim Rheticus (1514–1574) was the first reader of the full manuscript of Copernicus' treatise, and after the two years of common work in Frombork, became an unconditional enthusiast of his master's theory. However, he did not become a genuine successor of Copernicus, lacking sufficient recognition among the contemporary learned men, who stuck to Aristotle's and Ptolemy's ideas. One generation after Copernicus' death still only a few astronomers across Europe were convinced by heliocentrism.

Tycho Brahe, the most renowned astronomer of the next generation, owned a copy of the *Commentariolus*, and was quite enthousiastic about the new model of the Universe placing the Sun in the center. He eagerly engaged in observations that would prove its validity, trying to detect stellar parallaxes. Unfortunately, to no avail. All his attempts showed negative results, although his giant sextant was able to detect down to one arcminute variations of position between a given star and vernal and autumnal equinoxes. Unable to detect the slightest displacement of stars, he concluded that the Earth was the motionless center of the Universe, proposing a mixed system in which the Sun orbits around the Earth above the Moon, but in which all planets orbit around the Sun.

8.6.1 *Religious objections*

It seems that Copernicus was quite aware of a negative, to say the least, attitude of the Church authorities towards his model, depriving the Earth of its central position in the Universe. This is probably the reason for his long hesitation, or rather refusal, to make it public. When he finally gave consent to publish to his young friend Rheticus, he was already an old man in his 70-ties; and when in 1543 Rheticus brought a printed copy of "De Revolutionibus" to Frombork, his master Copernicus was on his deathbed.

The Universe in which the Earth was not central contradicted not only the Aristotelean worldview, but also Christian beliefs according to which the Hell is deep in the underworld, and heavenly Paradise is above. If the Earth is just a planet orbiting around the giant ball of fire, like the other five planets, then where is the Hell and where is the Paradise? The Scriptures were also cited as ultimate argument: Joshua could win the battle of Jericho because God ordered the Sun to halt, which means that the Sun is moving, not the Earth. Nevertheless the treaty was not put on the index of books prohibited by the Catholic Church (*"Index Librorum Prohibitorum"*) during next 80 years, when it was finally judged to be not only controversial, but subversive during the Galileo trial in Rome.

The Protestants were not seduced by the Copernican system either. Luther wrote that "Copernicus is a fool who wants to destroy astronomy". But the strongest opposition came from Luther's close collaborator Philip Melanchton (1497–1560), who used to teach Ptolemaic astronomy at the University of Wittenberg. Although he recommended Rheticus for the Deanship of the Faculty of Arts and Sciences at his University after this young disciple of Copernicus returned back from Poland, Melanchton became one of the most active enemies of Copernicanism. He went as far as to write a solemn letter calling the authorities to repress the Copernican theory by governmental force.

In 1549 he published *"Initia Doctrinae Physicae"*, a book in which he attacked Copernicanism from three different grounds, which were *"The evidence of the senses, the thousand-year old consensus of men of science, and the authority of the Bible"*.

The struggle between the partisans of the heliocentric Copernican system and their adversaries gained momentum until the end of the 16th century and led to very serious trouble for the first. *"De Revolutionibus"* was definitely put on the index of books prohibited by the Catholic Church in 1616. In 1633 Galileo Galilei was tried and condemned by the Inquisition for his stubborn defense of the Copernicanism. The Church recognised the error and rehabilitated Galilei only late in the 20th century.

8.6.2 *Giordano Bruno's ordeal*

On February 17, 1601, the Italian mathematician, poet and former Catholic friar Giordano Bruno was burned at the stake at Campo dei Fiori in Rome, after a painful and long trial by the Inquisition. In spite of terrible tortures inflicted on him, he did not deny adherence to his then unorthodox views, including the ideas that the Universe is infinite and that other solar systems exist. His name has become the synonym of martyrdom for science, although the main reason for the persecution was of religious nature: Bruno was considered as a heretic opposing the doctrines of the Catholic faith.

Born in 1548 near Naples, Bruno studied there and entered the Dominican Order in 1565. His vivid intellect and taste for science made his stay in the Monastery difficult. When the friars discovered prohibited books in his cell, facing danger Giordano shed the habit and fled. Giordano wandered across Europe trying to find a safe haven, but it soon became an impossible task. He stayed some time in Geneva, but local Protestants did not appreciate his teachings; similar problems arose in France in Lyon and Toulouse, then at the University of Oxford. He elaborated original methods improving memory and published a book on mnemotechnics, earning quite a notoriety in this domain. Some of his enemies accused him of practicing black magic and contacts with maleficious spirits.

Fig. 8.4   Left: Young Giordano Bruno; Right: The statue of Giordano Bruno by Ettore Ferrari, raised in 1899 at the Campo de Fiori in Rome.

Bruno not only promoted and defended the Copernican system; he propagated a view according to which we are not alone in the Universe, that stars are in fact other Suns which may have planets orbiting around them just like the planets of our Solar System, and some of them may be inhabited by intelligent creatures. Such a worldview was in direct contradiction with central and unique place human existence occupied in the religious teachings of the Catholic Church. Besides, Bruno did not subscribe to many Catholic dogmas, like Trinity or the immaculate conception. He also subscribed to the ideas of metampsychosis, and an eternal and infinite Universe without a center.

When Bruno was in Germany in 1591 , he was invited to teach in Venice, and made a fatal error accepting the invitation. In 1592 he was treacherously denounced to the Inquisition, and put to trial in Rome. Tortured during seven years, Bruno never recanted his views, and was accused of multiple heresies and condemned to death on the stake. Bruno bravely resisted until the end, and his last words were:

"Perhaps you pronounce this sentence against me with greater fear than I receive it".

Chapter 9

Tycho Brahe, the prince of astronomers

Tycho Brahe's biography - Early astronomical observations - The Uraniborg observatory - Great events: Supernova and the great comet - The Tychonic system - Prague and collaboration with Kepler - Brahe's legacy

9.1 Preamble

Tycho Brahe (1546–1601), recognised as the most prominent astronomer of the generation next after Copernicus, has played a crucial role in the development of celestial science that followed his untimely death when he was 55 years old. He raised the art of astronomical observation to an unprecedented level, reaching the natural limit of the resolution power of the human eye, close to one arcminute. His astronomical skills became famous, especially after Tycho built a luxurious home fit for astronomical studies, on the island of Hveen. Tycho's observatory, called "Uraniborg" by him, became an international center of astronomical study. Aware of the shortcomings of the Ptolemaic model of the Universe, Tycho was initially partisan of Copernicus' heliocentric system, and tried his best to detect stellar parallax due to the Earth's annual motion around the stationary Sun.

Having failed to observe any trace of annual variations of stellar positions, in spite of the extraordinary precision of his measuring devices, Tycho rejected the heliocentric Copernican system. However, being conscient of accumulating errors of the Ptolemaic system, he proposed a kind of compromise: a system of his own, based on a mixed geocentric-heliocentric model, in which the Earth remained the center of the Universe, the Moon its closest satellite, the Sun orbiting the Earth, but all planets orbiting the Sun.

Other important accomplishments of Tycho were the observation and description of the supernova of 1572 and of a giant comet in 1577. But perhaps his most important achievement was the invitation extended to the young Johannes Kepler to become his assistant in Prague.

9.2   Tycho Brahe's life

Tycho Brahe was born in 1546 to the noble and powerful family of Otte and Beate Brahe, belonging to the small circle of direct friends and supporters of the Danish King Frederick II. When he was still a child aged 2, he was taken by his uncle Jorgen Brahe and his spouse Inger Oxe to their home, with his parents' consent. This event played a crucial role in Tycho's life, because his uncle and aunt were much keener on science and education than his own parents. Tycho stayed in his uncle's castle until 1559, when his foster parents sent him to the University of Copenhagen. Following the wish of his uncle, Tycho studied law and medicine, but soon was attracted by mathematical sciences and astronomy, especially after he saw a solar eclipse in 1560, which was predicted by astronomers.

In 1562 Tycho left Denmark for Leipzig, to gain more experience abroad. He started his own astronomical observations in 1563. His first doubts concerning both the Ptolemaic and Copernican systems arose when he observed the conjunction of Jupiter and Saturn. Predictions based on the Ptolemaic system were off by more than a month; those based on the Copernican system were much better, but still off by a few days. After his uncle perished in 1565 rescuing his King from drawning, Tycho stayed for a while with his parents in Copenhagen, then took off to continue his studies at the University of Rostock. During his stay there, in 1566, he had parts of his nose cut off in a duel with another Danish student. After his return home in 1567 he got an artificial nose made for him of gold and silver, which he wore constantly, soon deserving the nickname of "the astronomer with a golden nose".

Tycho spent the next few years wandering across Europe with the goal of acquiring more astronomical skills. He went to Rostock, revisiting its university, then traveled to Basel, Freiburg and Augsburg, where he stayed for a longer time. Having obtained some help from a local patron, Tycho constructed a huge quadrant. It was very accurate, but so massive that it required many servants to turn it in the desired direction.

In 1572 Tycho met Kirsten Jorgensdatter, a young girl from his home town, and fell in love with her. The Danish law prohibited marriages between a noble (Tycho) and a commoner (Kirsten), so she remained his common law wife, not recognised as legally married. They remained together until Tycho's death. Kirsten gave him seven children, four daughters and three sons, and survived him only by three years. Both were buried in Prague.

In 1574 Tycho started lecturing on astronomy at the University of Copenhagen, but a year later he went to Kassel in Germany, where a huge observatory has been built 15 years earlier. Impressed by the improved methods of observation which were in use there, Tycho decided to create an observatory of his own. Looking for funding he visited Frankfurt, Basel and Venice, and chose Basel as the best place for construction.

However, when Tycho came back to Denmark to take his belongings and leave for Basel, King Frederick II offered him an entire island, Hven, located between Denmark and Sweden, in the middle of the Strait of Øresund. Such a generous gift was received with due gratitude; Tycho's dream came true, and he spent the next 20 years at the Observatory he created, endowed with excellent astronomical instruments of his own design.

Unfortunately, Tycho's character, cheerful and attentive to others when he was young, deteriorated during those years. He became irascible and despotic towards his servants, and megalomaniac in his relationship with other people, including the King. In 1588 Frederick II died, and his son Christian was still too young to reign, so a Regency was established, under which Tycho was still getting

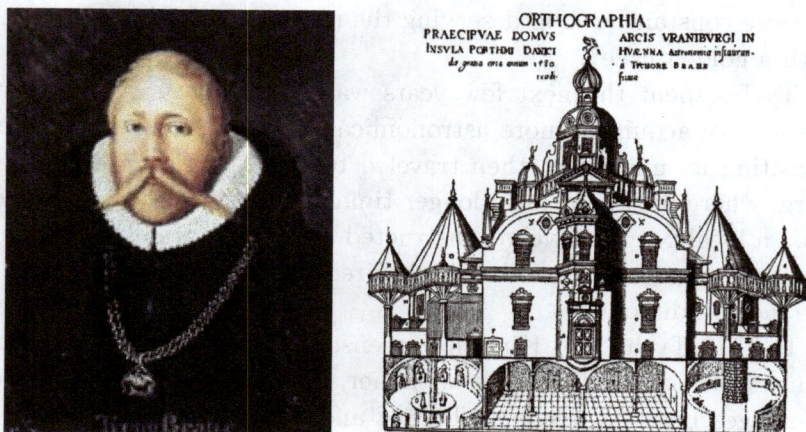

Fig. 9.1   Tycho Brahe, 1546–1601.

the financial support he needed. But when Christian became King, things turned rather sour for Tycho, and after some dispute over not respecting his obligations concerning the Royal Chapel which should have been repaired by Tycho being on his grounds, the King made it clear that his financial support came to an end.

Tycho moved to Copenhagen first, but soon discovered that the perspectives there were as grim, and decided to leave Denmark looking for another sponsor for his astronomical work. He was heartily invited to Prague by the German Emperor Rudolf II, and in 1697 moved there with his family and all his astronomical tools. Soon after he invited the young German mathematician Kepler to become his assistant in Prague. At first their collaboration was uneasy, but after a year Tycho gave Kepler access to most of his notes.

Unfortunately Tycho died in October 1601 of blood poisoning resulting from bladder burst. During an official dinner in the palace of a local baron, Tycho drank a lot, but being seated to the right of his host, could not leave the table obeying the etiquette of the day, with tragic consequences. He died 10 days later, but convinced the Emperor to name Kepler as his successor, which profoundly changed the history of astronomy for centuries to come.

9.3 The Uraniborg

King Frederick II became so admirative of Tycho Brahe's astronomical skills that he decided to help Tycho build a huge observatory endowed with many new instruments, unprecedented in Europe. He offered Tycho a part of the Island of Hven, between the Danish island Zealand and the Southern Swedish province of Scania, in the strait of Øresund. The construction started in 1576; the main building was erected during the next year. A few years later the observatory with a big library, an alchemical laboratory and an enormous mural quadrant were completed, and given the name "Uraniborg" — a citadel dedicated to Urania, the Muse of astronomy.

Uraniborg possessed a rich library and its own paper factory and printing office, where Tycho could edit his books, marked with inscription *"Uraniburgi Ex Officinae Typographicae Authoris"*, 1596. One of the most remarkable books was entitled *"Epistolarum astronomicarum libri"* ("the books of astronomers' letters"), containing the exchange of comments and opinions between Tycho, the Landgraf Willem IV of Hessen and his court astronomer Christoph Rothman, concerning the construction of astronomical tools of observation, the comet observed in 1585, Copernicus' heliocentric system, and many other topics.

The most impressive astronomical tool was the giant mural quadrant with radius about 1.94 meters, permitting the observations with precision less than one arcminute (Figure 9.2).

Tycho was extremely inventive in constructing new instruments in order to improve the quality of his observations. The first important observational device produced in 1576 according to Tycho's plans was a huge brass azimuthal quadrant, which could be turned in any direction around its vertical axis, and measure angular positions along any vertical great circle in the sky. With a radius of 65 cm, it was able to resolve angular distances corresponding to one arcminute.

Another of Tycho's undertakings was to construct a giant globe serving as stellar map. It took almost ten years in the making, and became operational in 1580. It was a hollow sphere of 1.6 meter in radius, made of fine wood and covered with brass sheets. It served

Fig. 9.2   Left: The great mural sextant built by Tycho in Uraniborg; Right: the revolving azimuthal steel quadrant.

as an exact map of the celestial sphere. Each time Tycho measured the position of a yet unclassified faint star, he would mark it on the surface of the globe with a small silver nail. Tycho had about 1000 accurately observed stars by 1595, all inscribed on his giant globe.

The same globe was originally intended as a computational device. By means of auxiliary circles, the local coordinates, azimuth and altitude could be converted into the conventional celestial coordinates used to record stellar and planetary positions.

A triangular sextant with a 1.6 meter radius was constructed in 1582. Supported by an ingenious spherical mounting its versatility matched that of smaller instruments of the same kind, but its resolving power was much greater, only slightly above half an arcminute. It was with this precision that Tycho determined the angle between the ecliptic and the equator (in fact, the Earth's axial tilt). His definition of the tropical year was only one second shorter than the correct value.

In 1585 the construction of Tycho's great equatorial armillary, 3 meters in diameter, was completed. Being essentially two-dimensional, it was closer to an astrolabe, the armillary sphere being reduced to its bare essentials, and became one of Tycho's workhorse instruments. It has an estimated accuracy of 38.6 arcseconds.

In 1586 the revolving wooden quadrant of radius 1.6 meters was added. It could be directed towards any meridian, and its accuracy, based on measurements of the positions of eight reference stars, was close to half an arcminute.

9.4 Two extraordinary events: the nova of 1572 and the comet of 1577

• **The new star** — On November 11, 1572, a star appeared in the constellation of Cassiopea. Its brilliance was more intense than that of Venus at its maximum brightness. A new celestial object appeared among the fixed stars, where nothing was previously observed except the apparent void. The new star was so brilliant that it could be seen with the unaided eye during the daytime. At the end of 1572 its apparent luminosity was still comparable with Venus, but a few months later it started to decline, and it became invisible to the unaided human eye after March 1574.

Tycho Brahe was not only among the first discoverers of this phenomenon, but was also able to perform regular observations of its position and intensity variation. From his precise measurements he concluded that it belongs to the sphere of fixed stars, far beyond the orbits of known planets. By the same token, it was the first demonstration that the sky beyond the spheres of the Moon and the planets is not perfectly and definitely fixed. Having observed that the new celestial object did not present the slightest parallax during months of scrutiny, Tycho Brahe concluded that it cannot be anything else but a new star — a daring opinion for that time.

Brahe published a small book in Latin, entitled *De Stella Nova* ("On the New Star"), exposing his observations and conclusions. The book was printed at his own expense in Uraniborg. Later on, the content of this book was included in another publication by Tycho, "*De mundi aetheri recentioribus phaenomenis*", printed in 1588 in Uraniborg.

Today the remnants of that supernova explosion, whose official name is *SN* 1572, or *B Cassiopeiae* can be seen in Cassiopaea only with a telescope, as a small nebula. It remains a strong source of X-rays.

• **The great comet** — The next important event happened in 1577, when Tycho could perform observations at the Uraniborg observatory. On November 13 that year he spotted a comet, which soon grew into one of the brightest celestial objects, almost as bright as the Moon, and clearly visible even at sunrise. It was observed all over the world, and described also by the Turkish astronomer Taqi ad-Din (1526 − 1585).

Tycho noted that during its travel through the sky the comet's tail was always directed away from the Sun. He also observed how the comet's brightness diminished as its distance from the Sun grew, and finally proved that the comet was not an atmospheric phenomenon, as was believed since Aristotle and Ptolemy, but was a cosmic body much farther than the Moon.

He did this by comparing the position of the comet in the night sky where he observed it (the island Hven, near Copenhagen) with the position observed by the Bohemian astronomer Thadaeus Hagecius (Tadeàs Hàjek) (1525−1600) in Prague at the same time, taking into account the motion of the Moon. It was discovered that, while the Moon seemed displaced with respect to a fixed star nearby due to the parallax between Uraniborg and Prague, as shown in Figure 9.3, the comet was apparently in the same place in the sky for both of them.

It is not difficult to evaluate the lunar parallax between Uraniborg, i.e. the Island Hven (55°54′ N, 12°42′ E) and Prague (50°05′ N, 14°25′ E). The meridians of these two places are very close, the difference of their longitudes being about 1°44′ (Prague is just a little bit farther to the East). The difference of latitudes is dominant: it is equal to 5°51′. Recalling that one angular degree along a meridian is worth $40\,000/360 = 111.11$ kilometers, the distance between Uraniborg and Prague would be 650 km. Taking into account the difference in longitude, the distance along the great circle is slightly bigger, 663 km, as it follows from Pythagoras' theorem applied to the rectangular triangle formed by 5°51′ segment along the meridian passing through Uraniborg and hitting the parallel of Prague's

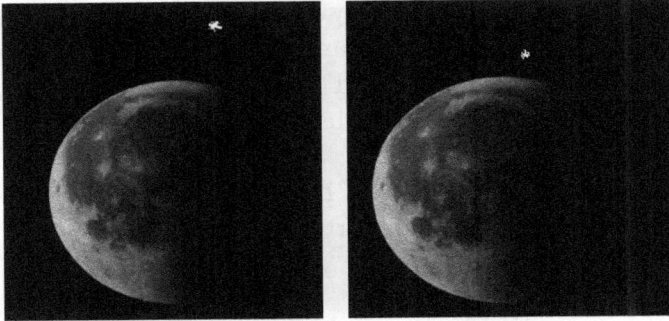

Fig. 9.3 Moon seen at the same time in Uraniborg (left) and in Prague (right). The parallax can be observed comparing Moon's position with respect to a fixed star which happened to be close in the sky.

latitude $50°05'$; the orthogonal segment along the $50°05'$ parallel until it reaches Prague — and this is only 132 km long, because one degree along the $50°$ is equivalent to 111.11 km $\times \cos 50° \simeq 111.11 \times 0.6 = 66.67$ km; finally, the geodesic distance between Uraniborg and Prague (i.e. along the unique great circle passing through these two cities) is $\sqrt{132^2 + 650^2} = 663$ km.

With this in mind, one can find the parallax of an object (in this case the Moon) which is at distance of $D_M = 384\,000$ km. Dividing 663 km by D_M we get $\delta = 1.727 \cdot 10^{-4}$ radians. Multiplying δ by $57.3°$ per radian and by $60'$ angular minutes per degree, we get the value of the lunar parallax between Island of Hven and Prague equal to $\delta = 5.94' \simeq 6'$ angular minutes. Such a displacement with respect to nearby fixed stars can be easily noticed with the naked eye, and measured with an average quality sextant or quadrant. Tycho used his new azimuthal brass quadrant with which he was able to distinguish angular distances slightly smaller than one arcminute.

In the absence of any observable (more than one arcminute) parallax of the great comet Tycho concluded that it must orbit the Earth (which he considered to be the center of the Universe) more than three times farther than our Moon. From dimming and brightening, and from its displacements among the stars and relative to the Sun Tycho drew a yet more radical conclusion: the trajectory of the comet was crossing several Ptolemaic crystalline spheres to which the planets were supposedly fixed. During its closest approach to the Sun it was nearer to it than Venus, and before disappearing the comet was as far as Saturn! This fact contradicted the physical reality of the spheres described in officially accepted astronomical science as contiguous real, hard, and transparent spherical shells. The alternative notion of a "fluid heaven", defended by Copernicus, gained thereafter a serious support.

9.5 The Tychonic system

Already in his *"Commentariolus"* Copernicus suggested that the distance between the Earth and the Sun is extremely small compared to the distance separating us from the stars, which is so large that the annual stellar parallax must be far below the observational possibilities (as the typical resolution at his time was a few arcminutes). Such enormous distances were beyond imagination for Copernicus' contemporaries, who rejected his model as unrealistic, although its practical predictive power was recognized — but not more than a clever mathematical device.

As we said before, Tycho Brahe owned a copy of the *Commentariolus*, and was quite enthusiastic about the new model of the Solar System presented within. In order to prove its validity, he engaged in detecting stellar parallaxes — unfortunately, to no avail. All his attempts showed negative results, although his giant sextant was able to detect down to one arcminute variations of stellar and planetary positions with respect to vernal and autumnal equinoxes. Although Copernicus suggested that the absence of stellar parallaxes is due to their tremendously large distance from us, the observational limit set by Tycho pointed towards hypothetical distances so mind-boggling, that Tycho rejected this possibility.

Let us follow Tycho's reasoning. Supposing that the heliocentric Copernican system describes astronomical reality, hypothetical stellar parallax must be less than $1'$, one arcminute. Translated into radians, this gives

$$\Delta\alpha \leq 1' = \frac{1}{60} \text{ degree} = \frac{1}{60} \times \frac{1}{57.3} = 2.91 \cdot 10^{-4} \text{ rad}. \qquad (9.1)$$

According to the well known relation between object's size d, distance from the observer D and the angle of vision $\Delta\alpha$ expressed in radians, $d = D \cdot \Delta\alpha$, with the baseline given by the diameter of the Earth's orbit the distance to stars (supposed to be found on the same sphere) would be no less that 3437 times the radius of Earth's orbit — incredibly far away, especially as compared to the distance to Saturn, the last and the farthest of the known planets, whose orbit was just 10 times larger than the orbit of Earth as proposed by Copernicus.

Another problem Tycho considered as a serious argument against the heliocentric system was the stars' own size implied by their distance. Assuming that a typical angular size of a star as seen from the Earth (or from the Sun for that matter, so far away stars were supposed to be) is one arcsecond.

The radius of the Earth's orbit was estimated by Copernicus as close to about 1500 Earth's radii — which was still more than 15 times lower than the actual distance to the Sun, but represented already a mind-boggling distance of $\simeq 10\,000\,000$ kilometers. This distance should be multiplied by 3437 in order to get the average distance to the stars, which should be $3437 \cdot 10^7 \simeq 3.5 \cdot 10^{10}$ km. According to the formula valid for very small angles, the angular size δ (in radians) is equal to the ratio d/D, the diameter of the object divided by its distance from the observer.

When Tycho applied this equation to the distant stars assuming that their angular diameter as seen from the Earth is equal to at least one arcsecond, he arrived to the conclusion which he deemed totally absurd: the stars must have diameters many times greater than the Sun! Today we know that this is true, some stars are so huge that if they were in our Sun's place, they would extend beyond Earth's orbit. Table 9.1 shows the angular sizes, diameters and distances from the Sun of several of the brightest stars.

What Tycho proposed as a compromise between the Ptolemaic and Copernican systems was the following hierarchy: the Earth remained the center of the Universe, the Moon its closest satellite, the Sun orbits the Earth, and the planets orbit around it. The inferior planets, Mercury and Venus, were orbiting close enough as to

Table 9.1　Angular sizes, diameters and distances of the nearest stars.

Star	δ (arcsec)	d (in km)	D
The Sun	1 930″	$1,4 \cdot 10^6$	$1496 \cdot 10^8$ km
Betelgeuse	0.049″	$1.23 \cdot 10^8$	700 l.y.
Sirius	0.006″	$2.1 \cdot 10^6$	8.6 l.y.
Canopus	0.006″	10^8	310 l.y.
α Centauri	0.007″	$1.6 \cdot 10^6$	4.37 l.y.
Altair	0.003″	$1.8 \cdot 10^6$	16.7 l.y.
Deneb	0.002″	$5.7 \cdot 10^8$	2620 l.y.

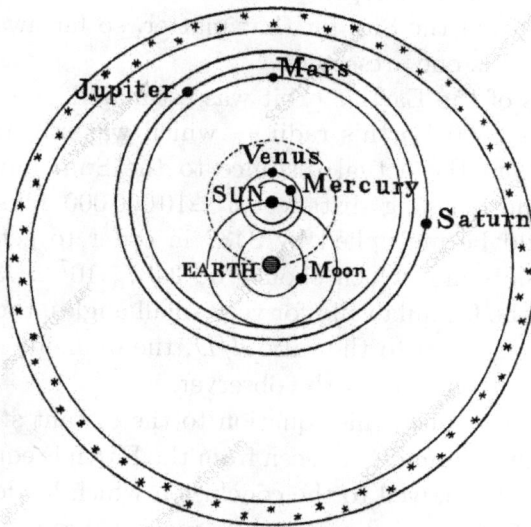

Fig. 9.4　Tycho Brahe's geo-heliocentric model, a compromise between Ptolemaic and Copernican systems.

ensure their observed maximal elongations, $\simeq 28°$ for Mercury and $\simeq 47°$ for Venus. In order to explain the difference between inferior and superior planets whose elongations can be arbitrary, and which never pass between us and the Sun, the orbits of Mars, Jupiter and Saturn were supposed to be large enough to include the Earth and the Moon's orbit, as it appears in Figure 9.4.

9.6 Prague and Kepler

When Tycho arrived in Prague in 1597, he was already acquainted with the young Kepler's *Mysterium Cosmographicum* edited in 1596, in which Kepler explained the relative sizes of the planetary orbits' sizes by inscribing into their "spheres" regular platonic polytopes with the Sun as common center. An intense exchange of letters between Tycho and Kepler, who lived in Graz, followed.

Tycho had almost all his astronomical tools brought to Prague, where the construction of new observatory in Benátky, a village 30 km south of Prague, started immediately. But Tycho needed an assistant with good mathematical skills in order to analyze his 38-years long observations of Mars, and Kepler was an ideal candidate. Their collaboration was rather bumpy at first, but soon Tycho became more confident and gave Kepler access to his notes. Tycho introduced his assistant to the Emperor Rudolph II, and convinced him to name Kepler his successor as the Imperial Mathematician.

As a faithful and respectful pupil, Kepler published his observations of Mars in the special volume after Tycho's death; but he did not subscribe to the Tychonic model, still popular among the astronomers until Galileo's astronomical discoveries after 1610. Using Tycho's vast observational material Kepler interpreted them from the point of view of the Copernican heliocentric system, Kepler's deep analysis and interpretation of Tycho's observations of Mars led to a radically new vision of the solar system and the laws of planetary motion. Without Kepler's discovery that planetary orbits are elliptical and not circular, and that the areal velocity is constant, Newton would not have been able to derive his theory of universal gravity, in which the constant areal velocity (Kepler's second law) is implied by *central forces* indifferent to their dependence on the distance r from the attracting center, while the ellipticity of orbits (Kepler's first law) and the relation $a^3 \sim T^2$ (Kepler's third law) result from the inverse square law of the gravitational attraction.

9.7 Tycho's legacy

Tycho's was the last great astronomer, observing by the naked eye before the invention of the telescope. His astronomical observations were exceptional both for their accuracy, quantity and systematization. His measures of stellar and planetary celestial positions were much more accurate than those of his contemporaries, not to mention his predecessors. He was the first astronomer to include systematic corrections of celestial objects' latitude taking atmospheric refraction into account.

After Tycho's death Kepler took possession of his astronomical instruments and manuscripts, including hundreds of pages of observations of Mars, which Tycho wanted to publish under the title *"Tabulae Rodolphinae"* in homage to his new sponsor, the Emperor Rudolf II. A legal battle with Tycho's widow followed, and Kepler had to give all astronomical instruments back; but he was allowed to keep all Tycho's manuscripts, including the Martian tables which he secretly took shortly before Tycho's death.

The mural quadrant, along with many other instruments of the observatory, was depicted and described in detail in Brahe's 1598 book *"Astronomiae instaurata mechanica"*, which remains the sole source of information about Tycho's instruments. None of them survived wars and plundering. The giant globe found its way to Copenhagen after Tycho's death, but was destroyed in the great fire that consumed the Danish Library where it was conserved.

It took almost a full century until Tycho's extraordinary precision was reattained and surpassed when better telescopes started to be used by astronomers in the second half of 17th century: Johannes Hevelius, John Flamsteed, Christiaan Huygens, Edmund Halley and Giovanni Domenico Cassini.

Chapter 10

Galileo and the new physics

Galileo's life: education and career - Discovering the law of free fall - Galileo's new mechanics - The experimental approach to science - Critics and admirers: Descartes and Kepler - Construction of the telescope - Astronomical discoveries - Defense of the Copernican system - Troubles with the Inquisition - Trial and home arrest - The last years

10.1 Galileo Galilei's life and achievements

Galileo Galilei was born on February 15, 1564 in Pisa as the eldest son of a renowned musician Vincenzo Galilei. In spite of Galileo's manifest musical capacities his father decided that a medical career would be the most appropriate and ensuring a steady income in the future. At that time the family lived in Florence, but young Galileo initially studied philosophy and medicine in Pisa; in 1585, without a doctor's diploma yet, he came back to Florence, where he continued studies in mathematics and physics, which included learning almost by heart most of Aristotle's writings. He became so knowledgeable in mathematical sciences that his skills were noticed by several mature scientists. In 1589, thanks to a recommendation by the marquis Guido Ubaldo del Monte he was appointed professor at the University of Pisa by Ferdinando Medici, the Count of Tuscany.

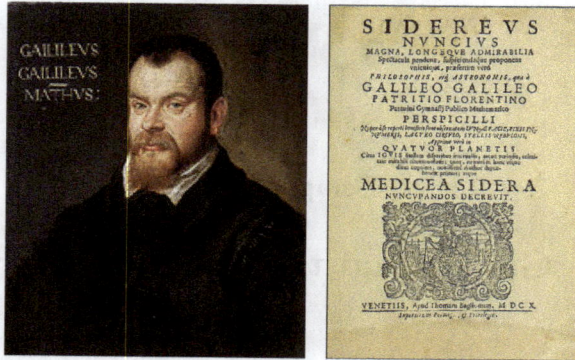

Fig. 10.1   Left: One of the rare portraits of Galileo Galilei in his earlier years. Many portraits were painted after he became famous, so his most widely known image is that of an old man. Right: Galilei's book "The Starry Messenger" (1610).

It was in Pisa that Galileo performed experiments and formulated the law of the free fall.[1]

Galileo served as professor of mathematics in Pisa until 1592, when he moved to Padova, still as professor; he stayed in Padova for the next 19 years, until 1610, which he considered the happiest and the most fruitful period of his life. In 1609 he constructed his first telescope, with which he made a series of fundamental astronomical discoveries in 1610. He published most of his findings in a book entitled "*Sidereus Nuncius*" — "The Starry Messenger", which he dedicated to his protectors, the Medici family of Florence. He also presented the book to the Pope Urban VIII, seeking papal protection. In 1611 he became a member of the *Academia dei Lincei*, the "Lynx Academy" founded in Rome by Federico Cesi (1586–1630) in 1603 with the intention to provide an opportunity to the best minds of Italy to discuss science and philosophy.

Galileo's telescope presenting a straight image was also appreciated as an exceptionally helpful tool for naval and military use. Galileo could sell many telescopes manufactured by himself to Venetian merchants, becoming rich enough to become more independent.

---

[1]Although Galileo's family name was "Galilei", he tended to refer to himself by his first name, "Galileo". We respect his choice in what follows.

He exchanged many letters with Kepler, with whom he shared a strong belief in the correctness of the Copernican system. In 1632, not without Kepler's influence, he published a book "Dialogue Concerning the Two Chief World Systems".

Three characters, Simplicio, Salviati and Sagredo, exchange their views on the Ptolemaic and Copernican systems. Simplicio is a naive defender of Ptolemy, but his arguments against the Copernican system are shown to be inconsistent. Although the book was initially approved by the Inquisition, a year later Galileo, accused of heresy, was summoned to Rome, underwent a trial and was condemned to retract his views publicly. After that he was forced to spent the rest of his life in a kind of house arrest in Arcetri near Florence, until his death in 1642. In spite of blindness that struck him at the end of his life, Galileo remained active and conceived a project of a pendulum clock that was realized only after his death.

10.2   Galilean mechanics

10.2.1   *The free fall*

Galileo, who diligently studied Aristotelean physics, became skeptical about the laws of free fall as they were presented by Aristotle. It seemed that for Aristotle velocity of the free fall was constant, different for bodies of different densities, but the same from the beginning of the fall until the end point, the contact with the ground. Moreover, bodies of similar shape having different weights should fall with different speeds, the heavier body with greater speed than the lighter one. Finally, the speed was supposed to be inversely proportional to the density of the surrouding medium: in fact, the same stone falls quickly in the air, and much less rapidly in water.

As early as 1554 the Venetian mathematician and philosopher Giovanni Battista Benedetti published a book *"Demonstratio"* refuting most of Aristotelean postulates concerning the free fall. The most important argument was quite simple: take two bodies A and B, with average densities ρ_A and ρ_B, respectively. If $\rho_A > \rho_B$, according to Aristotle the body A should be falling more rapidly than the body B. Now let us glue the two bodies together so that they would fall as one. The average density of the combined body will be ρ_C, with $\rho_A > \rho_C > \rho_B$, and the agglomerate should fall down less rapidly then the body A, although more rapidly than the body B. However, this is not what is observed: the conglomerate of A and B falls down *at least* as rapidly as A itself, if not even faster.

Many common experiences pointed towards the Aristotelian hypothesis — a feather would fall down slowly while an iron ball would fall very quickly. However, Galileo justly attributed the reason for that difference to the friction of the surrounding air. In his time he was unable to carry an experiment shown frequently in contemporary high schools, in which falling objects are placed in a glass tube from which the air was evacuated. Then a feather and a metallic coin fall down at the same pace.

The experiment Galileo was supposed to perform in order to show that masses fall down in the same manner independently of their weight and density provided both were heavy enough, was realized using the famous Tower of Pisa, is most probably a myth invented by his pupil Viviani after Galileo's death.

One of the consequences of universality of free fall discussed by Galileo was that a freely falling body is in a state of weightlessness. He illustrated this conclusion by considering a man carrying a quarter of beef on his shoulders.[2]

To assume that the velocity of a body in free fall is greater at the end of the fall if the initial position was higher was in agreement with everyday life. Galileo argued by evoking how piles are driven into the ground by heavy stones falling from some height. With diminishing height of initial position, the force of the falling stone decreases and becomes close to zero if the stone starts to fall from only a few inches above the pile. The relationship between the force and velocity was not very clear at that time, but for sure, increasing the initial height

---

[2] A similar experiment took place some years earlier (1585–1586) in Delft in the Netherlands, where Simon Stevin and Jan Cornets de Groot conducted the experiment from the top of the Nieuwe Kerk church. The experiment, showing that two lead balls, one ten times heavier than another, fall to the ground at the same time with the same speed, is described in Simon Stevin's 1586 book "The Principles of Statics". In the 6th century B.C.E. Philoponus (490–570 C.E.) performed a similar experiment in Alexandria.

resulted in greater velocity at the end of the free fall, and much stronger physical action on the pile.

Initially Galileo supposed that the velocity v of a freely falling body is proportional to the distance s it covered since the beginning of fall: $v = Cs$. To his astonishment, Galileo discovered that this hypothesis leads to an absurd conclusion: the free fall obeying such rule must take infinitely long time. Just as Achilles' run in Zeno's paradox! Here is Galileo's reasoning. Let the total distance traveled by the freely falling mass be h, and let the final velocity when it hits the ground be $V = Ch$, supposing that the initial velocity was 0. Linear dependence of velocity on distance s is shown in Figure 10.2.

Let us consider the last half of the total distance covered by the falling mass, i.e. from the altitude $h/2$ till the ground level. According to the linear law, the velocity of the falling object was equal to $\frac{Ch}{2}$ when the object was half way to the ground, and Ch at the end. Therefore the average velocity between the positions $\frac{h}{2}$ and h — the last half of the course — is

$$\langle v \rangle_1 = \frac{1}{2}\left(\frac{Ch}{2} + Ch\right) = \frac{3}{4}Ch. \tag{10.1}$$

According to the relation between the distance Δs covered during time Δt and the average velocity $\langle w \rangle$, $\frac{\Delta s}{\Delta t} = \langle w \rangle$, we conclude that the time Δt_1 during which the body fell from the height $h/2$ down to the ground is equal to

$$\Delta t_1 = \frac{\Delta s}{\langle v \rangle_1} = \frac{\frac{h}{2}}{\frac{3Ch}{4}} = \frac{2C}{3}. \tag{10.2}$$

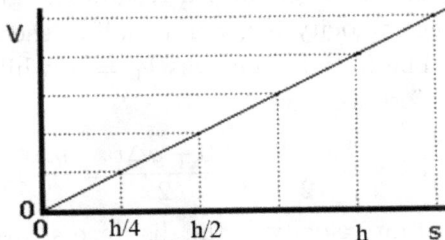

Fig. 10.2   Linearly dependent velocity $V = Cs$ versus covered distance s.

Let us consider now the part of the free fall starting from the altitude h above the ground contained between $\frac{3h}{4}$ and $\frac{h}{2}$, so that the distance between these points is equal to $h/4$. The average velocity on this part of the course is

$$\langle v \rangle_2 = C\frac{1}{2}\left(\frac{h}{4} + \frac{h}{2}\right) = \frac{3C}{8}. \qquad (10.3)$$

Dividing the length of the run $h/4$ by the corresponding average velocity $\langle v \rangle_2$ we get the time Δt_2 which is necessary to travel the distance $h/4$ separating the points $3h/4$ and $h/2$ above the ground equal to the time Δt_1 that was necessary to travel the distance separating the points $h/2$; indeed,

$$\Delta t_2 = \frac{h/4}{\frac{3Ch}{8}} = \frac{2C}{3} = \Delta t_1. \qquad (10.4)$$

It is not necessary to repeat this exercise further: the reader can easily conclude, as Galileo did in his time, that continuing to divide the course into smaller parts, each time dividing the previous distance by 2, the free fall from the height $y = H$ to the ground level $y = 0$ will take infinite time, being an infinite sum of equal contributions, each of them lasting $\Delta t_i = 2C/3$, for $i = 1, 2, 3, \ldots, \infty$. In fact, this was another version of Zeno's paradox.

Galileo turned then to the alternative possibility, which was to assume that the velocity of freely falling bodies grows with time. By simple reasoning using the average velocity concept, Galileo derived the mathematical law expressing the dependence of velocity on time during the free fall. Let us suppose that a body starts falling at $t = 0$ from the height h; if its velocity is proportional to the duration according to the law $V = gt$, and if it hits the ground after time Δt, then its average velocity is given as half of the sum of its initial and final values. The initial value was $V_0 = 0$, while the final value is $V_f = g\Delta t$; therefore we have

$$\langle V \rangle = \frac{V_0 + V_f}{2} = \frac{0 + g\Delta t}{2} = \frac{g\Delta t}{2}. \qquad (10.5)$$

In the case of constant velocity V the distance s covered after time t is equal to the product of velocity by time, i.e. $s = Vt$. Applying the same formula to the mean velocity — which is quite an audacious extrapolation — Galileo obtained the quadratic law, with the right coefficient[3]:

$$s = \frac{gt^2}{2}. \qquad (10.6)$$

This was Galileo's luck, because the same extrapolation based on the notion of mean velocity would give an erroneous answer for any other dependence of velocity on time. Suppose that the law would have been $v = gt^\alpha$, with $\alpha \neq 1$. Let us choose $\alpha = 2$ for simplcity. Then the same trick would give the following result:

$$V_0 = 0, \quad V_f = gt^2; \quad \langle V \rangle = \frac{V_0 + V_f}{2} = \frac{gt^2}{2}, \qquad (10.7)$$

and then, multiplying the mean velocity by total time of the fall, we would get $s = \frac{gt^3}{2}$, which is not true — the real result, as we know it now, is $s = \frac{gt^3}{3}$.

But Galileo was not satisfied with purely theoretical approach, no matter how logically consistent it might appear. The ultimate truth could be obtained only via direct experiments, reproducible and measurable. Galileo can be rightly called the father of experimental physics as we know it. In order to establish the law of free fall, he first replaced a freely falling body by a billiard ball rolling down an inclined plane. Of course the "plane" was in fact a pair of wooden rails keeping the ball on a straight trajectory. The vertical driving force of Earth's gravity was thus replaced by its tangential component, as shown in Figure 10.3. By choosing the inclination low enough, Galileo was able to slow down substantially the ball's downward motion. And he performed many tries, with different inclinations, velocities and accelerations, checking by the same token the validity of vector addition of forces.

There are many versions relating Galileo's experiment, one of which is shown in Figure 10.3 on the right. The main problem Galileo faced was the absence of time measuring devices with which it would

---

[3]This formula, was known to the Oxford Calculators of Merton College (ca. 1330); its graphical proof was given by Nicole Oresme (1325–1382) ca. 1360.

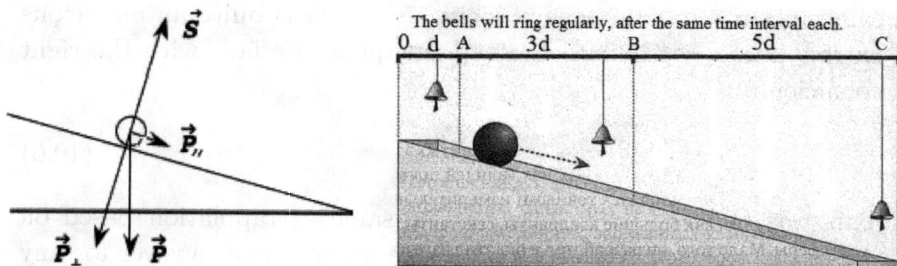

Fig. 10.3 Left: Decomposition of the gravitational force acting on a ball on an inclined plane into normal and tangent components; Right: the bells suspended over the rolling ball trajectory so that the times between each two sounds is equal.

be possible to evaluate correctly the time elapsed between different distances traveled by the rolling ball with accelerating speed. The best device available was the water clock, with drops pouring regularly through a little hole in a tank. But the additional difficulty consisted in determining the exact distance at which the ball was to be found at a given time, because even on an inclined plane the motion was too rapid to make the fixing of the particular instantaneous positions by human eye possible.

It was perhaps Galileo's musical knowledge inherited from his father, and of musical instruments in particular, that suggested him a brilliant solution. Installing thin strings perpendicular to the rail on which the ball rolled down, (an alternative version of this experiment consists in attaching small bells above the passage of the ball), it was not the eye, but the ear that was sollicited: when the rolling ball touched the string (or the bell), a characteristic sound was emitted. By trial and error Galileo was able to arrange the strings at increasingly long distances from the top in such manner that consecutive sound signals were heard at equal time intervals. A good musical ear is very sensitive to rhythm, and the distances corresponding to equal time intervals were very well defined.

It turned out that in equal successive periods of time, the distances traveled by a free-falling body were proportional to the succession of odd numbers: $1, 3, 5, 7$, etc. The total distance traveled from the starting point was therefore equal to the sums of the above, which means that if after the first second the distance was equal to $1 \cdot \Delta d$, after two seconds the distance from the starting point is $(1+3) \cdot \Delta d = 4 \cdot \Delta d$, and after the third second the ball will arrive at the location $(1+3+5) \cdot \Delta d = 9\Delta d$, i.e. the squares of consecutive integers: $1^2 = 1$, $2^2 = 4$, $3^2 = 9$, and so on. The readers acquainted with the mathematical induction method can easily prove that this is not only a guess, but an exact formula:

$$\Sigma_{k=1}^{N}(2k-1) = 1+3+5+7+\cdots+(2N-1) = N^2. \qquad (10.8)$$

[* Here is the rigorous proof of formula (10.8):
First let us check that formula works for a given value of N. Indeed, the result for the first three values, $N = 1, 2$ and 3, was found above. What remains to be proved is the following assertion: if the formula (10.8) is valid for N, it is also valid for $N+1$. Supposing that $\Sigma_{k=1}^{N}(2k-1) = N^2$, let us evaluate the next sum. We have:

$$\Sigma_{k=1}^{N+1}(2k-1) = N^2 + [2(N+1)-1] = N^2 + 2N + 1 = (N+1)^2,$$

$$(10.9)$$

which is the desired result, so the theorem is proved by the mathematical induction method.*]

Galileo found another regularity concerning the laws of free fall. Comparing the time of arrival at the bottom for a ball rolling without friction (i.e. with negligible friction in practice) along various chords of the same circle, he observed that the time was always the same.

Galileo checked this assertion experimentally as well as theoretically. In Figure 10.4 we see a circle of radius R, centered at C, and touching the ground at its lowest point O. The free vertical fall from the highest point A towards O will take, according to Galilei's law of free fall, time $t = 2\sqrt{R/g}$. Now consider the chord $A'O$, materialized so that a ball can roll down on it towards the same point O. Denoting by α the angle between the segment OA' and the horizontal plane, the length of $A'O$ is equal to $2R\sin\alpha$. On the other hand, the vertical gravitational pull g splits in two components, the tangent to the segment being $g\sin\alpha$ and normal to the segment $A'O$ equal to $g\cos\alpha$, the latter being canceled by resistance of the support. Therefore the law of free fall along the inclined segment $A'O$ takes on the

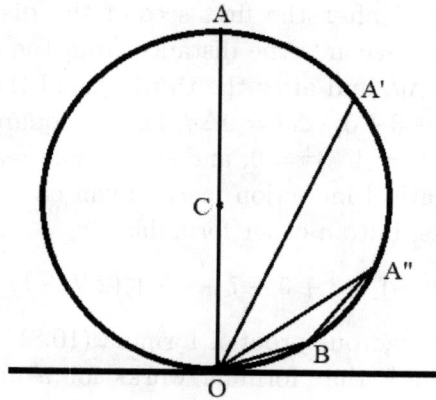

Fig. 10.4   A ball rolling down along two or more chords subtended between two given points on a circle arrives at the bottom more quickly than along single chord.

following form:

$$2R \sin \alpha = \frac{(g \sin \alpha)t^2}{2} \;\rightarrow\; t = 2\sqrt{\frac{R}{g}}, \qquad (10.10)$$

so the time of the fall along the chord $A'O$ is the same as the time of vertical free fall from A down to O.

   Using quite complicated reasoning Galileo proved that free fall along any chord takes longer than the sequence of two free falls along two shorter chords joining the same points on the circle via an intermediary point B: one always has $t_{A''O} > t_{A''B} + t_{BO}$. In order to prove this, Galileo took into account what we recognize today as the principle of energy conservation: the ball that started to roll down at A'' arrives to the intermediary "station" at B with non-zero speed acquired during its travel from A'' to B, and it is with the same velocity it starts its travel along the next segment BO. Without this assumption the time of free fall (or rather "free rolling down") along the chord BO would take exactly the time of the free fall from A to O, as shown before.

   However, Galileo's conclusion that rolling down along the circle from A'' to O is the quickest was wrong. Only half a century later

Huygens showed that the most rapid rolling down with no friction happens along a curve of cycloidal shape.

10.2.2 *Relativity of motion*

Galileo's ingenuity did not stop at this stage. He checked not only how heavy balls roll down an inclined planar surface, but observed also how high they could climb upwards along an ascending part of curved wooden rail. The result was independent of inclination, as shown in Figure 10.5. The conclusion Galileo drew from this experimental fact was that in the absence of inclination, motion would continue indefinitely with constant speed. Galileo also experimented with pendulums, showing that the period depends only on the length, not on the weight, and that a pendulum can move upwards to the same height it started from, even if one shortens its length (by putting a pin halfway forcing only the lower part of the pendulum to continue).

One of the consequences of the law of free fall formulated by Galileo was the proof that material bodies ejected horizontally with initial speed V follow the motion which is a combination of a uniform rectilinear motion with constant velocity V and the free fall with constant acceleration g directed vertically downwards. The resulting trajectory is a parabola. Figure 10.6 shows Galileo's original drawing explaining the parabolic character of motion with initial horizontal velocity. The drawing comes from Galileo's last book *Dialogue Concerning Two New Sciences*, a scientific testament written in last years of his life under the house arrest ordered by the Inquisition [Galilei (1638)].

This was a major blow to Aristotelean mechanics, in which motions could follow only rectilinear or circular paths, or combinations of rectilinear and circular segments.

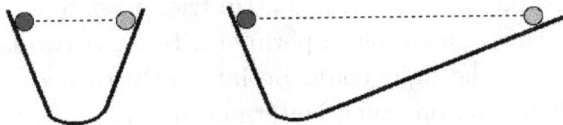

Fig. 10.5   The height reached by a ball that starts to roll down and then must roll up depends only on the initial altitude.

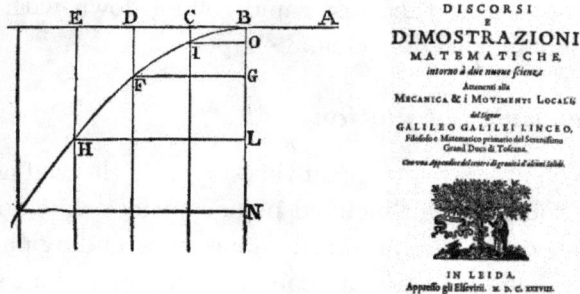

Fig. 10.6 Left: Galileo's sketch explaining the parabolic shape of trajectory motion with initial horizontal velocity; Right: Galileo's "Dialogue on Two New Sciences" (1638).

One of the most important of Galileo's ideas concerned the relativity of rectilinear motion with constant velocity: two observers, one of whom is at rest while another moves along a straight line with a constant speed, will see the same results performing the same experiments: the water will pour downwards, the balls will bounce from the floor in the same manner, etc. Today we call the reference frames moving with constant velocity with respect to each other *Galilean*, or *inertial* frames.

Another fine observation by Galileo was the cancellation of the gravitational force during a free fall. A man carrying heavy load on his shoulders will not feel any pressure if he happens to fall down, because the load falls with the same speed and acceleration, and would have remained with no contact if they were separated by an infinitesimal distance. This fact was called the *principle of equivalence* by Albert Einstein in the early 20th century, and became the cornerstone of his Theory of Relativity.

Besides mechanics and problems of motion, Galileo was interested in many other physical phenomena. He tried to determine the speed of light by means of a simple experiment. He took two lanterns with shutters, so that the light could be intstantly turned off or on. He and his student took one such lantern each, and when the night fell, climbed two distant hills, but close enough to make the lanterns' light visible. The idea was to open the shutter abruptly while the shutter of the other lantern was closed. The student should open his shutter the very instant he saw the light of the first lantern. Galileo intended to use his pulse to measure the time between opening his shutter and the student's reaction, but it was so swift that it seemed instantaneous. Galileo's conclusion was that the speed of light must be so great that human senses cannot detect the time it takes to travel even along a great distance on Earth, and to determine its value.

Another of Galileo's inventions was an ingenious thermometer based on thermal expansion of water. In a long vertical vessel filled with water floated several glass balls of different average density. With changing temperature the density of water also changed slightly, making less dense balls go up to the surface while the denser ones stayed at the bottom. One or two were in the middle, and when calibrated properly, they indicated temperature variations. Galileo's thermometers are still for sale today and used in some homes for their decorative value.

10.3 The telescope and astronomical discoveries

In 1608, still in Padova, Galileo constructed his first telescope, with two lenses, one convex serving as objective, the other one concave, serving as eye piece. Similar devices, of which Galileo had heard, were made by opticians in Holland, with very low magnifying power, but Galileo arrived at the right combination of lenses by himself. After many trials and errors he produced his first telescope, with only three times, and soon later with eight times magnifying power. The next year he was able to improve the magnifying power to 10 and even 14 with lenses which he learned to grind by himself.

The field of vision of Galileo's telescopes was very narrow — when he directed his telescope to look at the Moon, he could see its disc only partly. But what he saw was already a major surprise — the first one in a series of revolutionary discoveries that followed.

• **1.** Mountains, craters and dark planes on the Moon: On the Moon's surface Galileo saw mountains casting sharp shadows, many circular formations resembling volcanic craters on Earth, but also

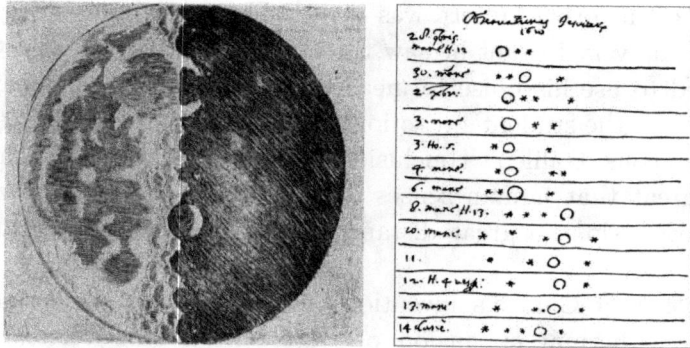

Fig. 10.7   Left: Galileo's drawing of Moon's surface; Right: First sketch of four Jupiter's satellites observed by Galileo on January 7, 1610.

very wide ones, of apparently non-volcanic origin. The mountainous regions were bright, while great flatlands were rather dark, and much less mountaineous. Galileo called them the "seas", although it was not clear whether they contained water or not.

• **2.** The Milky Way full of stars: Although Galileo's first telescopes' field of vision was so narrow that he could see only partly the disc of the Moon, they concentrated incoming light rays, thus enabling the observation of faint stars normally not accessible to the naked eye. When Galileo directed his telescope towards the Milky Way, he saw myriads of stars instead of what was considered by astronomers a hazy nebula of continuous nature. Nevertheless it remained unclear to Galileo how far away all those stars were, and how many they could be.

• **3.** The four moons of Jupiter: A major astronomical discovery, the four satellites orbiting at different distances and with corresponding periods provided a miniature model of the solar system. Galileo noticed that the moons periodically disappeared behind Jupiter and appeared regularly after a period of time depending on their orbital velocity. The major impact of this descovery was the fact that no matter what system one adopts, geocentric or heliocentric, circular motions could have many centers, and not the unique and universal one. Galileo proposed to use the regular eclipses of the satellite closest to Jupiter (today its name is Io) as celestial clock marking universal time, which compared to local time anywhere on Earth, can determine the longitude of the place. Ingenious as it was, the method revealed itself to be too complicated and impractical.

An interesting event happened with the names of Jupiter's moons. Galileo named them *"Medicean stars"* to honor the four Medici brothers, who ruled in Tuscany at that time. This helped him to get the professorship in Padova. However, the German astronomer Simon Marius, Tycho Brahe's former student, claimed priority of observing Jupiter's moons before Galileo, and considered himself to be entitled to give names to Jupiter's satellites. Taking inspiration from Greco-Roman mythology, he chose the names of four of Jupiter's divine love conquests: the three nymphs Io, Europa and Callisto, and a young boy Ganymede. Marius' claim was disputed by Galileo, whose first observation of Jupiter's satellites dated from January 7, 1610. Marius' claim for priority was based on the date of his discovery of Jupiter's moons, which was December 29, 1609. However, at that time everyone in Italy was using new Gregorian calendar which was 10 days ahead of the Julian calendar still used by the protestants in Germany. When transposed from Gregorian calendar to the Julian one, the date January 7, 1610 turned out to be December 28, 1609, one day before Marius. In spite of this, although Galileo did indeed discover Jupiter's four moons before Marius, the names the latter gave to Jupiter's satellites have become officially *Galilei's four moons: Io, Europa, Ganymede*, and *Callisto*.

• 4. The phases of Venus: When Galileo saw Venus as a "horned star", he immediately undestood that what appeared before his eyes was the best confirmation of the validity of Copernican heliocentric system. Not only did Venus appear as a Moon in miniature, developing the same phases as if it were illuminated by the Sun from different angles, but also its angular size varied as its distance from the Earth grew or decreased, accordingly to the assumption that it followed a circular orbit with a radius about 0.7 of the radius of Earth's orbit. The planet appeared much greater when it was in the "new Venus" phase, very close to the Sun, and much smaller when it was close to its "full Venus" phase, as shown in Figure 10.8.

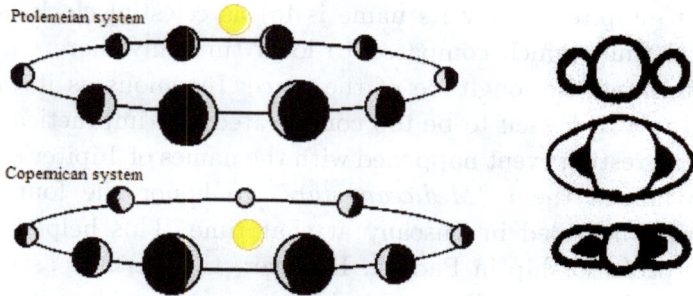

Fig. 10.8   Left: Phases of Venus according to Ptolemeian versus Copernican system; Right: Sketches of Saturn made by Galileo.

● **5. Saturn's unusual shape:** After the discovery of Jupiter's satellites Galileo pointed his telescope to Saturn. What he saw was so strange that he hesitated whether the image was not distorted by some atmospheric phenomenon. Instead of one spherical shape, like in the case of Venus, Mars and Jupiter, Saturn seemed to possess "ears" on each side, or perhaps it were two giant moons? Galileo's conclusion was that what he saw was a triple planet. This is what he sent in the form of an anagram to his friends and fellow astronomers:

   smaismrmilmepoetaleumibunenugttauiras

whose secret meaning was the following sentence: "Altissimum planetam tergeminum observavi" — which gives in English "I have observed the highest planet to be triple-bodied".

Johannes Kepler, who was among the recipients of the encoded information, began diligently trying various combinations, and finally found (as he thought) the right answer: "Salve, umbistineum geminatum Martia proles", meaning "Be greeted, double-knob, children of Mars", thinking that Galileo observed two satellites of Mars. Seeking numerical coincidences everywhere, Kepler was convinced that if Venus has no moon at all, Earth has one, and Jupiter four, then it would be logical if Mars had two moons, to fit a geometric progression. Unfortunately, the two moons of Mars are so small that they have been discovered only after very huge telescopes were built in the 19th century. Kepler received the correct answer from Galileo at the end of the year 1610.

But this is not the end of Galileo's encounters with Saturn. The two "companions" he discovered did not move, so they were not likely to be Saturn's moons, and Galileo lost interest in them for a couple of years. When he looked at Saturn again through his telescope, he discovered to his great dismay that Saturn had recovered a spherical shape, like the other planets. Galileo predicted that the two companions will reappear soon. As a matter of fact, the two bulges reappeared, but not in the way Galileo expected. Later on, in 1616, he saw those two strange shapes again, but this time they looked like handles of a cup. The sketches Galileo made to represent what he saw are shown in Figure 10.8 on the right. It took more than 70 years until the strange shapes on Saturn's sides were identified as rings periodically disappearing from sight when they become parallel to the plane containing the line of vision. The rings of Saturn are so thin that they become invisible. The rings were definitely identified by Huygens and Cassini at the end of 17th century.

- **6.** Spots on the Sun: Observing the Sun with his telescope, but through a dimmed glass, Galileo discovered dark spots on the solar disc. They could not be an illusion or an atmospheric phenomenon, because they persisted during many days. Moreover, the ones that were close to the edge displayed flattened form, as it should be the case for spots painted on spherical surface. Some of Galileo's contemporaries believed the sunspots could be small planets between Mercury and the Sun, but Galileo refuted this guess observing that the spots changed their size and shape, and sometimes disappeared behind Sun's disc. The nature of the phenomenon remained mysterious; however, their very existence was another blow at the Aristotelian worldview in which the Sun must be perfect, like all heavenly objects.

10.4 The "Dialogo" and its consequences

Galileo's astronomical discoveries struck an ultimate blow to Ptolemy's geocentric system. The fact that Mercury and Venus orbited the Sun and not the Earth became obvious by the simple

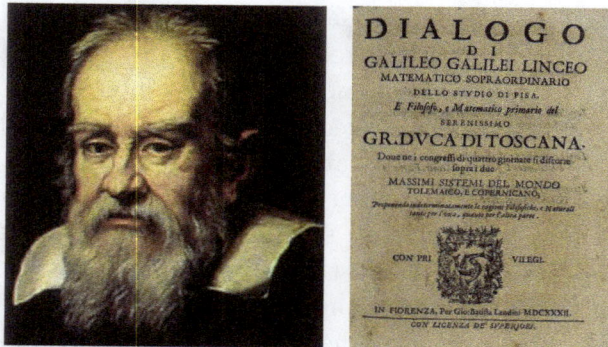

Fig. 10.9 Left: Galileo in his elderly years. Right: Galileo's book, the "Dialogo", published in 1632 in Leiden.

observation how the phases of Venus changed depending on its position between the Sun and Earth (similar phases of Mercury were not easy to observe with the imperfect telescopes of that time). Other dicoveries contradicted most of the Aristotelian assumptions about the heavens in general, especially the perfect and unchangeable nature of heavenly bodies and the entire structure of the Universe. New stars, like the Nova observed by Tycho Brahe in 1577 who proved that it belongs to the most distant stellar realm, the spots on the Sun, the mountains on the Moon — all that pointed towards a physical unity between the terrestrial, "sub-lunar" world and the Cosmos at large. Galileo's conviction that the Copernican system describes the objective truth was strenghtened by his discoveries, and the long correspondence with Kepler reinforced it. In one of his letters Kepler exhorts Galileo to defend the Copernican system publicly — "be bold, Galileo, step forward", Kepler wrote. For quite a long time Galileo was hesitant to "step forward", fearing the reactions of the Church and Inquisition.

10.4.1 *The trial*

In 1616 a Catholic priest Francesco Ingoli (1578–1649) wrote an essay in which he attacked the Copernican theory, and sent Galileo a copy. In his argumentation Ingoli used a number of physical and mathematical arguments, reinforced by theological ones. Most of the physical and mathematical arguments were taken from Tycho Brahe's writings, among which was the fact that projectiles sent vertically upward should fall down west from the point from which they were shot, and why there was no strong wind opposing the Earth's motion. Another argument repeated after Brahe was the absence of stellar parallaxes, and the conclusion that if the Copernican model was real, then the stars must be incredibly large, perhaps even larger than the Sun — which in Galileo's time was considered as being totally absurd. Theological arguments included citations from the Holy Scriptures implying that *"God placed lights in the firmament of the heavens to divide the day from the night"*, and that the Hell is in the farthest place from Heaven, therefore in the center of the Earth. However, at the end of his essay Ingoli summons Galileo to respond exclusively to scientific arguments, not to theological ones.

Under the influence of Ingoli the Catholic Church put Copernicus' *"De Revolutionibus"* on the index, arguing that it was a "false Pythagorean doctrine, altogether contrary to the Holy Scripture". On the orders of Pope Paul V, Cardinal Robert Bellarmine gave Galileo prior notice that the decree was about to be issued, and warned him that he could not "hold or defend" the Copernican doctrine. Galileo answered Ingoli in a long "Message to Ingoli" written in 1624, but it remained unpublished. It did not change the opinion of learned ecclesiastical authorities.

Galileo decided to write an apology of the Copernican system and heliocentricity in popular form, accessible not only to the scientists, so he chose to express his views in the commonly spoken Italian and not in Latin. The form was original, too: not a narrative, but a dialogue between three characters, Sagredo, Simplicio and Salviati, under the title *"Dialogue Concerning the Two Chief World Systems"*. The work on the book lasted four years; it appeared in 1632 and was presented with a dedication to Ferdinando Medici of Florence.

Although it is not entirely proven, it seems that the Pope Urban VIII felt offended by Galileo's book, finding that the character "Simplicio" defending the Aristitelean geocentric system and painted as a man of little understanding, was describing him personally. In spite of

Medicis' protection, the Papal Court pronounced a ban on Galileo's
book, and asked the Inquisition to conduct the investigation. Galileo
was summoned to present himself before the Inquisition court in
Rome, and after a humiliating trial was convicted in 1633 of grave
suspicion of heresy for "following the position of Copernicus, which
is contrary to the true sense and authority of the Holy Scripture".
At the age of 69 he was forced to kneel and to read aloud a repu-
diation of his views on heliocentricity and the moving Earth. The
legend about Galileo saying aloud *"Eppur si muove!"* ("And yet it
moves!", meaning the Earth) at the end of his trial is most probably
apocryphal: it was just too dangerous to pronounce such a phrase
aloud. But he certainly *thought* along these lines.

10.4.2 *The last years*

The trial imposed by the Inquisition was severe and humiliating, but
the verdict was not too harsh, especially as compared with Giordano
Bruno's fate: Galileo was placed under house arrest for the rest of
his life. He had to stop his astronomical investigations definitely, but
was permitted to work on other subjects. He spent the rest of his life
in his house in Arcetri, in the company of his loving son Vincenzo
and his faithful disciple Vincenzo Viviani (1622–1703) who helped
him as he was progressively losing sight. Also his devoted disciples
Torricelli (1608–1647) and Cavalieri (1598–1647) visited him many
times, comforting their master and helping him to write his last book
resuming Galileo's views on motion and mechanics.

Although the Church prohibited the print of any Galileo's works,
he wrote another book in form of a dialogue, involving the three characters of his first *Dialogue*, Sagredo, Salviati and Simplicio, exposing
in a popular and vivid manner all his previous findings in mechanics
and physics. The book was printed in Leiden, and despite the Inquisition's ban on Galileo's writings, was smuggled to Rome; fifty copies
were sold almost immediately.

Galileo's curious mind remained active during his home arrest.
He proposed a method of determining geographical longitude using
the periodic (every 42.5 hours) occultations of the Jovian satellite Io

as a celestial clock marking universal time. It consisted in comparing the local time with the universal time defined by Io's occultations. Theoretically perfect, this method was hardly realizable, especially at sea; this is why the Dutch Admiralty, to whom Galileo proposed his method, politely declined it, however recognizing his ingenuity by a generous gift (a gold chain and a medal).

Creativity did not abandon Galileo until his last days. Being aware that longitude determination using the Jovian satellites was too sophisticated to become feasible, he concentrated his efforts in constructing a reliable clock that could be carried along by travellers, keeping local time of the point of departure. Despite total blindness since 1636, with the help of his son Vincenzo he conceived a pendulum clock with anchor mechanism similar to the one constructed by Huygens several decades later. Unfortunately it was realized only two years after Galileo's death in 1642.

Chapter 11

Kepler and the new astronomy

Kepler: scientist and mystic - from Graz to Prague - Tycho Brahe's legacy - The analysis of Brahe's observations of Mars - Laws of planetary motion - Correspondence with Galilei - hexagonal snowflakes - "Somnium", the first science fiction

11.1 Johannes Kepler' life and achievements

Johannes Kepler (1571–1630, Holy German Empire) was one of the greatest astronomers of all times. Born to an empoverished protestant family, as a young child he was rather sickly and weak, however endowed with curious mind and great intellectual capacities. Kepler's father was a mercenary, and perished in the Netherlands when Johannes was five years old. Kepler's mother had to sustain her three children earning a living as healer and herbalist.

As a young boy, Kepler saw the great comet in 1577, which was also observed and described by Tycho Brahe. In 1580 Kepler observed a full lunar eclipse and noticed the unusual red color of our satellite.

Kepler studied theology in Tübingen to become a pastor; but soon he became interested in mathematics, impressed by his excellent teacher Michael Maestlin, with whom he maintained a long-lasting friendship. Maestlin acquainted young Kepler with Copernicus' book "De Revolutionibus", which played a major role in Kepler's scientific convictions — he became a staunch defender of the Copernican system. In 1594 Kepler got a position as mathematics teacher in Graz (today's Austria). The city was mostly Catholic, and Kepler did not

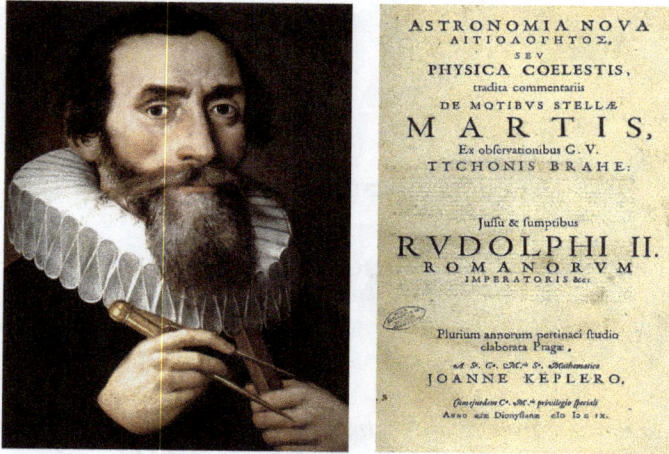

Fig. 11.1   Johannes Kepler portrayed by an unknown artist in 1610. On the right, his most important book, *"Astronomia Nova"*, published in 1609.

feel at ease. At home he continued his efforts to apply mathematics to discover the harmony of the celestial world, being convinced that the Sun kept planets orbiting it by a force he called "magnetism".

Kepler was convinced that our world, being God's creation, must possess internal beauty and harmony which human beings have to discover. Looking for regularities in planetary orbits, Kepler combined the five Platonic perfect polytopes interposing them between the spheres of the six planets (including the Earth, according to the Copernican model), finding a reasonable agreement with the observed orbital radii.

In 1586 he published his findings in a book *"Mysterium Cosmographicum"*, where he defended the Copernican heliocentric system. He sent a copy to Tycho Brahe, who a few years later invited Kepler to become his assistant in Prague. This was a major turn in Kepler's life, because Tycho was the best teacher of observational astronomy he could ever dream of.

Unfortunately, less than two years later Tycho suddenly died; but before, he asked Rudolph II to make Johannes Kepler the imperial mathematician. Tycho's family fought legal battles with Kepler for years. They were able to win possession of Tycho's instruments, but Kepler kept Tycho's books. Having inherited the entire archive of his master's astronomical data gathered during more than twenty years of continuous observations, he used them while elaborating his laws of planetary motion based above all on Tycho's meticulous observations of Mars. Kepler's most important book *"Astronomia Nova ΑΙΤΙΟΛΟΓΙΤΟΣ seu physica coelestis, tradita commentariis de motibus stellae Martis ex observationibus G.V. Tychonis Brahe"* was published in 1609; later on, another important book, *Harmonices Mundi*, was published in 1619.

However, Mars was not the only object Kepler was interested in, especially as his astrological activities obliged him to keep track of other planets, too. Observing in 1603 the conjunction of Jupiter with Saturn, a rare astronomical event occurring once in 20 years, Kepler discovered a new star, which turned out to be the supernova of 1604. He published the description of the phenomenon in a booklet *"De Stella Nova"* edited in 1606.

During many years Kepler exchanged letters with Galileo, who shared his findings with certain reticence, informing Kepler about certain discoveries he made with his newly constructed telescope, but without explaining its optical construction.

In 1611 Kepler constructed his own telescope, with two convex lenses. It was more powerful than Galileo's one, but it gave an inversed image. This was a shortcoming only for terrestrial observations, but was insignificant for the astronomical ones; and when astronomers started to use telescopes more widely, Kepler's construction prevailed. Nowadays Galileo's scheme is used in military or theater binoculars. With his new device Kepler was able to confirm Galileo's observations of Jupiter's moons, earning Galileo's praise expressed in a special letter.

The year 1611 turned out to be unfortunate for Kepler: his wife Barbara died, as well as one of their sons. The new German Emperor Matthias succeeding Rudolf II was less friendly to Kepler, who had to leave Prague and move to Linz. Although he formally kept his title of imperial astronomer, he was offered a position of mathematics professor in Linz. In 1613 he remarried with Susanna Rettinger, with who gave birth to seven children, but only two of them survived.

The times were marked by strong tensions between protestants and catholics. After the defenestration of catholics by protestants in the city hall in Prague in 1617, a terrible war started in Europe; it lasted 30 years and took away more than 5 million lives. In 1625 the Counter-Reformation happened, and catholic authorities temporarily removed Kepler's library. Other problems marred Kepler's last ten years: his mother was accused of witchcraft, and he had to spend a lot of energy and almost an entire year fighting for her honor and freedom. She was exonerated in 1621, but died a few months later.

During his stay in Linz, which lasted until 1626, Kepler remained extremely active, in spite of all the vicissitudes that fell on him. He continued to analyze Tycho's Martian tables, and published the supplement (*"Epitome"*) to the *Tabulae Rudolphinae*. In 1619 he published his major book *Harmonices Mundi*, in which he formulated his third law of planetary motion, stating that the squares of the periods of planets are proportional to the cubes of the semi-major axes of their orbits.

In 1625, Kepler's library in Linz was placed under seal by the agents of the Catholic Counter-Reformation, and in 1626 the city of Linz was besieged in the course of the 30-years war. Kepler was forced to move to Ulm, where he arranged for the printing of the *Rudolphine Tables* at his own expense.

In 1628 Kepler became an official advisor to Wallenstein, a famous general of the Emperor Ferdinand II. He made astronomical calculations for Wallenstein's astrologers and occasionally wrote horoscopes himself. In his final years, Kepler often changed places, traveling from Prague to Linz and Ulm, and finally to Regensburg. Soon after arriving in Regensburg, Kepler fell ill. He died on 15 November 1630, and was buried there; his burial site was lost during the Swedish occupation of the city; only Kepler's self-authored poetic epitaph survived the merciless time:

> *I used to measure the heavens,*
> *now I shall measure the shadows of the earth.*
> *Although my soul was from heaven,*
> *the shadow of my body lies here.*

11.2 Quest for harmony

From his earliest years Kepler was convinced that God created the Universe endowing it with perfect beauty and harmony, the ultimate goal of science being gradually uncovering God's intelligent design. The apparent imperfections and irregularities being due to the limitations of human vision and intelligence, which can be overcome with patient effort and deepened insight. According to the heliocentric Copernican system, six planets — including our own, the Earth — orbited the Sun, supposed to be the motionless center of the Universe. The relative sizes of their orbits (supposed circular yet) were well established by Copernicus more than fifty years earlier. Kepler was looking for some logical pattern, a kind of divine architectural plan that must have been founded on geometry. The model of concentric spheres bearing the planets was still present in his mind, although the spheres were no more believed to be like material bodies we know from everyday's experience. Then Kepler had a revelation: six spheres meant five interstitial spaces between them, and there were exactly five perfect Platonic solids known in Euclidean geometry.

Quite remarkably, comparing the radii of the spheres inscribed in the Platonic solids with the planetary orbits, he found agreement within 5% , with the exception of Jupiter, of which he said: "no one will wonder, considering such a great distance". How much Kepler was impressed by the coincidence of the platonic solids' sizes with the orbits of planets can be appreciated by the letter he wrote to Maestlin: *"I wanted to become a theologian; for a long time I was restless. Now, however, behold how through my effort God is being celebrated in astronomy."*

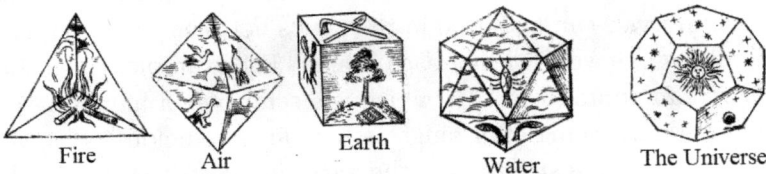

Fig. 11.2   Five Platonic solids: tetrahedron, octahedron, cube, isocahedron and dodecahedron.

Table 11.1   A comparison between the planets' average distances from the Sun and the inner and outer radii of five Platonic polyhedra.

Planet	Symbol	a (in A.U.)	Solid	R_{out}	r_{in}
Mercury	☿	0.395	-	–	–
Venus	♀	0.723	Octahedron	0.707	0.408
Earth	⊕	1.000	Icosahedron	0.951	0.756
Mars	♂	1.530	Dodecahedron	1.401	1.114
Jupiter	♃	5.209	Tetrahedron	6.12	2.04
Saturn	♄	9.551	Cube	8.66	5.00

Table 11.1 contains the average radii of orbits of six planets (including the Earth) as they appear in the heliocentric Copernican model, and the radii of spheres circumscribed on and inscribed in the five platonic solids.

Kepler exposed his geometrical model in the book entitled *"Mysterium Cosmographicum"* published in 1596. The great astronomer Tycho Brahe upon receiving the book from Kepler, was so impressed by the young man's talent, that he offered Kepler to join him as an assistant in his Prague observatory. Kepler accepted, especially as being a protestant he felt uneasy in catholic Graz. This was a stroke of destiny, because without Brahe's collection of data on Mars Kepler would never have found the laws of planetary motion.

The Pythagorean influence can be clearly traced in Kepler's efforts to detect not only a geometrical, but also a musical harmony in the celestial order. At some stage of his investigations, Kepler thought that the heavens are filled with harmony in its literal sense, with some kind of music being constantly played, although not audible for the human ears. According to the views explained in *Mysterium*, musical intervals can be found in the ratios between the maximal and minimal angular velocities of each planet. For example, this ratio for the Earth amounts to 16 : 15, which is a semitone, a half of an entire tone. Therefore the Earth sings: a *mi, fa, mi* melody. For Venus, whose orbit is almost circular, the ratio between the maximal and minimal angular velocities is fairly approached by the fraction 25 : 24, which can be identified with a quartertone (half of a semitone). Also

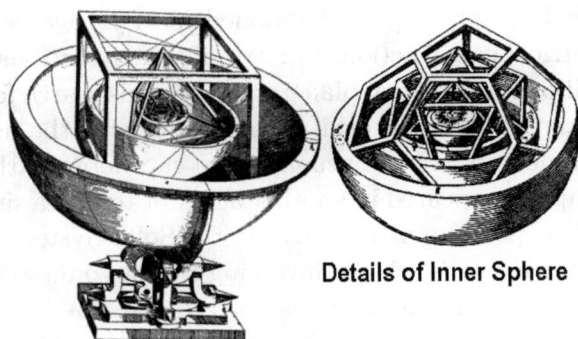

Details of Inner Sphere

Fig. 11.3 The model of the Solar System with spheres inscribed and circumscribed around the five Platonic solids. From Kepler's "Mysterium Cosmographicum" book (1596).

the ratios of average orbital radii fell into a musical harmonic scheme. The voices Kepler attributed to planets of the Solar System are as follows: Mercury is a soprano, Venus and Earth are altos, Mars is a tenor, and Jupiter and Saturn are bass voices. In order to take into account the variation of angular velocities in the case of eccentric orbits, an addition of small epicycles was needed even in the Copernican system. In musical terms this was equivalent to the phenomenon of harmonics observed in musical instruments using chords. For Mercury, the planet with greatest eccentricity in the Solar System, as many as five harmonics were added; for Mars, two harmonics, while Venus "sang" with one pure note, with no harmonics needed.

There was one noticeable exception which Kepler was unable to explain. While all neighboring orbits radii produced ratios close to simple fractions between notes of musical scale within the 25 : 24 margin of precision (corresponding to less than one "sharp", or quarter of musical tone), the ratio between the average radii of Jupiter and Mars escaped the rule, being off the closest musical interval by an anharmonic ratio 18 : 19 which cannot appear in a natural scale. The too large interval between Mars and Jupiter remained unexplained for more that two hundred years, until the first asteroid was observed, named Ceres by its discoverer, the Italian astronomer Piazzi (1746–1826).

Although later on Kepler abandoned his Pythagorean and Platonic geometrical constructions in favor of the revolutionary hypothesis of the elliptical nature of planetary orbits, his search for some law underlying the sequence of orbital radii influenced the next generations of astronomers. The unusually huge gap between the orbits of Mars and Jupiter appeared in another attempt to find a simple mathematical rule explaining orbital sizes in the Solar System. More than 150 years after Kepler's scheme involving five platonic solids, in 1766 the German astronomer Johann Daniel Titius (1729–1796) proposed a simple formula reproducing the ratios between the orbital radii of planets with one noticeable exception, which was again the gap between the orbits of Mars and Jupiter. The formula has been popularized by published works of another German astronomer, Johann Elert Bode (1747–1826), so that the formula is known as the Bode-Titius law.

The regularity described by the Bode-Titius law is based on a very simple arithmetic involving a geometrical series with basic ratio 2. The construction is as follows:

a) Consider the series of integers $0, 1, 2, 4, 8, 16, 32, 64, 128, \ldots$
b) Multiply these numbers by 0.3 to get $0, 0.3, 0.6, 1.2, 2.4, 4.8, 9.6, \ldots$
c) Add 0.4 to each of the above, and get $0.4, 0.7, 1.0, 1.6, 2.8, 5.2, 10.0, \ldots$

Table 11.2 shows how close are the average radii of planets of our Solar System to the values predicted by the Bode-Titius law.

The apparent empty space between Mars and Jupiter, first noticed by Kepler, corresponds to the asteroid belt discovered in the 19th century. The biggest of its components are like small planets with regular spherical shape: the biggest one, discovered in 1801 by Italian astronomer Giuseppe Piazzi (1746–1826), was named *Ceres* after Roman goddess of harvest, and has a diameter of about 956 km; soon after followed *Vesta* and *Pallas*, with diameters 525 and 512 km, respectively. As of today, the asteroid belt is estimated to contain between 1 and 1.8 million asteroids larger than 1 kilometer in diameter, and myriads of smaller ones, with irregular shapes. They

Table 11.2   The average planetary distances to the Sun compared with
the predictions of the Bose-Titius law.

Mercury	Venus	Earth	Mars	—	Jupiter	Saturn	Uranus
0.4	0.7	1.0	1.6	2.8	5.2	10.0	19.6
0.387	0.723	1.000	1.524	—	5.2	9.539	19.18

are supposed to be the remnants of a proto-planet which failed to
agglomerate due to Jupiter's destructive gravitational influence.

The comparison with the planet's orbits expressed in Earth orbital
radii (i.e. in astronomical units, AU) is quite satisfactory and indicates that it is not just a coincidence — there must be something
deeper concerning the stability of certain orbits and instability of the
other ones, outside the stability zones. In spite of numerous efforts,
until today no satisfactory theory was found explaining this particular pattern in planetary systems.

11.3   The limits of the Copernican system

The most appreciable feature of the Copernican heliocentric system
was the fact that the retrograde motion was explained by the difference between the angular velocities of a moving terrestrial observer
and the real angular velocities of planets with respect to the stars.

However, the discrepancies between the predicted and the
observed planetary motions and positions were greater than those
given by the sophisticated versions of the Ptolemaic geocentric system with equants, deferents and many epicycles, as well as the predictions of the Tychonic system, in which the inner planets (Mercury
and Venus) rotated around the Sun which orbited around the static
Earth, together with the Moon and the outer planets (Mars, Jupiter
and Saturn).

Among all planets, Mars seemed to be the most difficult to
be understood. Its motion among the stars, as well as the periodic changes of its brightness, eluded explanation even along the
lines of the Copernican system. Its retrograde motion was the most
complicated of all, with variations in angular velocity needing the

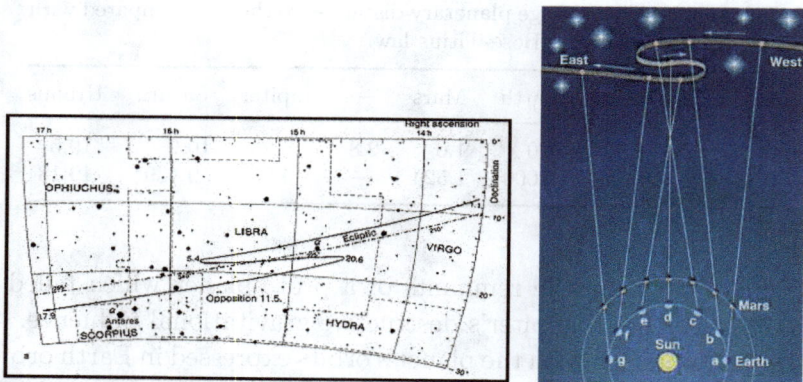

Fig. 11.4   Mars' trajectory among distant stars, as seen from the Earth, and its explanation according to the Copernican system.

addition of extra epicycles, with various angular velocities each (see Figure 11.4).

After a thorough analysis of a great amount of observations left by Tycho Brahe, Kepler concluded that neither of the three available astronomical systems, the Ptolemaic, the Tychonic as well as the Copernican one, was able to provide predictions agreeing with observations. Kepler was totally confident in the accuracy of Tycho's results, who was able not only to determine positions of planets with up to $1'$ precision, but who also systematically substracted corrections due to the atmospheric refraction, varying with different celestial altitudes. Kepler presented his criticisms of the three systems in his book. One of its pages with three models compared can be seen in Figure 11.5.

Kepler tried in vain to describe correctly consecutive positions of Mars using the Copernican model with eccentric circular orbits, he found that his attempts always presented discrepancies with Tycho's observations. The differences were not huge, but amounted to 6 to 8 arcminutes. Knowing that Tycho reached the precision of less than $2'$, and very often just $1'$, so great was Kepler's confidence in Tycho's data, that he arrived to the conclusion that it was the circular shape of orbits that was, in fact, an erroneous assumption. In a sudden revelation, the idea of replacement of the circular orbits by elliptic

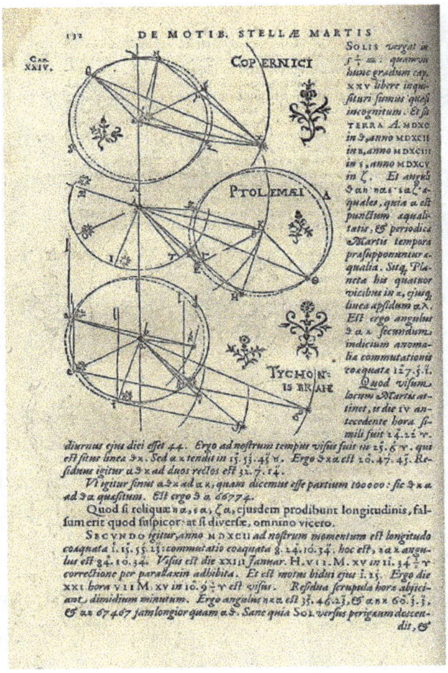

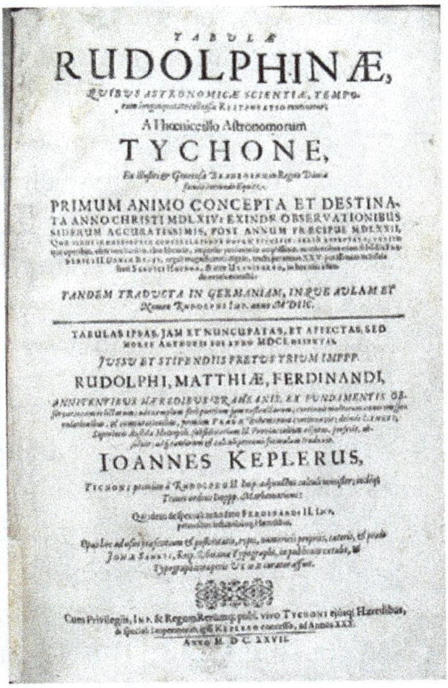

Fig. 11.5 Left: The page of Kepler's "Astronomia Nova" in which he critically compares the three systems. Right: Tycho Brahe's observations of Mars edited by Kepler.

ones came to Kepler's mind. He spent several years on geometrical analysis of Tycho's Martian observations, and using consecutive triangulations, was able to reproduce the elliptical shapes of both orbits, Martian and terrestrial alike.

11.4 How Kepler determined the shape of the Martian orbit

As we already underlined many times before, to a first approximation, Kepler could safely admit that the Earth was moving around the Sun along a circular orbit. In fact, this orbit is slightly elliptical, but the eccentricity is very low, $e = 0.0167$, which corresponds to the ratio between the aphelion and perihelion distances of the order of

$$\frac{\text{aphelion}}{\text{perihelion}} = \frac{r_a}{r_p} = 1.034. \tag{11.1}$$

On the contrary, as we presently know, the Martian orbit is elliptical with an eccentricity $e = 0.0934$, i.e. near to 10%. This explains
why the orbit and motion of Mars show certain anomalies that the
Copernican system was unable to explain. Kepler had at his disposal
a huge amount of high precision observations of Mars' positions in
the sky which he inherited from his master Tycho Brahe.

Using meticulously systematic data of Mars' positions obtained
by Brahe during almost twenty years, Kepler was able to reconstruct
the orbit of the Red Planet in a way that was as simple as it was
ingenious. He did it in three steps:

• **Step 1.** First of all, Kepler needed to determine the sidereal
period of Mars' rotation around the Sun. In principle, this could be
done by precise measurements of Mars' position among the stars,
and noting the time between successive positions of this planet coinciding with some well defined zone in the vicinity of characteristic
stars or constellations. But the trajectory of the Red Planet seems
quite erratic, due to its ellipticity and inclination (close to 5° with
respect to the ecliptic). To arrive to his first goal, instead of analyzing consecutive positions of Mars with respect to the stars, Kepler
scrupulously noted the periods between consecutive Martian oppositions, and found that on the average they were separated by 780
days (or 2 years and 50 days), although each time they occurred in
a different celestial region.

According to the Copernican heliocentric system, the oppositions
of Mars occur when the planet is found on the straight line joining
the centers of the Sun and the Earth, i.e. exactly at midnight, on
the southern meridian. This kind of measurement is much more precise than tracing Mars' position among the stars, especially when
the bright and well-known stars are many degrees apart from it. So,
the establishment of the *synodic period* of Mars was made with as
little error as possible, with the aforementioned result, 780 days.

Comparing this period with one year, i.e. the period of full rotation
of Earth around the Sun, i.e. 365 days, we obtain the fraction

$$\frac{780}{365} = 2.13698 \quad \text{or} \quad \frac{780}{365} \simeq \frac{62}{29}. \tag{11.2}$$

The last fraction is the result of a Diophantine expansion of the ratio 780/365, whose successive approximations are given by the series of fractions 15/7, 62/29, 139/65, etc., of which the second one gives already the precision up to 1 part in 10 000.

If Mars is in the next opposition after the Earth has completed 2.137 revolutions around the Sun, Kepler inferred, then Mars must have made during the same time $n + 0.137$ revolutions, n being an integer number. Obviously enough, n cannot be zero, because if such was the case, Mars' angular velocity would have been so slow (only about 49° per year), that more than seven opposition of Mars would occur in one terrestrial year. Also, $n = 2$ must be excluded, because in that case Mars would turn around the Sun with the same angular velocity as the Earth does, so that as seen from Earth it would occupy the same position in the sky all the time. And obviously, $n = 3$ or more is unacceptable, because if Mars turned around the Sun more rapidly than Earth, we would observe more than one opposition during one year. This gives the unique solution $n = 1$, which means that during 780 days Mars covers $1.137 \times 2\pi = 1.137 \times 360° = 409°19'$. Therefore, one Martian revolution, or Martian year, lasts

$$T_M = \frac{780}{1.137} = 686 \, \text{days.} \qquad (11.3)$$

(As we showed when analyzing the discrepancy between the Moon's sidereal and synodic months, the same result can be obtained by comparing average angular velocities, or the inverse periods of rotation. The time between two consecutive oppositions T_2 would play the role of a synodic month, and our sideral year T_E is well known. Therefore the Martian year T_1 should satisfy the following equation: $\frac{1}{T_2} = \frac{1}{T_E} - \frac{1}{T_1}$, which gives the same result as in (11.3).)

Now Kepler could use Mars as a stable beacon in space, knowing that it will come back to the same position after 686 days. This new geometric tool, added to the observations of consecutive oppositions, became the key to the reconstruction of the real shape of orbits, not only the Martian one, but also our own, reconstructing the motion of Earth around the Sun to yet unknown precision.

• **Step 2.** At first, Kepler took for granted the Copernican system, supposing that the orbit of Earth is a perfect circle. But how to explain the variation of Earth's orbital velocity? The Ptolemaic system gave the answer by constructing an imaginary circular orbit called *deferent* whose center was shifted with respect to the Earth, which was believed to be motionless. Adapting the Copernican point of view, Kepler used the huge amount of Tycho's obsevations in order to trace the Earth's path in its motion around the Sun.

In order to determine positions of the Earth with respect to both the Sun and Mars at various times of the year, he contructed two triangles with the same base, which was the straight line joining the Sun and Mars. Knowing the exact direction towards Mars and the direction towards the Sun on the same day, and having at his disposal similar data taken by Tycho Brahe 686 days later, Kepler was able to construct two triangles with one common summit coinciding with Mars' position, identical after exactly one Martian year, and one common side joining the Sun with Mars. The angles between the directions Earth-Sun and Earth-Mars were found in both cases in Tycho's Martian observations. The resulting construction is shown in Figure 11.6.

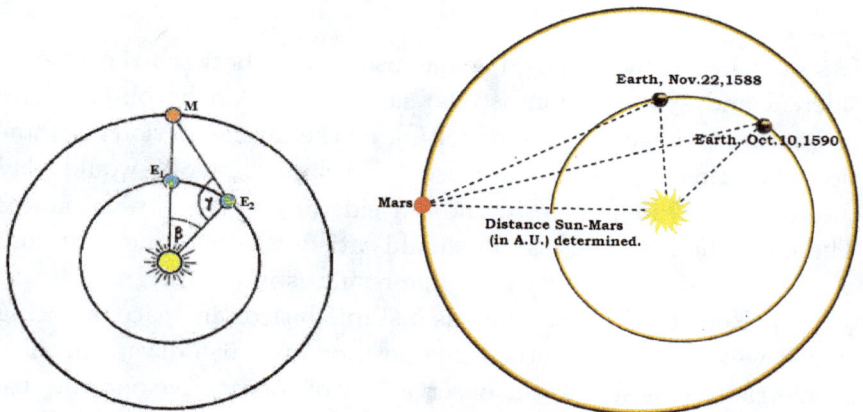

Fig. 11.6   Left: Two of Tycho Brahe's observations of Mars separated by 686 days, opposition first, and at different angle one Martian year later; Right: Two observations separated by 686 days, at an arbitrarily chosen position.

• **Step 3.** After the shape of the Earth's orbit was fixed, the reconstruction of the Martian orbit was possible due to the same triangulation principle, as shown by the right diagram in Figure 11.6.

Kepler performed similar constructions with many pairs of data giving precise positions of Mars separated by 686 days, like in Figure 11.6, but during different seasons. His findings were in contradiction with all the three models of Solar System available in his time: the Ptolemaic with equants and epicycles; the Copernican model with circular orbits; and the "mixed" model proposed by Tycho Brahe, with the Earth at the center, the Moon and the Sun revolving around it, and planets revolving around the Sun. His conclusion was that neither of the three could explain and forecast the observed motions of Mars, and also of other planets with precision matching Tycho Brahe's observations.

The discrepancies accumulated in all these systems, which should be either rejected, or improved. Kepler was a staunch partisan of the Copernican heliocentric system, so his efforts concentrated on modifications and improvements to be implemented in order to explain the differences between the observations and positions of planets predicted by the Copernican system with circular orbits and constant angular velocities. The second assertion could not be true, and was taken into account in both Ptolemaic and Copernican systems by *deferents* and *equants*.

11.5　The ellipse and its properties

The ellipse is one of *conical sections* obtained when a plane cuts through a cone. When the plane happens to be orthogonal to the cone's axis, the curve resulting from the intersection is a circle with its radius increasing linearly when the plane is shifted down from the summit. When the plane is inclined with respect to the cone's axis, the resulting intersection is an ellipse. When the plane is parallel to the cone's axis, the intersection is no more a closed curve. It is a parabola. Finally, increasing the angle of intersection further,

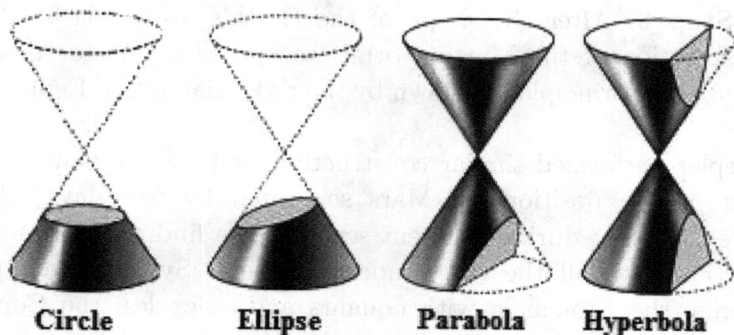

| Circle | Ellipse | Parabola | Hyperbola |

Fig. 11.7   Conical sections: circle, ellipse, parabola and hyperbola.

we get a hyperbola. All types of conical sections are represented in Figure 11.7.

The knowledge of the geometrical properties of ellipses enabled Kepler to identify the shape of planetary orbits after an extremely patient and deep analysis of astronomical observations provided by Tycho Brahe. The most important were the data on successive positions of Mars gathered during more than 20 years.

Let us review the most important properties of ellipses. First, the construction: it is slightly more complicated than drawing a circle with compass which marks the set of all points at equal distance R (the radius of the circle) from a given center O. To draw an ellipse, one has to choose TWO points, F_1 and F_2, called *the foci*, or *focal points* of the ellipse to be created. The distance between F_1 and F_2 is called *the focal distance*. Then fix two pins at the focal points and fix a string with length $2a$, greater than the focal, and inserting a pencil so as to keep the string always tense, draw the upper half of the ellipse. The pins prevent drawing the entire closed curve, so the lower half should be drawn separately.

As seen in Figure 11.8, the sum of two distances $\mid F_1 P_2 \mid + \mid P_2 F_2 \mid$ is equal to the sum of distances $\mid F_1 P_1 \mid + \mid P_1 F_2 \mid$, as well as for any other randomly chosen point P of the ellipse. It is also obvious that the total length of the string is equal to the major axis $2a$. This is easily understood if the point P approaches the horizontal axis of symmetry. Then the shorter segment of the string covers the distance

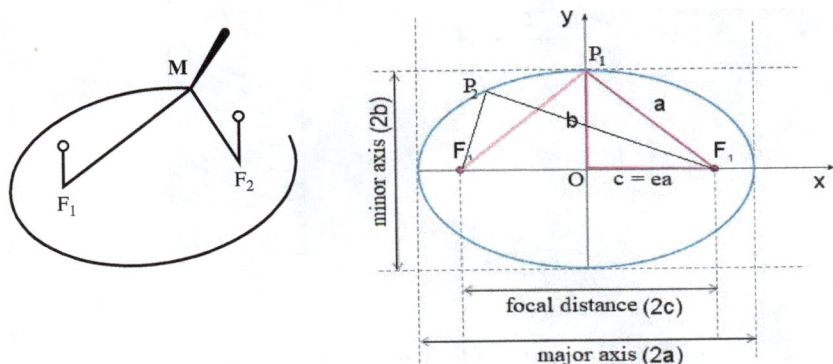

Fig. 11.8   Construction of an ellipse (left) and its basic properties (right).

between ellipse's left extremity and the left focus F_1, while the longer segment will cover the same distance plus the distance between the foci $\mid F_1F_2 \mid$, altogether covering the distance between left and right extremities, which is by definition the major axis.

From this we infer that the distance P_1F_2 is equal to a, the semi-major axis. From Pythagoras' theorem applied to the rectangular triangle P_1OF_2 we get the length of the segment $c = ea$; the factor e is called the *eccentricity* of the ellipse.

$$a^2 = b^2 + c^2 = b^2 + (ea)^2 \;\to\; e^2a^2 = a^2 - b^2 \;\to\; e = \sqrt{1 - \frac{b^2}{a^2}}.$$
$$(11.4)$$

The area of an ellipse is given by a simple formula generalizing the formula for the area of a circle:

$$A = \pi ab. \qquad (11.5)$$

This formula was found in third century B.C.E. by Archimedes, and independently by the Japanese mathematician Takakazu Seki in the 17th century, whose proof is quite original because it starts from the volume of a cylinder, which is obliquely sliced and divided into a huge number of thin elliptical plates. Consider a cylinder of radius b and of height h; its volume V is equal to the surface of its circular basis πb^2 multiplied by its height h, which yields $V = \pi b^2 h$. Let

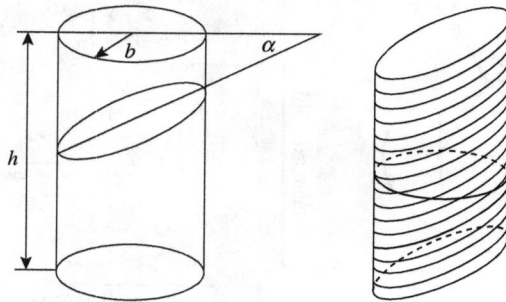

Fig. 11.9   Derivation of formula for the area of an ellipse.

us cut it in two parts with a plane forming an angle α with the horizontal plane. The intersection is an ellipse with the semi minor axis b and the semi-major axis $a = \frac{b}{\cos \alpha}$ (see Figure 11.9 left).[1] Next, take off the upper part and glue it to the lower part by its circular side. The volume of the resulting solid did not change, it is still the same V of the cylinder, but now its upper and lower sides are elliptical. Moreover, we can slice the skew cylinder into N identical elliptically-shaped slices, the thickness of each slice being equal to $\frac{h}{N \cos \alpha}$. When the angle α tends to zero, the thickness of a slice tends to h/N.

[* Let us denote by S the surface of single elliptic slice, its volume is then equal to $S \times \frac{h}{N \cos \alpha}$. The total volume of N such slices must be equal to the initial cylinder's volume V, which leads to the following identity:

$$V = N \times \left(S \frac{h}{N \cos \alpha} \right) = \pi b^2 h. \qquad (11.6)$$

From the above, simplifying by h and by N, and recalling that $a = b \cos \alpha$, we get the desired result:

$$S = \pi b^2 \cos \alpha = \pi a b. \qquad (11.7)$$

---

[1] A cylinder can be considered as an extremely elongated version of a cone. The intersections with oblique planes yield ellipses, but when the plane becomes parallel to the cylinder, instead of a parabola one gets two parallel lines, and when the plane's angle continues to grow, instead of hyperbolas one gets ellipses again.

Curiously enough, no concise formula exists for ellipse's perimeter length, generalizing the simple formula for the perimeter of a circle of radius R, $L = 2\pi R$. The length of the perimeter of an ellipse can be computed by integration (which gives a so-called elliptic integral),

$$L = 4a \int_0^{\frac{\pi}{2}} \sqrt{1 - e^2 \sin^2 \theta}\, d\theta. \tag{11.8}$$

In practical applications an approximate formula can be used, giving a very good agreement with the actual perimeter length of an ellipse with given semi-axes a and b:

$$L \simeq \pi \left(\frac{3}{2}(a + b) - \sqrt{ab} \right). \tag{11.9}$$

Luckily enough, the perimeter of an ellipse does not play any essential role in the laws of planetary motion discovered by Kepler, while the area of an ellipse does. *]

11.6 Kepler's laws

Establishing the three laws of planetary motion in our Solar System represents one of the greatest triumphs of scientific thought applied to astronomical observations. Only after acknowledgement of the elliptical form of planetary orbits the Copernican system started to give better predictions of planetary motions than the Ptolemaic geocentric model. Kepler's genius did not stop at just correcting the shape of orbits from circular to elliptic; he also formulated the laws of motion along those orbits, eliminating once and for all, the complicated combinations of deferents, equants and epicycles, necessary also in the Copernican model in order to describe planetary motions. Here are three laws established by Kepler:

• **First Law:** Planets move along elliptical orbits, the Sun being placed in one of the focal points.

• **Second Law:** During their revolutions, the line connecting a planet with the Sun sweeps equal areas in equal times.

• **Third Law:** The squares of the periods of revolution are proportional to the cubes of great axes of planet's elliptical orbits.

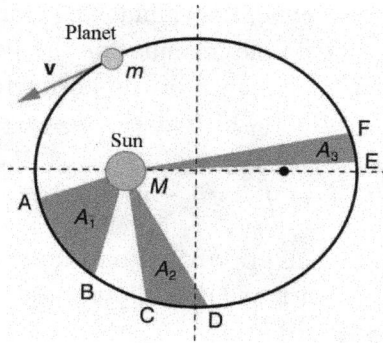

Fig. 11.10   Illustration of Kepler's second law. The areas swept by a planet in equal times are equal.

The first two laws are illustrated by Figure 11.10, in which a planet orbits around the Sun whose fixed position coincides with one of the foci. Equal areas swept by the planet during equal time intervals are represented by hatched sectors, thinner when the planet is close to the aphelion, and wider close to the perihelion.

In fact, the second law is the expression of conservation of planet's angular momentum (with the Sun supposed to be motionless, at the center of a Galilean reference frame). In Figure 11.11 *six* parameters completely describing planetary motion are shown. These are:

The two elements defining the orientation of the orbital plane:

- **1)** *Inclination i*: Vertical tilt with respect to the ecliptic reference plane (x, y), measured at the ascending node upward through the ecliptic plane;

- **2)** *Longitude* of the ascending node Ω with respect to the vernal point x;

The two elements defining the shape and the size of the ellipse:

- **3)** *Eccentricity e* of the ellipse;

- **4)** *Semi-major axis a*: The sum of the *periapsis* and *apoapsis* distance divided by two. In the case of Solar System planets the denominations *perihelion* and *aphelion* are used by the astronomers.

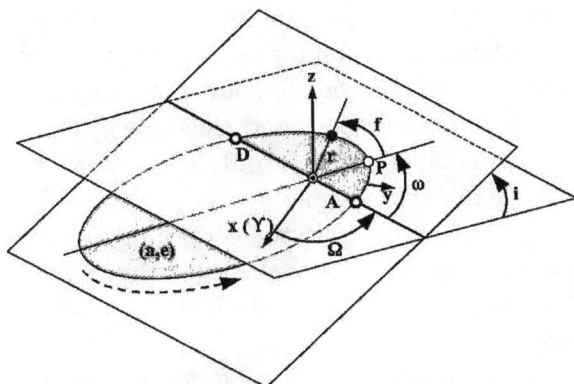

Fig. 11.11 Six Keplerian parameters defining planetary motion in the Solar System.

The two elements defining the orientation of the ellipse and planet's position on it at a given time:

- **5)** *Periapsis argument* ω defines the internal orientation of the ellipse in its orbital plane, or the direction of the *line of the apsids*;

- **6)** *True anomaly f* at the *"epoch"* t_0 defines the planet's position on the elliptic trajectory at a given time, usually counted from the moment of crossing the line of the nodes at the ascending node. The true anomaly is also denoted by ν or θ, depending on established tradition in a given country.

Not all Keplerian parameters defining the shape and position of planetary orbits remain constant over very long periods of time. Our Solar System exhibits long-term stability, but the planets' mutual gravitational attraction influences their orbits which may undergo certain secular changes. This is true in particular when one considers Jupiter's gravitational influence perturbing the Martian orbit, or Venus' influence on Mercury, to name a few. The influence of the Sun on our Moon is so strong that it causes the precession of Moon's orbit with a period of 18.3 years. This phenomenon is called the *motion of the line of apsids* and was explained by Newton in his *"Principia"*. Another important example is Mercury's perihelion precession.

Kepler's second law in its purely geometric version given by Figure 11.10 explains in a qualitative manner why orbital motions of planets orbiting around the Sun are not uniform: their velocities increase near the perihelion, and slow down near the aphelion. However, the astronomers need not only a qualitative description of motion, but above all, the possibility to predict the planet's position at any moment by giving two celestial coordinates as function of time $t - t_0$, where t_0 is usually chosen as the time when the planet is found at its orbit's perihelion, but it can also be fixed arbitrarily according to current needs.

The position of a planet following an elliptical path can be described in different ways, with different geometrical parameters. In Figure 11.12 we show the ingenious choice made by Kepler in order to derive a simple equation relating time t and the angle called *eccentric anomaly* from which the angle accessible to observation, called *true anomaly*, identical with the angle φ used in a polar coordinate representation of an ellipse, can be found.

In Figure 11.12 the elliptical orbit centered at O, with semimajor axis a and eccentricity e, is circumscribed by a circle with radius a and centered at O, so that it touches the ellipse at two extremal points representing perihelion (on the right) and aphelion

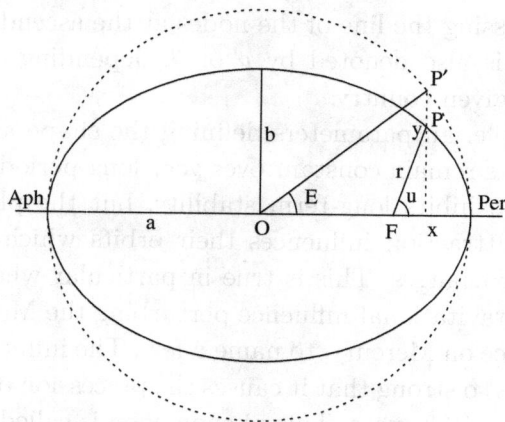

Fig. 11.12   Real elliptic orbit of a planet and imaginary circular orbit, defining angular variables known as true anomaly u and eccentric anomaly E.

(on the left). The Sun is placed in the focal point F, and the planet P is supposed to follow the elliptical orbit counter-clockwise. The origin of the orthogonal coordinate system is also at O, while the polar coordinate system is centered at the focal point F. The current position of planet P can be given either by the polar coordinates r and its true anomaly u, which coincides with the angular polar coordinate φ used in geometry, or by the pair of its Cartesian coordinates x, y.

[* An alternative way to spot a planet is to use the angle between the x-axis passing through the perihelion, and the point P' right on the vertical line passing through P and the circle of radius a circumscribing the ellipse. The angle between the horizontal axis and the radius OP' is called *eccentric anomaly*, denoted by E. Finally, one can describe the motion of the planet by the angle $M(t)$ called *mean anomaly*, proportional to the area swept by planet's radius on the elliptical orbit, which increases linearly with time according to the second law. The angle M is not represented in the figure, but it is very easily defined by a simple formula. Knowing that the surface area of an ellipse is given by

$$S = \pi ab = \pi a^2 \sqrt{1 - e^2} \quad \text{(because} \quad b = \sqrt{1 - e^2}\, a), \qquad (11.10)$$

the area swept by planet's radius vector during time $t - t_0$ is given by

$$S(t) = \frac{\pi a^2 \sqrt{1 - e^2}}{T}(t - t_0). \qquad (11.11)$$

Without loss in generality, we can set the origin of time at $t_0 = 0$, with t_0 corresponding to the moment when the planet is at the perihelion.

The *true anomaly u* is the angle between the direction towards the perihelion (along the *line of apsids* coinciding with the x-axis, and the radius-vector pointing from Sun (focal point F) to the planet P. To find the relation between $S(t)$ and u, Kepler made use of the eccentric anomaly E. Choosing the origin of a Cartesian coordinate system of coordinates (x, y) at F, the coordinate x of the planet at point P is

$$x = a \cos E - aE, \quad \text{with } E \text{ given in radians.} \qquad (11.12)$$

The point P belongs to the ellipse with semi-major axis a and semi-minor axis b, therefore the coordinates of any point of the ellipse satisfy the equation

$$\frac{x^2}{a^2} + \frac{y^2}{b^2} = 1, \qquad (11.13)$$

which is solved by $x = a\cos E$, $y = b\sin E$. Let us express the coordinates (x, y) using the angle E, semi-major axis a and eccentricity e

$$x = a(\cos E - e), \quad y = b\sin E = a\sqrt{1 - e^2}\sin E. \qquad (11.14)$$

On the other hand, using polar coordinates (r, u) we have

$$x = r\cos u = a(\cos E - e), \quad y = r\sin u = a\sqrt{1 - e^2}\sin E. \quad (11.15)$$

In order to find the area S of the elliptical sector P-F-Per (see Figure 11.12) swept by the planet after time t, we shall compare the areas of the following geometric figures:

i) The area S_1 of sector P'-O-Per is proportional to its angle E:

$$S_1 = \pi a^2 \cdot \frac{E}{2\pi} = \frac{a^2 E}{2}. \qquad (11.16)$$

(Note that E is measured in radians, which are parts of the full circle 2π.)

ii) The area of the corresponding elliptical sector, which is the image by projection of the circular sector S_1. Just as the total area of the ellipse is obtained by replacing a^2 by ab in the formula for the area of a circle with radius a, the same recipe applies to any sector of ellipse as compared to the sector of the circumscribed circle cut out by the same vertical line:

$$S_2 = \pi ab \cdot \frac{E}{2\pi} = \frac{abE}{2} = \frac{a^2\sqrt{1 - e^2}E}{2}, \qquad (11.17)$$

so that we can write $S_2 = \sqrt{1 - e^2}S_1$.

iii) The area S_3 of the triangle O-P-F is easily found to be

$$S_3 = \frac{1}{2}(ea) \cdot y = \frac{1}{2}(ea) \cdot b\sin E = \frac{1}{2}a^2 e\sqrt{1 - e^2}\sin E. \qquad (11.18)$$

The area of S of the elliptical sector *P-F-Per* swept by the planet since it passed the perihelion *Per* is readily obtained by substracting the area of the triangle S_3 from the area of elliptical sector S_2 (see Figure 11.12):

$$S = S_2 - S_3 = \frac{1}{2}a^2\sqrt{1-e^2}(E - e\sin E). \qquad (11.19)$$

Remembering that the area $S(t)$ is proportional to the angle $M(t)$ called *mean anomaly* (expressed in radians), we get from (11.17) and (11.18) the famous Kepler's equation

$$M(t) = \frac{2\pi t}{T} = E(t) - e\sin E(t). \qquad (11.20)$$

In spite of its apparent simplicity, this equation cannot be solved algebraically, because it involves the angular variable E with its sine functon $\sin E$. Such equations are called *transcendental,* and some of the best mathematicians devoted their efforts in order to find a way of solving Kepler's equation, at least approximately. One of the most ingenious methods was invented by German astronomer Friedrich Wilhelm Bessel (1764–1846) in the 19th century.

However, even if we have at our disposal precise values of eccentric anomaly E, we still have to transform these data in order to predict the true anomaly u which is obtained via direct astronomical observations. This part of the job can be reduced to solving trigonometric identities and using trigonometric tables.

Let us start by recalling that from (11.14) and using the polar parametrisation $x = r\cos u$, $y = r\sin u$ we have

$$\cos u = \frac{\cos E - e}{1 - e\cos E} \quad \text{and} \quad \sin u = \frac{\sqrt{1-e^2}\sin E}{1 - e\cos E} \qquad (11.21)$$

from which we get the expression for $\tan u$:

$$\tan u = \frac{\sin u}{\cos u} = \frac{\sqrt{1-e^2}\sin E}{\cos E - e}. \qquad (11.22)$$

But even in this form it is not easy to express u as function of E directly; luckily enough, things become much simpler if we use trigonomic identities for half-angle values — for a given angle $0 \le \alpha < \frac{\pi}{2}$

we have

$$\tan \frac{\alpha}{2} = \frac{\sin \alpha}{1 + \cos \alpha}, \tag{11.23}$$

from which we get, using the expressions from (11.21) and expressing $1 + \cos u$ with common denominator,

$$1 + \cos u = 1 + \frac{\cos E - e}{1 - e \cos E} = \frac{1 - e \cos E + \cos E - e}{1 - e \cos E}, \tag{11.24}$$

the following expression:

$$\tan \frac{u}{2} = \frac{\sqrt{1 - e^2} \sin E}{1 - e \cos E + \cos E - e} = \frac{\sqrt{1 - e^2} \sin E}{(1 - e)(1 + \cos E)}, \tag{11.25}$$

and finally,

$$\tan \frac{u}{2} = \frac{\sqrt{1 - e^2}}{1 - e} \frac{\sin E}{1 + \cos E} = \sqrt{\frac{1 + e}{1 - e}} \tan \frac{E}{2}, \tag{11.26}$$

from which we can extract the explicit expression for the true anomaly u as function of eccentric anomaly E:

$$u = 2 \arctan \left(\sqrt{\frac{1 + e}{1 - e}} \tan \frac{E}{2} \right). \tag{11.27}$$

Obviously, when $e \to 0$, the angles $M(t)$, $E(t)$ and $u(t)$ grow closer to each other, and for circular orbits they become identical. *]

After establishing the value of the true anomaly $u(t)$ and knowing the exact position of the Earth at the same time t, astronomers can compute the position of a given planet with respect to fixed stars as seen from the Earth. In the case of Mars, Kepler was able to compare the results of his calculations with precise observations left by Tycho Brahe, and got agreement up to 1 arcminute.

While applying the Ptolemaic, Tychonic or Copernican system with circular orbits, Kepler found differences up to 8 arcminutes, but being confident in Tycho's observational results, he looked for a radically different explanation and rejected circular orbits in favor of elliptical ones, which along with the correct time dependence given by the law of constant areal velocity, led to the perfect agreement with observation.

The third law could be verified still during Kepler's lifetime not only by the five planets of the Solar system known at that time, but also by the four satellites of Jupiter, freshly discovered by Galileo. We can check the validity of the third law using precise contemporary values of the semi-major axes and sidereal periods of the six planets of Solar System, known at the time, expressed in astronomical units (A.U.), days, and terrestrial years. Then for the Earth we get $a = 1$ and $\tau = 1$ in dimensionless units, so that for our planet the ratio a^3/τ^2 will be 1. Table 11.3 shows how close to 1 is the same ratio evaluated for other planets of the Solar System.

Table 11.4 shows how accurately Kepler's third law is verified by the four biggest of Jupiter's moons.

It is also worthwhile to note that the average distances from the central body in this miniature of solar system fit quite well with the Bode-Titius formula for planets orbiting around our Sun. Let us recall that for the first four planets, Mercury, Venus, Earth and Mars, Bode-Titius law is $0.4 + 0.3 \times 2^{n-1}$ (for Mercury one should put $n = -\infty$ to get the mean radius of the orbit equal to 0.4 A.U.);

Table 11.3   Checking the validity of Kepler's third law in the Solar System.

Planet	Sign	a (in A.U.)	T (in days)	τ (in years)	$\frac{a^3}{\tau^2}$
Mercury	☿	0.387	87.97	0.241	0.9976
Venus	♀	0.723	224.7	0.615	0.9992
Earth	⊕	1.000	365.24	1.000	1.000
Mars	♂	1.524	687	1.881	0.9992
Jupiter	♃	5.204	4328.85	11.852	1.008
Saturn	♄	9.583	10.759	29.46	1.013

Table 11.4   Checking the validity of Kepler's third law for the moons of Jupiter.

Satellite	a (in km)	$\alpha = \frac{a}{a_{Io}}$	T (in hours)	$\tau = \frac{T}{T_{Io}}$	$\frac{a^3}{\tau^2}$
Io	421700	1.000	42.46	1.000	1.000
Europa	670900	1.591	85.23	2.01	0.9991
Ganymede	1070400	2.538	171.71	4.044	1.00
Callisto	1882700	4.465	400.54	9.43	1.001

then we get 0.7 for Venus, 1.0 for Earth and 1.6 for Mars. When expressed in millions of kilometres, the four moons of Jupiter display a very similar pattern: 0.421 for Io, 0.671 for Europa, 1.07 for Ganymede and 1.89 for Callisto. Still curiously close to the sequence 04, 0.7, 1.0, 1.6! In spite of many attempts, a credible model for the stability of certain orbits during the primary accretion from dust orbiting around a massive central body has yet to be produced.

11.6.1   *Explaining the epicycles*

Kepler's laws of planetary motion solved once for all the complicated problems that marred former systems, both Ptolemaic and Copernican. Suddenly there was no need for the epicycles, great or small. However, introduction of epicyclic model based on the works of Apollonius was not due to a whim or to a chance. They were nothing else but successive approximation of Kepler's laws.

Caution: the following exercise requires the knowledge of derivation rules in Newton's notations, with the first derivative of a function denoted by a dot over it, and the second derivative by two dots.

[* Let us recall the equations defining the elliptic trajectory and the Kepler's second law of equal areas expressed in polar coordinates r and φ:

$$r(t) = \frac{R}{1 + e \cos \varphi(t)}, \quad r^2(t)\frac{d\varphi}{dt} = l = \text{Constant.} \tag{11.28}$$

(We make a tacit assumption that the motion remains strictly planar, which enables us to eliminate the polar variable θ, posing $\theta = \pi/2 = $ Constant.)

The obvious particular solution is given by a perfect circle with radius R, corresponding to zero eccentricity $e = 0$; the second equation reduces then to $R^2\dot{\varphi} = l = $ Constant, whose solution is $\varphi(t) = \omega t$, with $R^2\omega = l$, and which expresses Kepler's second law of constant areal speed in this particular case.

Astronomical observations lead to the conclusion that a circular orbit with constant angular speed cannot describe planetary motions, although the departures from such behavior are not very big. Therefore it is natural to assume that these departures can be described by periodic functions of time, small when compared to orbit's average radius R. One can even suppose that there exists small parameter ε whose consecutive powers can serve to make better approximations:

$$r(t) = R + \varepsilon n_r(t) + \varepsilon^2 b_r(t) + \cdots ,$$

$$\varphi(t) = \omega t + \varepsilon n_\varphi(t) + \varepsilon^2 b_\varphi(t) + \cdots \qquad (11.29)$$

Concerning the radial coordinate r, for elliptical orbits such development results from the formula $(1+x)^{-1} = 1 - x + x^2 - x^3 + \cdots$ for $|x| < 1$. The eccentricity e is smaller than 1, and so is the absolute value of $\cos\varphi$; therefore

$$r(t) = \frac{R}{1 + e \cos\varphi} = R\,(1 - e \cos\varphi + e^2 \cos^2\varphi - e^3 \cos^3\varphi + \cdots).$$

$$(11.30)$$

We can identify the small parameter ε of the development (11.29) with the eccentricity e, the functions n_r and b_r being equal to $n_r(t) = R\cos\varphi(t)$, $b_r(t) = R\cos^2\varphi(t)$. Now we can find the time dependence of functions $n_\varphi(t)$ and $b_\varphi(t)$ inserting their derivatives in the Kepler's second law: as we have now $\dot\varphi = \omega + e\,\dot n_\varphi + e^2\,\dot b_\varphi + \cdots$ and $r(t)$ given by (11.30), we can write

$$r^2\dot\varphi \simeq R^2\,(1 - 2e\cos\varphi + 3e^2\cos^2\varphi \cdots)(\omega + e\dot n_\varphi + e^2\dot b_\varphi + \cdots),$$

$$(11.31)$$

which up to the first order in e reduces to:

$$r^2\dot\varphi = l \simeq R^2\omega + e(\dot n_\varphi - 2e\omega\cos\varphi), \qquad (11.32)$$

assuming that the constant areal velocity is still the same, i.e. putting $l = R^2\omega$, we get the equation for $\dot n_\varphi$, replacing ω by $\dot\varphi$, which can be done at this stage of approximation, when only terms linear in e count.

$$\dot n_\varphi - 2\omega\cos\varphi = \dot n_\varphi - 2\dot\varphi\cos\varphi = 0 \quad \text{or} \quad \dot n_\varphi = 2\dot\varphi\cos\varphi. \quad (11.33)$$

The last equation can be easily integrated, knowing that $\frac{d}{dt}(\sin\varphi(t)) = \dot\varphi\cos\varphi$, therefore $n_\varphi(t) = 2\sin\varphi(t)$. Orbital displacement corresponding to $\Delta\varphi$ equals $r\Delta\varphi$ and is perpendicular to the radius-vector. Therefore the departure from circular orbit is, up to the terms linear in small parameter e.

$$r \simeq R - eR\cos\varphi, \quad r\varphi \simeq R\omega t + 2eR\sin\varphi. \qquad (11.34)$$

The deviations from circular orbit behave like an epicycle;

$$n_r(t) = eR\cos\varphi(t), \quad eRn_\varphi = 2eR\sin\varphi(t), \quad \frac{n_r^2}{(eR)^2} + \frac{n_\varphi^2}{4(eR)^2} = 1.$$
$$(11.35)$$

However, this epicycle is not a circle, but an ellipse with semi-major axis $2eR$ and semi-minor axis eR, the semi-major axis oriented along the circular orbit ("the deferent" of the Ptolemaic system), and the semi-minor axis aligned on the radius-vector. Of course, using an elliptical epicycle instead of a circular one would have improved the predictive power of both Ptolemaic and Copernican systems, but neither of their authors was ready to abandon the Aristotelian axiom of exclusiveness of uniform circular motion to be applied to all bodies in the Heavens.

We leave to the reader the task of determining the explicit expressions for the next order of approximation taking into account the terms proportional to ε^2, $b_r(t)$ and $b_\varphi(t)$; we give here only the hint: both expressions contain the cosine and sine functions with doubled frequency 2ω, $b_r(t) = A + B\cos(2\omega t)$, $b_\varphi = C + D\sin(2\omega t)$. These corrections can be interpreted as a second epicycle. They also modify the initial semi-major axis value R and the initial circular frequency ω.

The analytic definition of elliptical orbit combined with Kepler's second law lead directly to the inverse square law of gravitational force. Here is one of the shortest mathematical demonstrations. Let us take the time derivative of $r(t)$ describing an elliptical orbit in the parametric representation:

$$\dot{r} = \frac{d}{dt}\left[\frac{p}{1 + e\cos\varphi(t)}\right] = \frac{pe\dot{\varphi}\sin\varphi}{(1 + e\cos\varphi)^2}. \qquad (11.36)$$

Substituting $(1 + e\cos\varphi)^{-2} = p^{-2}r^2$, we get $\dot{r} = ep^{-1}r^2\dot{\varphi}\sin\varphi$ and using the Kepler's second law $r^2\dot{\varphi} = l = \text{Const.}$ we can write the radial velocity as:

$$\dot{r} = \frac{el}{p} \sin \varphi. \tag{11.37}$$

With constants p, l and e the only time dependent function in this expression is $\varphi(t)$, which makes the time derivation of $\dot{r}$ particularly easy:

$$\frac{d}{dt}(\dot{r}) = \ddot{r} = \frac{el}{p} \dot{\varphi} \cos \varphi. \tag{11.38}$$

Now, $e \cos \varphi$ can be expressed as a function of r using the constitutive equation of the ellipse:

$$e \cos \varphi = \frac{p}{r} - 1, \tag{11.39}$$

which we insert now in (11.38):

$$\ddot{r} = \frac{el}{p} \dot{\varphi} \cos \varphi = \frac{l}{p} \dot{\varphi} \left(\frac{p}{r} - 1 \right). \tag{11.40}$$

Recalling that $l = r^2 \dot{\varphi}$ we can substitute l in the first term, and express $\dot{\varphi}$ as function of r in the second, $\dot{\varphi} = lr^{-2}$ to obtain

$$\ddot{r} = r\dot{\varphi}^2 - \frac{l^2}{p} \frac{1}{r^2}, \tag{11.41}$$

where we recognize the two essential terms on the right side of the equation for radial acceleration: the centrifugal acceleration $r\dot{\varphi}^2$ and acceleration directed towards the focus (i.e. the center of the coordinate system), proportional to the inverse square of radius-vector's length r.

There is an extra bonus in formula (11.41): it is valid for *any* value of eccentricity e. This means that the inverse square law for central attracting force causing centripetal acceleration keeps it validity not only for closed elliptical orbits, but also the parabolic ($e = 1$) and hyperbolic ($e > 1$) ones. The converse is also true, i.e. the motion of a point-like mass under the action of central force proportional to the inverse square of distance follows necessarily one of the conic curves.

The proof was given by Newton in his *Principia*, using purely geometrical tools and the "calculus of fluxions" invented on this occasion. Of course Kepler did not have at his disposal Newton's mathematical apparatus and Newton's laws of motion and gravity; nevertheless his views on what causes planetary motion anticipated Newton's approach. Kepler wrote that the Sun, being the center of the Universe, rules the planets and their motions through a mysterious force of attraction which he called "magnetism". He also foresaw that such a force should decrease with distance, although erroneously ascribed to it a simple inverse law $\simeq r^{-1}$. *]

11.7   Ahead of his time

11.7.1   *Science beyond astronomy*

Johannes Kepler was a universal genius whose scientific interests covered a wide scope of natural phenomena. Beside being the first to correctly explain planetary motion, thus founding celestial mechanics and its first natural laws in the modern sense (precise, universal and verifiable), he laid down the principles of modern optical science in his book *Astronomia Pars Optica*. These included the following fundamental findings:

— explaining the formation of pictures with a pin hole camera obscura;
— explaining the process of vision by refraction within the eye;
— formulating eyeglass design for nearsightedness and farsightedness;
— explaining our depth perception by parallax due to the use of both eyes.

Kepler continued further his investigations in optics. In his book *Dioptrice* he defined the basic concepts, such as real, virtual, upright and inverted images and magnification. He also described the principle of a telescope and laws of total internal reflection.

Kepler correctly attributed the main cause of the tides to the Moon (in contrast to Galileo). The last point was beautifully commented in an essay published in 2017 by Ivan T. Todorov [Todorov

(2017)] analyzing the paradoxical controversy on the nature of tides opposing Kepler and Galileo:

Galileo was right affirming that the tides provide one of the strong proofs of Earth's rotation. But he dismissed the idea of direct or indirect influence of the Moon and Sun as tides' primary cause, while Kepler was strongly convinced that this was indeed the case. In this sense, Kepler foresaw the law of universal gravitation, whereas for Galileo the idea of action at a distance through empty space seemed irrational. Kepler was a mystic, and Galileo a down-to-the-Earth realist; Kepler believed in certain influence planets might have on what is happening on Earth and even on human behavior, and was an adept of Astrology. His horoscopes were highly appreciated, and they were one of sources of his earnings before he got sponsored by the Emperor Rudolph II. Surprisingly enough it was Kepler the mystic who got closer to Newton's theory of gravitation, whereas the materialist Galileo, who with his ideas on relative motion and the role of acceleration introduced the solid basis for Newtonian mechanics, was unable to imagine gravitational attraction acting at a distance through the empty space.

Another domain in which Kepler was ahead of his time concerned the structure of matter. In his book *De Nive Sexangula* he explains the macroscopic properties of material bodies, including crystalline symmetries, from the point of view of atomistic theory. On the example of snowflakes he explains their hexagonal symmetry comparing symmetries of close packing of cannonballs leading to regular hexagonal structures, with close packing of "atoms of water" which he supposed to be of perfect spherical form. By doing so, Kepler came very close to the bases of modern crystallography.

11.7.2 *Somnium, the first science-fiction novel*

Johannes Kepler can be also considered as the precursor of science fiction. *Somnium* (Latin for "The Dream") was perhaps the first attempt ever at using fiction to illustrate scientific principles. This fantasy was written between 1620 and 1630, and edited posthumously in 1634 by his son Ludwig. Initially, it was conceived as a dissertation written in defense of the Copernican system, describing how an observer on the Moon would perceive our own planet's movements. Conscious of possible dire consequences of its publication, Kepler chose to disguise the argument in the fantastic form of a dream, in which an Icelander named Duracotus traveling to Tycho Brahe's observatory is taken by daemons to a strange place called *Levania* (the Moon), orbiting around a huge globe called *Volva* (the Earth).

Levania is divided in two hemispheres, *Privolva* and *Subvolva*, the first one never seeing *Volva*, the second always directed towards it. Although the descriptions of plants, animals and intelligent inhabitants of Levania are pure fantasy, the explanation of eclipses of the Sun by Volva and shadow Levania casts upon Volva, the monthly variation of Volva's phases, annual cycles and timing of the eclipses, of how the planets look like as seen from Levania — all this represents an excellent handbook of lunar astronomy.

The original literary form combining fiction with scientific content inspired a number of similar texts by modern authors, from Cyrano de Bergerac to C. Flammarion and H.G. Wells.

Chapter 12

Newton and universal gravity

Natural science after Galileo and Kepler - Great philosophers of
17th century - Newton's life - Mathematical inventions - Newton's
mechanics - Theory of gravitation - "Principia" - Kepler's laws
derived - Newton's contributions to astronomy - Critical assessment

12.1 Preamble: The glorious 17th century

With Isaac Newton (1642–1727) science has performed an enormous
qualitative leap forward — it may be said that it passed from a young
and adolescent age into the age of maturity. Newton was rightfully
praised as a founder of the rigorous mathematical approach to physics
and astronomy. Nevertheless he paid tribute to his predecessors, having written in a letter to Robert Hooke in 1676, when they were still
on friendly terms: *"If I have seen further it is by standing on the*
shoulders of giants".[1]

In this letter Newton puts an end to a bitter controversy with
Hooke concerning optics, due to the latter's strong criticism of the
presentation of his views on the physics of light and optics Newton
made before the Royal Society. Hooke claimed also the paternity of
the inverse square law for gravity. In the end, Newton and Hooke

---

[1]This metaphor can be traced to the medieval scholar Bernard de Chartres (? –
ca. 1124) who was the first who used it comparing his contemporaries to the
ancient scholars of Greece and Rome.

were friends no more, and after Hooke's death Newton tried to erase the memory of Hooke's contributions to science.

Whom Newton meant in his famous statement remains open to hypotheses; it might be that he suggested Hooke to be one of the "giants". But we would certainly make no mistake by citing Galileo and Kepler, and perhaps Descartes, whose writings Newton exhaustively studied, although his physics was totally opposite to Descartes' theories.

The first half of the 17th century in Europe was an era of terrible conflicts and wars, culminating in the English Civil War (1640–1660), "La Fronde" in France (1648–1653), the Dutch fight for independence (1558–1638) and the Thirty Years' War that started in the Holy German Empire (1618–1638). Yet, during the same period Europe's artistic, intellectual and scientific life thrived, engendering a new generation of philosophers, mathematicians and physicists who carried on the scientific revolution initiated by Galileo and Kepler. The generation that followed them prepared and determined the next decisive leap forward due to Newton's genius.

It is worthwhile to present briefly the main scientific and philosophical successors of Galileo and Kepler whose peak of creativity preceded Newton's start in science and whose writings and scientific personalities marked profoundly his worldview, even when he was not sharing it.

Galileo's pupils and disciples in Italy:

- **Bonaventura Cavalieri** (1598–1647) introduced logarithms, worked on optics and mechanics; in mathematics, invented the "calculus of indivisibles" enabling him to compute volumes and surfaces, thus anticipating the integral calculus.

- **Evangelista Torricelli** (1608–1647), Galileo's last disciple, who helped him to write the last dialogue, found the barycenter and the area of a cycloid, and invented the mercury filled barometer. Later on, Blaise Pascal (1623–1662) used Torricelli's invention to prove that a vacuum can exist, contrary to Aristotle's assertion that "Nature abhors vacuum".

- **Vincenzo Viviani** (1622–1703) who at the age of 17 became Galileo's assistant and stayed with him in Arcetri until the end, then became a pupil of Torricelli. He wrote an exhaustive biography of Galileo, and was first to measure the speed of sound in air.

In England:

- **John Wallis** (1616–1703), Professor of Geometry at the Oxford University, developed further Descartes' and Cavalieri's geometrical methods, and laid foundations of infinitesimal calculus, precursor of integration.
- **Isaac Barrow** (1630–1677), Newton's teacher at Trinity College in Cambridge, proved the fundamental theorem of calculus, stipulating that integration and differentiation are inverse operations to each other.
- **Robert Boyle** (1627–1691), who is often called the father of modern chemistry. Known also thanks to his experiments on gases and formulating the *Boyle's law* (called in France Boyle-Mariotte's law) relating the pressure of a gas to the inverse of the volume it occupies. Boyle can be also seen as one of the founders of the scientific experimental method in physics.
- **Robert Hooke** (1635–1703), was an assistant to Robert Boyle and conducted experiments on gases with vacuum pumps of his own construction. He built a telescope with which he observed the axial rotation of Mars and Jupiter. He foresaw the inverse square law of gravity, but was not able to prove it mathematically. Using a microscope Hooke observed the microfossils and announced the first hypothesis of biological evolution. The law of linear elasticity bears his name.

In Holland:

- **Willebrord Snell** (1581–1626), also known as *Snellius*, discovered the law of refraction of light, introduced the decimal notation with a comma after the integer part of a number, laid the bases of *triangulation* in geodesy and measured a part of the meridian between Alkmaar and Bergen.
- **Baruch Spinoza** (1632–1677) elaborated new philosophical system based on logic and reason applied not only to natural phenomena, but equally to ethics and faith.

- **Christiaan Huygens** (1629–1695) clarified the laws of elastic collision and gave the correct mathematical expression of the centrifugal force; found the isochronic property of the cycloid and geometry of evolving curves. His wave theory of light was rejected by Newton, unfortunately.

Beside all these influential predecessors, the works of the outstanding French mathematicians and philosophers of the first half of the 17th century played a particularly important role in forming Newton's worldview. Among these, Mersenne, Pascal, Fermat and Descartes should be mentioned in the first place.

- **Marin Mersenne** (1588–1648), an ordained priest, inventor of the *Mersenne primes* $2^p - 1$, of the laws of musical harmony, a staunch defender of Galileo, fighting against superstitions and prejudices in science. Meresenne maintained an abundant correspondence with many important thinkers of his time, including Galilei, Pascal, Hobbes, Huygens, Fermat and Descartes.
- **Pierre de Fermat** (1601–1665) made important contribiutions to mathematics in first essays on infinitesimal calculus, and probabilities. He formulated a mathematical principle bearing his name describing the propagation of light. Fermat's principle states that light propagating in different media chooses always the quickest path from one point to another. This gave rise to the calculus of variations.
- **Blaise Pascal** (1623–1662), a mathematical prodigy; legend says that at the age of 12, having read the first pages of Euclid's "Elements" exposing the five postulates, he closed the book and derived all the fundamental theorems of Euclidean geometry by himself. He made not only great contributions to mathematics — the probability theory, to combinatorics, to geometry — the magic theorem of conic curves (at the age of 19), used later by Newton; he also constructed the first computing machine. In physics, Pascal developed the bases of physics of fluids, and formulated the law of uniform pressure in liquids.
- **René Descartes** (1596–1650), was born and educated in France, but spent the half of his life in Holland and Sweden. He formulated and explained mathematically the laws of reflection and

refraction of light, previously proven experimentally by Snellius, and analyzed the foundations of mechanics, in particular the principle of inertial motion and the laws of elastic collisions. Descartes insisted on separating physics from metaphysics, concentrating on exclusively rational approachto physical phenomena. Descartes' major work, the *"Discourse on Method"* contains the principles of the scientific approach to reality. The gist of his rules for discovering truth can be expressed as follows:

— To accept nothing in one's judgements beyond what presents itself clearly and distinctly as in mathematics;
— To divide each difficulty in as many parts as possible and solve them one by one;
— Start from the simplest objects and advance the knowledge towards the more complex ones;
— Make careful enumerations and reviews so that nothing is left out.

Today these rules seem quite obvious, but they were far from being so in Descartes' times. Nevertheless Descartes' criticism of Galileo's approach to mechanics and laws of free fall was not fully justified. He claimed that Galileo's laws of free fall are not reliable unless he could explain "what is weight". Descartes identified matter with "extension" in space, and his model of Universe explaining the circular motion of planets around the Sun by vortices of ether pulling them along was obsolete already when he proposed it.

Descartes greatest gift to modern science was the invention of *analytic geometry*, without which it is hard to imagine the subsequent development of calculus by Newton and Leibniz, the description of mechanics in rigorous mathematical terms, and introduction (along with Viète (1540–1603)) of the modern algebraic notation of equations.

12.2 Newton's life and achievements

One of the greatest scientists of all times, Isaac Newton was born in 1642 (the year of Galileo's death) in Woolsthorpe Manor

(Lincolnshire, England). Newton's father died before the child's birth, and his mother remarried soon after, so that Isaac spent his youth with his maternal grandmother in the nearby Town Grantham. His teacher at the school he attended recognized the boy's extraordinary mathematical talent, and when after a second re-marriage Newton's mother wanted him to return to the family manor, convinced her to let her son prepare for entrance to the University. In 1661 Newton entered the Trinity College at the University of Cambridge, where he got his degree four years later under the tutelage of Isaac Barrow, the Lucasian Professor of Mathematics. Barrow (1630–1677) was not only an eminent linguist and theologian, but also a fine mathematician. His innovative calculus of tangents to planar curves paved the way to the *calculus of fluxions* and greatly influenced Newton's own mathematical discoveries.

Since his youngest years Newton showed not only mathematical talents, but also technical skills, constructing windmills and water clocks. At the age of 26 he presented a new type of telescope using a hemispherical mirror instead of lenses, and presented it to the Cambridge University. Later on he performed many experiments, in particular with fluids and pendulums. His discovery of the separation of colors by refraction of white light by a prism, and of diffraction rings between a lens and a flat piece of glass opened new vistas in optics, on which he wrote a book in his late years (*Opticks*, 1704).

In 1665 Newton retired to Woolsthorpe to escape the great bubonic plague that broke out in London, and worked there making great progress in his novel approach to calculus and performing optical experiments. During the eighteen months of this semi-voluntary reclusion Newton wrote almost the entire text of what appeared later as one of the greatest contributions to science ever made by a single author. Newton did not publish his text during the next twenty years. Back in Cambridge in 1667 he continued his scientific work, succeeding Barrow as the Lucasian Professor of Mathematics in 1669. In 1672 Newton was elected to the Royal Society.

Newton's most important work, entitled "*Philosophiae Naturalis Principia mathematica*", was published in 1686, the edition being financed by Newton's friend Edmund Halley. It was a masterpiece

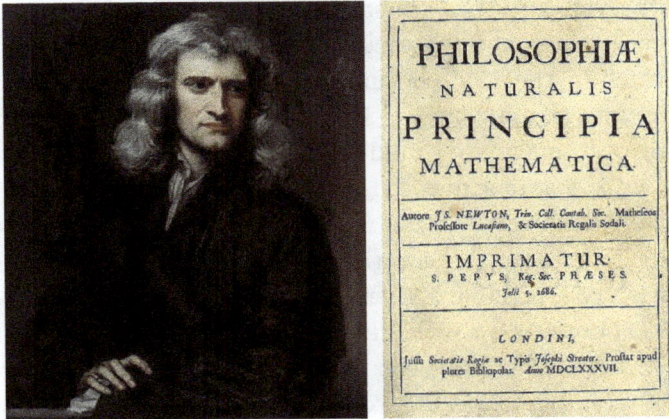

Fig. 12.1   Left: Isaac Newton when he was about 30 years old; Right: the cover page of "Principia", first edition of 1687.

containing most of Newton's findings in mechanics, physics and astronomy. It brought him fame and financial stability. In 1688 he was appointed the warden, and later the master of the mint, a very lucrative position. Newton was also a member of Parliament, and ennobled by Queen Anne in 1704. The *"Principia"* had two more editions during Newton's life, in 1713 and in 1726, one year before his death. Newton was also a devout exegete of biblical texts.

Newton maintained scientific contacts with numerous scientists of his time and kept himself informed about the progress of science on the continent, especially in France. The third edition of *"Principia"* contains the latest findings of the French astronomers Richer and Cassini, as well as Huygens' theory of the isochronous pendulum.

12.3   Mathematics

Newton's contributions to mathematics are almost always connected with concrete physical problems he proposed himself to solve. Newton hypothetised that the instantaneous acceleration of a body of mass m is proportional (as a vector) to the force acting upon it. When the force is constant, like what happens with gravity (a very good approximation on the surface of Earth), the solution was found by

Galileo: the velocity grows linearly with time. But how to solve the general case, when the force changes its value during the motion? Mathematics in the 17th century did not propose methods able to solve equations like $\mathbf{a}(t, \mathbf{r}) = \frac{1}{m}\mathbf{F}(t, \mathbf{r})$.

The revolutionary step made by Newton consisted in the use of the *calculus of fluxions* invented for this occasion, known today as the *differential calculus*. Newton's version of differential calculus was of practical rather than of theoretical nature. He applied it to evaluate the variation of a variable y given as a function of another parameter, e.g. time t, i.e. to find the value of $y(t + \Delta t)$. For a linear function, $f(t) = Ct$, the things are clear: we have $f(t + \Delta t) = C \cdot (t + \Delta t) = Ct + C\Delta t$. The variation is proportional to the time variation Δt, with the same coefficient C.

[* In the case of a uniformly accelerated motion the distance s is proportional to the square of time; for example, for the free fall one has, as established by Galileo,

$$s(t) = \frac{gt^2}{2}, \quad \text{with } g = 9.81\,\text{m/sec}^2.$$

Let us evaluate $s(t + \Delta t)$:

$$s(t + \Delta t) = \frac{g}{2}(t + \Delta t)^2 = \frac{g}{2}(t^2 + 2t\Delta t + (\Delta t)^2)$$

$$= \frac{gt^2}{2} + gt\Delta t + \frac{g}{2}(\Delta t)^2. \tag{12.1}$$

The average velocity between the subsequent positions at t and $t + \Delta t$ can be evaluated as

$$\langle v \rangle = \frac{s(t + \Delta t) - s(t)}{\Delta t}.$$

This expression becomes as close as we want to the instantaneous value $v(t)$ if the time interval becomes so small that it can be neglected when compared to the result $\langle v \rangle$:

$$\langle v \rangle = \frac{s(t + \Delta t) - s(t)}{\Delta t} = gt + \frac{g}{2}\Delta t \simeq gt \tag{12.2}$$

in the limit $\Delta t \to 0$. The differential calculus, which Newton called the "Method of Fluxions" and exposed in 1665 in a special treatise with the same title, followed the same scheme as establishing the relationship between the velocity and acceleration during a free fall. Supposing that a certain quantity f varies with time following the power law $f(t) = At^n$, we can evaluate the momentaneous rate of change at time t as the difference between $f(t + \Delta t)$ and $f(t)$. In order to do it, we must know the general expression for the n-th power of a sum of two numbers, $(a + b)^n$. The formula is known as Newton's binomial theorem:

$$(a + b)^n = \Sigma_{k=0}^n C_k^n a^{n-k} b^k, \quad \text{with} \quad C_k^n = \frac{n!}{(n-k)!k!}, \qquad (12.3)$$

where the *factorial* $N!$ of an integer number N is equal to the product of all integers from 1 to N:

$$N! = 1 \cdot 2 \cdot 3 \cdot 4 \cdots \cdots (N-1) \cdot N,$$

and by definition the factorial of zero is declared to be equal to one, $0! = 1$. For example,

$$(a+b)^2 = a^2+2ab+b^2, \quad (a+b)^3 = a^3+3a^2b+3ab^2+b^3, \quad \text{and so forth.}$$

The coefficients C_k^n form the so-called *triangle of Pascal*, defined by the great French philosopher and mathematician Blaise Pascal (1623–1662); it was known long before in India and China, but Pascal re-discovered it independently. Here are the first few rows of Pascal's triangle:

$$1 \ \ 1$$
$$1 \ \ 2 \ \ 1$$
$$1 \ \ 3 \ \ 3 \ \ 1$$
$$1 \ \ 4 \ \ 6 \ \ 4 \ \ 1$$
$$1 \ \ 5 \ \ 10 \ \ 10 \ \ 5 \ \ 1$$
$$1 \ \ 6 \ \ 15 \ \ 20 \ \ 15 \ \ 6 \ \ 1$$

Each row starts and ends with 1; then each entry can be obtained as a sum of its immediate upper left and upper right neighbors, e.g. $6 = 3 + 3$, $15 = 10 + 5$, etc. On the other hand, it is easy to check that the k-th entry in the n-th row coincides with the Newton's coefficient C_k^n.

Therefore the n-th power of the sum of t and its "fluxion" Δt is given by the following expression:

$$A(t + \Delta t)^n - At^n = A\left[(t^n + nt^{n-1}\Delta t + \frac{n(n-1)}{2}t^{n-2}(\Delta t)^2\right]$$

$$+ A\left[\frac{n(n-1)(n-2)}{3}t^{n-3}(\Delta t)^3\right.$$

$$\left. + \cdots + nt(\Delta t)^{n-1} + (\Delta t)^n\right]. \qquad (12.4)$$

When Δt is very small, the difference between $A(t + \Delta t)^n$ and At^n ("fluxion" in Newton's terminology) becomes also very small, and the terms containing higher powers of Δt become negligible as compared with the linear term proportional to nt^{n-1}. This approximation describes with arbitrary precision the instantaneous rapidity of change of the function t^n, so that we can set

$$\Delta(t^n) = nt^{n-1}\Delta t. \qquad (12.5)$$

The same operation applied to a linear combination (with constant coefficients) of two different powers of t will yield the sum of corresponding "fluxions":

$$\Delta(At^n + Bt^s) = [nAt^{n-1} + sBt^{s-1}]\Delta t. \qquad (12.6)$$

This important *linearity* property leads to the conclusion that infinitesimal variations of any function can be fairly described by an approximation provided by a finite linear combination of different power functions with appropriate constant coefficients. Using successive approximations by polynomials, Newton invented a method of solving arbitrary equation of the form $f(x) = 0$, bearing his name. *]

Differential calculus was introduced independently by Newton's contemporary, the German philosopher and mathematician Gottfried Wilhelm Leibniz (1646–1716), with whom Newton entertained a bitter dispute over priority. The differential calculus invented by Leibniz was more rigorous from a purely mathematical point of view.

The widely used modern concepts and notations used in the differential calculus have been introduced by Leibniz, including the notions of *differential dx* of a variable x, a differential of a function $df(x)$ and the *derivative* of a function, df/dx, denoted also by $f'(x)$. With these concepts at hand, the iteration of derivation became natural, and second, third, and in general n-th derivatives of any function could be introduced, e.g.

$$f'(x) = \frac{df}{dx}, \quad f''(x) = \frac{d(f')}{dx} = \frac{d^2 f}{dx^2}, \quad f'''(x) = \frac{d^3 f}{dx^3}, \text{ etc. } (12.7)$$

Leibniz gave differential calculus a rigorous mathematical form, whith clear definitions and axioms: linearity, meaning that the derivative of any linear combinations of two functions is the same linear combination of their derivatives, and the *Leibniz rule*, which is the formula for the derivative of a product of two functions:

$$\frac{d}{dt}(f(x)g(x)) = \frac{df}{dt}(x)g(x) + f(x)\frac{dg}{dt}(x). \quad (12.8)$$

12.4 Physics

Newton's greatest and the most long-lasting achievement is undoubtedly his formulation of the laws of mechanics and his theory of universal gravity, which following Galileo's and Kepler's ideas definitely did away with Aristotle's separation between terrestrial and celestial laws of motion and physical behavior. The astronomical revolution due to the two "giants", as Newton called his famous predecessors, was given a solid mathematical description, based on new methods of calculus, also invented by Newton.

However, Newton was interested in the larger domain of physical phenomena, in particular in the nature of light and optics, and his findings had an important impact on astronomy, too. His first achievements in this domain were the construction of a new type of the telescope, using a system of mirrors instead of lenses. and the proof that what we see as white light is a superposition of different colors, which can be separated when the light beam passes through the glass prism. He also had the idea of "crucial experiment" by showing that a second prism can recombine the colors to give white light.

More than one century after Newton's death spectroscopy became the astronomers' preferred tool, enabling them to establish the relationship between a star's dominant color (the spectral type) and its surface temperature.

The telescope, called also *Newtonian reflector*, is a large tube with two mirrors, the primary concave mirror, and the secondary flat mirror sending the magnified image produced by the concave mirror to the side of the tube making the observation possible. Contrary to the refracting telescopes, the observer's place was on the side of the device, looking perpendicularly to the rays arriving from the celestial objects.

The two discoveries were intimately related, because the use of reflecting mirror instead of glass lenses could remove the *chromatic aberration* which marred the images of stars and planets as seen by refracting telescopes. To obtain a greater magnifications and higher resolution, the lenses in refracting telescopes had to be as large as possible; but lenses act as prisms, too, and the light passing through their outer parts is split into various colors just like when it passes through a prism. The resulting image is appears as "smeared", being a superposition of images in different colors, slightly shifted with respect to each other. The mirror of the first reflector telescope produced by Newton had spherical shape, which led to another aberration, due to the fact that only a parabolic mirror concentrates incoming parallel rays in one focal point, while a spherical mirror does not produce a point-like image of a distant star. Later on, the technique of mirror production was substantially improved, and perfect parabolic mirrors replaced the spherical ones.

Another fundamental contribution Newton made to optics was the discovery of the interference rings arising when a lens is posed on a mirror or on a flat piece of glass. Curiously enough, Newton did not believe in the wave theory of light in spite of very convincing experiments and mathematical models proposed by Huygens. Perhaps this was because Huygens' ondulatory theory supposed the existence of an aether in which optical waves could propagate, while Newton believed that space is a void with no material content, as it does not oppose any resistance to the eternal motion of planets. Newton's approach to optical phenomena was based on the analogy with the mechanical behavior of elastic bodies, whose trajectories were similar to the rays of light. At the end of Book I of the *Principia*, Section XIV, Newton derives the laws of reflection and refraction already discovered by Snell (Snellius, 1580–1626) and by Descartes, supposing that small bodies have different velocities in different media.

Newton was very careful in his conclusions. In the *scholium* (general comment summarizing previously proven theorems) he declares the following: *Therefore because of the analogy there is between the propagation of the rays of light and the motion of bodies, I thought it not amiss to add the following Propositions for optical uses; not at all considering the nature of the rays of light, or inquiring whether they are bodies or not; but only determining the trajectories of bodies which are extremely like the trajectories of the rays.*

12.5 Newtonian mechanics

In his *Principia Mathematica* Newton formulated with a great precision the fundamental laws of mechanics of material bodies. The motion with a constant velocity, according to Galileo, could not be detected by the observer confined inside his vehicle without being able to see the surrounding world move in the opposite direction. A uniform motion could not be dinstinguished from rest. Only an accelerated motion was immediately detectable by inertial forces opposing the exterior force which was the initial source of acceleration.

$$\Delta(m\mathbf{v}) = \mathbf{F}\Delta t. \tag{12.9}$$

Finally, the third law stipulated that if a body A acts on a body B with certain force $\mathbf{F}$, then the body B acts on the body A with an opposite force $-\mathbf{F}$ (this law is often called the action-reaction law).

The *Pricipia Mathematica* were written in Latin, as all scientific texts of that time. Here are the three fundamental laws of Newtonian mechanics as they were originally formulated:

LEX I *Corpus omne perseverare in statu suo quiescendi vel movendi uniformiter in directum, nisi quantenus a viribus impressis cogitur statum illum mutare.*

("Every object will remain at rest or in uniform motion in a straight line unless compelled to change its state by the action of an external forces.")

LEX II *Mutationem motus proportionalem esse vi motrici impressae et fieri secundum lineam rectam qua vis illa imprimitur.*

("The change in momentum is proportional to the force applied and occurs along the line of action of that force.")

LEX III *Actioni contrariam semper et aequalem esse reactionem: sive corporum duorum actiones in se mutuo semper esse aequales et in partes contrarias dirigi.*

("For every action there is always an equal reaction: or, the mutual actions of two bodies are equal and acting in opposite directions.")

The three laws of motion formulated in this concise manner by Newton were as fundamental for the subsequent development of mechanics as the five Euclid's postulates were for geometry, and not surprisingly, we refer to "Newtonian mechanics" as we refer to "Euclidean geometry". Newton's three laws of motion clarified the notion of *force*, still quite nebulous in writings of Kepler and Galileo, and the new important notion of *momentum*, or *quantity of motion* as the product of mass by velocity.[2] The second law contains implicitly the conservation of momentum in the absence of external forces. It can be also applied to the situations when the mass of an object is not conserved, like in the case of a rocket.

12.6 Universal gravity

Perhaps the most important impact of Newton on the philosophy of nature was the unification of celestial and terrestrial mechanics — in fact, the very idea that celestial bodies must obey the same rules and relationships between forces, masses, velocities and accelerations as any material bodies accessible to the direct observations and measures here on Earth.

---

[2]The concept of momentum was independently introduced by Huygens in his analysis of elastic collisions.

Newton's law of universal gravitational attraction can be formulated as follows: any massive body attracts another massive body with a force that is parallel to the line joining their centers of mass, proportional to the product of both masses, and decreasing as the inverse square of distance separating them. According to Newton's third law,

$$\mathbf{F}_{12} = -\frac{GM m \mathbf{r}_{12}}{|\,\mathbf{r}_{12}\,|^3} = -\mathbf{F}_{21}, \quad \text{with G a constant.} \tag{12.10}$$

In the case when the massive spherically symmetric body is placed at the center of coordinate system, the gravitational force it produces acting on a small mass m at a distance $\mathbf{r}$ can be written in simple form:[3]

$$\mathbf{F} = -\frac{GM_1 M_2 \mathbf{r}}{|\,\mathbf{r}\,|^3}. \tag{12.11}$$

The inverse square law was first based more on intuition than on a solid mathematical proof. When Newton arrived at the conviction that the Moon is maintained on its elliptical orbit around the Earth instead of continuing to move forward on with constant (including direction) velocity as it should follow from Galileo's assertion (clarified later by Descartes and Huygens) concerning inertial motion of massive bodies, he concluded that there is some force of attraction towards the center of the Earth that bends the Moon's trajectory and obliges it to remain in its orbit.

It seems that Robert Hooke (1635–1703) was aware of this independently of Newton, and at the same time, if not a little earlier. If the force Earth exerts on the Moon is of the same nature as the gravitational pull it exerts on massive bodies at its surface, Hooke reasoned, then such a force should act equally in all directions, the Earth being spherically symmetric. But then, as distance r from Earth grows, the force must act on all masses found at a given radius alike. But the surface of spherical shell growing as r^2, so the intensity of the gravitational influence should decrease like r^{-2}, just like

---

[3] Although G is often referred to as "Newton's constant", it was introduced late in the 19th century. For Newton, only the masses were important, not the units in which they were expressed.

the intensity of illumination of a sphere surrounding a light source decreases as r^{-2}, the amount of light emitted being the same, only dispatched on a larger surface area.

There was a dispute on the priority of this fundamental idea, and Newton contributed to obscuring Hooke's contribution. Nevertheless, the priority of genuine mathematical approach and explanation based on a brilliant use of the Kepler's third law combined with the centripetal acceleration, undoubtedly belongs to Newton.

Following the idea that heavenly bodies must obey the same mechanical laws as any other massive objects observed in our everyday life, the fact that the Moon does not fall down on the Earth should be explained by the velocity of its revolution, sufficient enough to compensate its tendency to fall down. According to the law of free fall established by Galileo, and the principle of inertia announced by Newton, in the absence of Earth's gravitational pull the Moon should follow a straight trajectory. Without any lateral velocity the Moon, as any other material object, would start falling towards the Earth, as an apple falling down from a tree. But when we throw a stone with some initial horizontal velocity, it will follow a parabolic trajectory, and fall down quite far from the place where the motion started; the farther the higher the initial speed was.

The law of parabolic motion contradicting Aristotle's assertions was established by Galileo; who also for short distances (cannon balls) concluded that maximal distance will be attained with an upward shot performed at the angle of 45°. The higher the initial velocity, the greater the distance covered by the cannon ball before it hits the ground.

Similar considerations can be applied to a shot with horizontal initial velocity, just little above the surface of our planet, as displayed in Figure 12.2 Let us evaluate the velocity at which an object will never fall down, continuing its way at the same initial height above the surface. According to the principle of additivity of velocities, the horizontal component of the initial velocity V should be conserved, while the vertical component is subjected to the law of the free fall. The downward acceleration at Earth's surface has the well known value $g = 9.81 \, \text{m/sec}^2$; according to the law of free fall established by

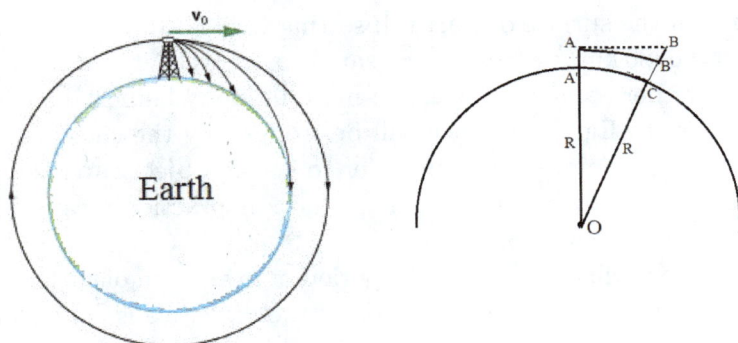

Fig. 12.2   A horizontal shot with sufficient initial velocity may end up by creating an artificial satellite. On the right: the scheme explaining the calculus of first cosmic velocity (satellization, i.e. launching a satellite near the Earth's surface).

Galileo, during the first second the falling body travels $h = 4.905 \, \text{m}$ towards the Earth's center. The picture on the right of Figure 12.2 shows (in an exaggerated manner) that in order to remain at the same distance from Earth's surface, the body must travel horizontally the distance denoted by x. All distances in the figure are exaggerated; in fact, the difference between the curved segment AB' and the straight segment AB is negligible, so small are these distances when compared to Earth's radius $R = 6380 \, \text{km}$. Similarly, the height AA' of the initial point A above the surface should be invisible on the drawing, so small it is with respect to Earth's radius, as well as h, the small loss of altitude after the first second of the free fall.

Applying Pythagoras' theorem to the triangle AOB, and supposing that the initial position A is so close to the surface (even if it were of the order of a few kilometers) that we can pose $OA = R$, we get the following identity:

$$(R+h)^2 = R^2 + x^2, \quad \text{or} \quad R^2 + 2Rh + h^2 = R^2 + x^2, \quad (12.12)$$

from which we get a simple fomula neglecting the term h^2, much smaller than Rh:

$$x^2 \simeq 2Rh, \quad \text{so that} \quad v_{sat} = \sqrt{2Rh} = \sqrt{gR}. \quad (12.13)$$

if we insert $h = gt^2/2$ with $t = 1$ second, so that the dimension of $\sqrt{gR}$ is that of velocity — here denoted by v_{sat}, the *satellization velocity* on the surface of Earth. Inserting in (12.13) $g = 9.81\,\mathrm{m/sec^2}$ and $R = 6380\,\mathrm{km}$, we obtain the satellization velocity equal to 7.91 kilometers per second. In reality a material body launched with this speed close to Earth's surface will be stopped by the enormous friction of the surrounding air. In order to get an object satellized, one has to get above the atmosphere, which in practice means higher than 250 km at least.

It is also useful to evaluate the period of revolution of such an artificial satellite: the full rotation around the big circle is equivalent to traveling the distance of 40 000 km; Earth's circumference. Dividing this distance by the satellization velocity, we get

$$T_{min} = \frac{40\,000}{7.91} = 5057\,\text{seconds} = 84\,\text{minutes}, \qquad (12.14)$$

close to an hour and a half. This period on a circular orbit with radius equal to 6380 km will be useful when we shall compare it with another satellite of the Earth, with greater orbital radius and with a longer period, according to Kepler's third law generalized for any satellite revolving about a heavy central body.

If mechanical laws apply equally well to heavenly bodies as to material objects near the surface of the Earth, then the motion of the Moon should be analyzed in the same terms as the imaginary artificial satellite close to Earth's surface. This means that the equation (12.13) can be used, too, but with different unknown quantity to be determined: the centripetal acceleration a due to Earth's gravity at a distance $D_M = 384\,000$ km. In contrast, the Moon's orbital velocity on its trajectory around the Earth was already well known at that time: it was enough to divide the circumference of the lunar orbit D_L by the sidereal month expressed in seconds. The result is

$$\langle v_L \rangle = \frac{2\pi \times 384\,000\,\mathrm{km}}{27.32 \times 24 \times 3600\,\text{seconds}} = 1.021\,\mathrm{km/sec}. \qquad (12.15)$$

Inserting this value along with the average radius of the Moon's orbit D_L into the formula (12.13), where the gravitational acceleration g on the surface of the Earth is replaced by the unknown value a of the gravitational acceleration on Moon's orbit, we get

$$\langle v_L \rangle = \sqrt{aD_L}, \quad a = \frac{\langle v_L \rangle^2}{D_L} = 2.714 \times 10^{-6} \frac{\text{km}}{\text{sec}^2} = 2.714 \,\text{mm/sec}^2,$$
(12.16)

an extremely low value, which means that the gravitational pull of the central mass (here the Earth) is considerably weakened with growing distance. But what is the exact mathematical expression describing the phenomenon? Well, let us compare the two accelerations first:

$$\frac{a}{g} = \frac{2.714 \times 10^{-3} \,\text{msec}^{-2}}{9.81 \,\text{m sec}^{-2}} = 2.766 \cdot 10^{-4}.$$
(12.17)

If the gravitational pull was inversely proportional to the distance from the attracting central mass, this ratio would be equal to the inverse ratio between the distance to the Moon and the Earth's radius; but that ratio is equal to $(D_L/R)^{-1} = R/D_L = (6380\,\text{km})/(384\,000\,\text{km}) = 1.66 \cdot 10^{-2}$, which is by two orders of magnitude greater than the value obtained in (12.17). But the square of this number gives a quasi-perfect fit:

$$\left(\frac{D_L}{R}\right)^2 = \frac{R^2}{D_L^2} = \frac{(6380\,\text{km})^2}{(384\,000\,\text{km})^2} = 2.761 \cdot 10^{-4}.$$
(12.18)

The coincidence is remarkable, taking into account that Newton was using a simplified version of the Moon's orbit, which in fact is not circular, but elliptic with eccentricity more than 5%; but even so, the accuracy of the law so obtained is of the order of less than 1%: to measure the discrepancy, it is enough to divide the first result by the second one, which gives

$$\frac{2.766}{2.761} = 1.001811\ldots,$$

an error less than two parts in a thousand! No wonder that Newton generalized this result as a universal law: gravitational pull created by massive bodies decreases as the inverse square of distance.

Let us check the validity of Kepler's third law applied to two circular orbits, the artificial satellite revolving close to the surface of Earth, the radius of its orbit being $R_T = 6380\,\text{km}$, with period of

$T_{sat} = 84$ minutes, and the Moon (making a simplifying assumption that its orbit is circular) with radius $D_{\leftmoon} = 384\,400$ km. The lunar sidereal period is $T_{Moon} = 27.32\,\text{days} = 39\,341$ minutes. Kepler's third law states that for satellites revolving around the same central body the ratio between the squares of their rotation periods is equal to the ratio between the cubes of corresponding orbits' radii (at this moment, we are considering circular orbits only). Let us compare:

$$\left(\frac{T_{\leftmoon}}{T_{sat}}\right)^2 = \left(\frac{39\,341}{84}\right)^2 = (468.35)^2 = 219\,347;$$

$$\left(\frac{D_{\leftmoon}}{R_T}\right)^3 = \left(\frac{384\,400}{6380}\right)^3 = (60.251)^3 = 218\,721. \tag{12.19}$$

Again, the coincidence is excellent: the ratio between the two numbers is 1.0029, the error less than one-third of 1%. However, Kepler's third law is not restricted to circular orbits: in general case closed orbits are ellipses, and circle's radius is replaced by the semi-great axis a. We shall derive the general formula later on, after proving Kepler's second and first laws.

12.7   Kepler's laws explained

The first major achievement presented in the first book of Newton's *Principia* was the proof of Kepler's second law of constant areal velocity. In Section II, entitled *Of the Invention of Centripetal Forces*, in just two Propositions and Theorems, I and II, Newton brilliantly proves that a body acted upon by a central force, i.e. the force directed towards one and the same central point, will follow a motion characterized by constant areal velocity, describing at equal times areas of equal surface.

To prove this statement, Newton considers the action of the central attractor as a series of intantaneous "kicks" it gives to the distant small body. The proof is explained in Section II, proposition I, Theorem I of the first book of *Principia*.

Suppose that the body is initially at A, with a given initial velocity. In the absence of gravitational attraction after a short while Δt the body would be found at the point B, the segment AB parallel to the initial velocity multiplied by the time interval Δt. Without any external force applied to it, the body would continue its straight-forward motion, arriving after the next time interval equal to the first one, Δt, with unchanged velocity at the point c, so that the distance Bc would be equal to the distance AB. Suppose now that a force directed towards the heavy central body placed in S is instantly applied to the body at B.

According to the Newton's second law, the momentum of small body will change by amount proportional to the force applied, so its direction will be modified. This modification, proportional to the applied gravitational pull, is represented in Figure 12.3 by the line Bv. The body will be deviated parallelly to Bv, and will follow the straight line BC instead of Bc, with the line cC parallel and equal in length to Bv.

It is also important to notice that the second triangle formed by the deviated velocity and two lines joining its ends with the central point S are contained in the same plane. This means that the central force, no matter how it depends on distance from the attracting point S, modifies the velocity of orbiting mass m bending it constantly towards S (supposing that the force is an attractive one),

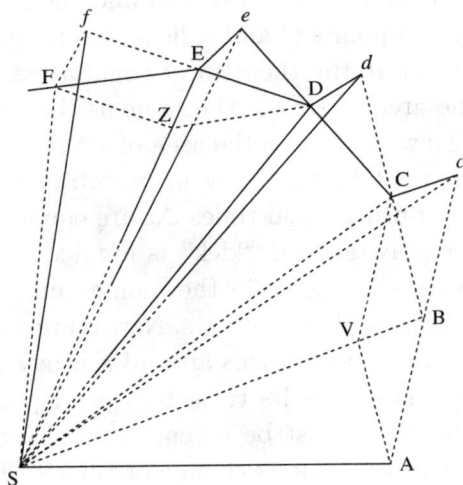

Fig. 12.3   Newton's purely geometrical proof of Kepler's second law.

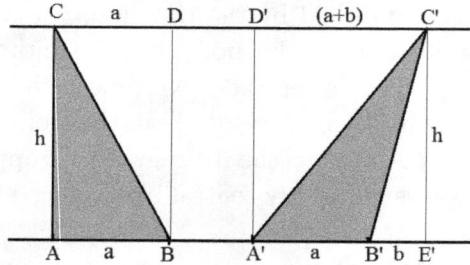

Fig. 12.4   Triangles with the same basis and height have the same surface area.

but the new velocity remains in the same plane: the motion is therefore planar.

To prove the theorem of equal areas swept in equal time intervals, Newton uses the well known geometrical fact: two triangles with the same basis and the same height have the same surface areas (Figure 12.4). This is also true for two triangles with equal sides placed on one common straight line, side by side, and sharing a common third vertex, as the triangles ASB and BSc do: these two triangles have the same surface area.

Now, Newton writes, consider two triangles SBc and SBC: they share one side in common, which is SB, and their vertices opposite to SB, which are the points C and c, lie on a straight line parallel to the base SB. Therefore the theorem of equal areas applies to these triangles, too: the area of BSc is the same as the area of BSC, and as the area of BSc was equal to the area of ASB, we get the desired result: the areas swept by the body in its orbit around the central attracting body S during equal times Δt are equal.

At C, another gravitational "kick" is applied , directed towards the central body, modifying again the momentum, and so forth. In Proposition II, Theorem II that follows, Newton proves the reciprocal statement, namely, if a body turns around a heavy center in such a way that the areas swept by its trajectory are equal for equal time intervals, then the force must be a central one. Both theorems are then extrapolated to continuous change of velocity by the action of the central force making the time intervals and the space segments covered by the orbiting body infinitesimal at will.

[* A more rigorous proof of Kepler's second law is as follows:

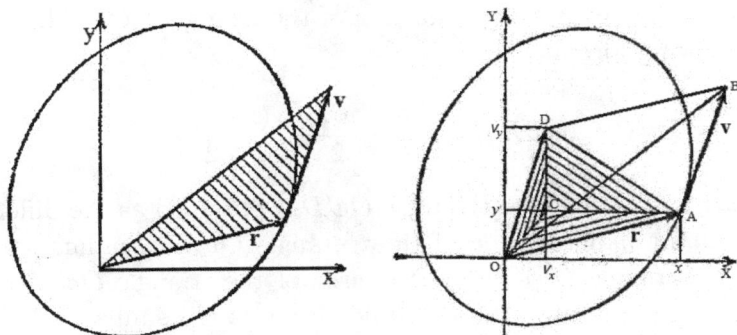

Fig. 12.5 The analytical proof of Kepler's second law.

In Figure 12.5, the areal velocity is symbolized by the triangle OAB. It is proportional to the area the radius-vector of the planet sweeps per unit of time. On the left, this area is represented by the triangle formed by the radius-vector $\mathbf{r}$ and the instant velocity $\mathbf{v}$. The radius-vector $\mathbf{r}$ has its origin at the focus of the ellipse representing the planet's orbit. The area of this triangle is clearly half of the area of the parallelogram $OABD$ formed by $\mathbf{r} = \vec{OA}$ and $\mathbf{v} = \vec{AB}$ and two parallel vectors, $\vec{OD}$ parallel to $\mathbf{v}$ and $\vec{DB}$ parallel to $\mathbf{r}$.

Now, it is clear that the area of the hatched triangle AOD on the right is also half of the area of the same parallelogram, therefore equal to the area swept out in unit time, i.e. the triangle OAB on the left.

The triangle AOD can be divided into three triangles hatched in different directions: OCA, OCD and ACD. The sum of their areas is the area of triangle AOD, thus also the area of triangle OAB representing the areal velocity.

Let us project the two vectors $\mathbf{r}$ and the vector $\vec{OD}$ parallel to the velocity $\mathbf{v}$ on the the axis OX and OY. Let us denote the projections (coordinates) of $\mathbf{r}$ by x and y, and those of vector $\vec{AD}$ parallel to the velocity $\mathbf{v}$ respectively by v_x and v_y. The point C is at the intersection of horizontal line passing through the point y and the vertical line passing through the point v_x.

The area of the triangle OCA is equal to half the area of the rectangle $OyAx$ minus half the area of rectangle Ov_xCy, i.e. the area of the triangle OAy minus the area of the triangle OCy. This yields the following formula!

$$\text{Area of } \Delta(OAy) = \frac{xy}{2} - \frac{v_xy}{2} = \frac{(x-v_x)y}{2}. \tag{12.20}$$

Similarly, the area of the triangle OCD is obtained as the difference between the half the area of the rectangle Ov_xDv_y minus half the area of rectangle Ov_xCy, i.e. the area of the triangle Ov_xD minus the area of the triangle Ov_xC, which leads to the similar formula:

$$\text{Area of } \Delta(OCD) = \frac{v_xv_y}{2} - \frac{v_xy}{2} = \frac{(v_y-y)}{2}. \tag{12.21}$$

Finally, the area of the triangle ACD, with right angle at C, is simply half of the product of its two sides AC and CD, given by the formula

$$\text{Area of } \Delta(ACD) = \frac{(v_y-y)(x-v_x)}{2}. \tag{12.22}$$

Summing up the three areas, we get

$$\text{Area of } \Delta(OCD) = \frac{xy}{2} - \frac{v_xy}{2} + \frac{v_xv_y}{2} - \frac{v_xy}{2} + \frac{(v_y-y)(x-v_x)}{2}$$
$$= \frac{(xv_y - yv_x)}{2}. \tag{12.23}$$

(The readers acquainted with vector calculus will easily recognize the result of the vector product of $\mathbf{r}$ with $\mathbf{v}$; but Newton had no such notion yet, and relied on purely geometric reasoning.)

We should notice at this point that the orbit's shape is not necessarily elliptic; all we have to assume is that it is a convex curve, and the center of gravitational attraction is at the point O identified as the origin of the Cartesian coordinate system in which the central body is fixed.

Let us evaluate the variation of the areal velocity after a short time Δt. The changes of the radius-vector $\mathbf{r}$ and the velocity $\mathbf{v}$ are given respectively by:

$$\mathbf{r} \to \mathbf{r} + \Delta\mathbf{r} = \mathbf{r} + \mathbf{v}\Delta t; \quad \mathbf{v} \to \mathbf{v} + \Delta\mathbf{v}. \qquad (12.24)$$

According to Newton's second law of dynamics (12.9), $\Delta(m\mathbf{v}) = \mathbf{F}\Delta t$. In the case of a planet rotating around the Sun we can safely assume $m =$ Constant, so that $\Delta(m\mathbf{v}) = m\Delta\mathbf{v}$. In the case of a two-dimensional motion the velocity $\mathbf{v}$ has only two components, v_x and v_y; the gravitational force $\mathbf{F}$ acting on the planet is also contained in the same plane. We can write therefore, as now $\mathbf{r} = [x, y, 0]$:

$$\Delta\mathbf{v} = \frac{1}{m}\mathbf{F}\Delta t \to (\Delta v_x, \Delta v_y) = \frac{1}{m}(F_x, F_y)\Delta t. \qquad (12.25)$$

On the other hand, the gravitational force is directed towards the central body at 0,

$$F_x = -\frac{GMmx}{r^3}, \quad F_y = -\frac{GMmy}{r^3}, \quad F_z = 0,$$

i.e. it is parallel to the radius-vector $\mathbf{r} = [x, y, 0]$. Inserting the new values radius-vector $\mathbf{r}$ and velocity $\mathbf{v}$, $\mathbf{r} + \Delta\mathbf{r}$ and $\mathbf{v} + \Delta\mathbf{v}$ as defined in (12.24) and (12.25) into the expression obtained for areal velocity in (12.21), we get

$$\frac{1}{2}((x + \Delta x)(v_y + \Delta v_y) - (y + \Delta y)(v_x + \Delta v_x))$$

$$= \frac{1}{2}(xv_Y - yv_x) + \frac{1}{2}(x\Delta v_y + \Delta x v_y - y\Delta v_x - \Delta y v_x)$$

$$+ \frac{1}{2}(\Delta x \Delta v_y - \Delta y \Delta v_x).$$

The last term containing quadratic expressions in small *fluxions*, $\Delta x \Delta v_y$ and $\Delta y \Delta v_x$ becomes negligible as compared with the linear terms when these quantities become very small, so that with any required precision of measurement we can decrease the magnitude of infinitesimal variations so as to make all quadratic expressions become small beyond the observational limit. What counts only is the linear part, which is in modern terms the *derivative* of areal velocity with respect to time. Substituting

$$\Delta v_x = F_x \Delta t \sim x \Delta t, \quad \Delta v_y = F_y \Delta t \sim y \Delta t$$

we find that

$$(x\Delta v_y + \Delta x v_y - y \Delta v_x - \Delta y v_x)$$
$$\sim (xy\Delta t + v_x \Delta t v_y - yx\Delta t - v_y \Delta t v_x) = 0, \qquad (12.26)$$

which proves that under the action of any *central force* the areal velocity remains constant. Kepler's second law is the direct consequence of Newton's gravitational force, which is aligned along the radial vector, $\mathbf{F} \sim -\mathbf{r}$. Any central force will display the same property — the constancy of the areal velocity. But only the force decreasing as r^{-2} produces closed *elliptical* trajectories. *]

The proof of the last statement using exclusively geometrical properties of forces, velocities and accelerations covers dozens of pages of Newton's *Principia*. Here we shall give a proof based on the geometrical properties of the ellipse, but with a short-cut using the notion of energy conservation and gravitational potential, yet unknown to Newton when he was working on his *"Principia"*.

[* Among many ways to parametrize a curve on a plane let us choose the $r - \Phi$ representation, where r is the distance of a given point on a curve from the chosen center O, in other words the absolute value of its radius-vector $\mathbf{r}$, $r = |\mathbf{r}|$, and Φ is the angle between the radius-vector and the tangent to the curve. The simplest example of such a parametrization is given by a circle: starting from a given point P at distance r from the chosen center of coordinates O, the law relating Φ to r is $\Phi = 90° =$ Constant. More complicated relations between Φ and r result in other curves, not necessarily closed ones.

In the case of elliptical motion the center of coordinates will coincide with one of the foci of the ellipse, say F_1, r is the distance of a given point from F_1, and Φ is the angle between $\mathbf{r}$ and the instant velocity $\mathbf{V}$, co-linear with the tangent line to the ellipse (straight line t in Figure 12.6). In Figure 12.6 the point K on the ellipse is parametrized by r and $\Phi = \Phi_1$.

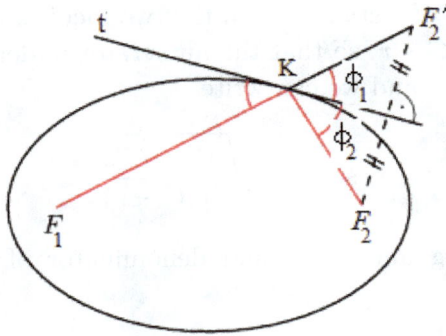

Fig. 12.6 An ellipse and the straight line t tangent to it, co-linear with instant velocity. We shall note $|F_1 K| = p$, $|F_2 K| = q$.

Our first step is to prove that at any point of the ellipse the two angles Φ_1 and Φ_2 are equal. This fact results from the very definition of the ellipse, according to which $p + q =$ Constant. If we move the contact point K slightly to the right, by infinitesimal amount δ, then the distance r will grow by $\delta \cos \Phi_1$ while at the same time the distance q will shrink by $\delta \cos \Phi_2$. By construction of the ellipse, the sum of new distances to the foci should remain constant, that is,

$$(r + \delta \cos \Phi_1) + (q - \delta \cos \Phi_2) = (r + q) \rightarrow \delta(\cos \Phi_1 - \cos \Phi_2) = 0,$$

therefore we must have $\cos \Phi_1 = \cos \Phi_2$ and finally, $\Phi_1 = \Phi_2$.

Let us find out the form of the equation relating the variables r and Φ when the curve is an ellipse. Consider the triangle $F_1 K F_2$. The law of the cosines gives

$$d^2 = r^2 + q^2 - 2rq \cos \alpha. \tag{12.27}$$

Because of $\Phi_1 = \Phi_2 = \Phi$ we have $\alpha = 180° - 2\Phi$; on the other hand, $\cos(180° - 2\Phi) = -\cos(2\Phi)$. Finally, using the formula for the cosine of a doubled angle, $\cos 2\Phi = 1 - 2\sin^2 \Phi$, we arrive at the following relation:

$$d^2 = r^2 + q^2 + 2rq(1 - \sin^2 \Phi), \tag{12.28}$$

and from this we separate the $\sin \Phi$ expressing it as function of r:

$$\sin \Phi = \sqrt{\frac{d^2 + (r + q)^2}{2rq}}. \tag{12.29}$$

Now, as the sum of distances from the two foci is a constant, $r + q = C$, we have $q = C - r$, so that the numerator under the square root is also a constant, and we can write

$$\sin \Phi = \sqrt{\frac{d^2 + C^2}{2r(C - r)}},$$

or, after dividing numerator and denominator of the fraction by $(d^2 + C^2)$,

$$\sin \Phi = \frac{1}{\sqrt{-Ar^2 + Br}}, \quad \text{with } A = \frac{2}{d^2 + C^2}, \quad B = \frac{2C}{d^2 + C^2}. \tag{12.30}$$

In order to prove that the inverse square law results in the parametric curve having the same form we must use another conserved quantity, the total energy given in this case by the sum of kinetic and potential energies:

$$\frac{1}{2}mv^2 - \frac{GMm}{r} = E < 0 \tag{12.31}$$

The total energy must be negative for the trajectory to be closed; for $E \geq 0$ the mass m will leave the vicinity of the attracting center M and never come back, going to infinity. This formulation was yet unknown to Newton, at least in the aforementioned terms; but we show this method for the sake of its elegance.

The conserved angular momentum is defined in modern terms as a vectorial product of the radius vector and the instantaneous velocity, multiplied by the mass of the orbiting body:

$$\mathbf{J} = m\mathbf{r} \wedge \mathbf{v}, \quad |\mathbf{J}| = J = mrv_\perp = mrv \sin \Phi,$$

and is constant according to Kepler's second law. Now we can express the velocity in terms of r and Φ and insert it into the energy expression (12.31): $v = \frac{J}{mr \sin \Phi}$, so

$$\frac{mv^2}{2} = \frac{mJ^2}{2m^2r^2 \sin^2 \Phi} = E + \frac{GMm}{r}, \tag{12.32}$$

so that we have

$$\frac{1}{\sin^2 \Phi} = \frac{2mr^2}{J^2}\left[E + \frac{GMm}{r}\right] \rightarrow \sin \Phi = \frac{1}{\sqrt{-Ar^2 + Br}}, \quad (12.33)$$

with $-A = \frac{2mE}{J^2}$ (we remind that $E < 0$, so that $A > 0$), and $B = \frac{2GMm^2}{J^2}$, this is exactly the parametric equation of an ellipse given in (12.30) a while ago. The two constants $d = 2ea$ and C appearing in the former expression can be determined by identifying the constants A and B in both formulas. *]

The first two of Kepler's laws are proven; what remains is the proof of the third one. We have already seen the proof of its simplified version valid only for circular orbits. Let us now follow a simplified version of Newton's derivation of Kepler's third law of planetary motion.

Let us consider two bodies of masses m_1 and m_2, orbiting their (stationary) center of mass at distances r_1 and r_2, as shown in Figure 12.7. The two bodies have different velocities, respectively v_1 and v_2, but because gravitational forces act along the line joining m_1 and m_2, they must complete a full rotation around C after the same time T. Let us suppose now that the situation shown in Figure 12.7 corresponds to the moment when the two bodies are at their maximal distance $a = r_1 + r_2$, when the velocities are perpendicular to the vector joining m_1 with m_2. For elliptic motion this corresponds to periastron or apastron points. In such a case the centripetal force can be computed like for a circular orbit. If so, the centripetal forces acting on m_1 and m_2 are equal to the gravitational forces taken with

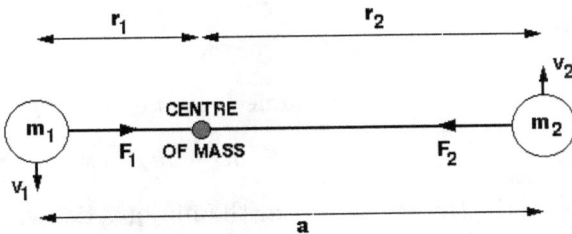

Fig. 12.7   Two bodies orbiting around their common center of mass C.

opposite sign; their absolute values are

$$F_1 = \frac{m_1 v_1^2}{r_1}, \quad F_1 = \frac{m_2 v_2^2}{r_2}. \tag{12.34}$$

When velocity is perpendicular to the radius vector, the sectorial velocity can be evaluated as the area of a triangle swept by the radius vector per unit of time, which is equivalent to

$$\frac{dA}{dt} = \frac{rv}{2}. \tag{12.35}$$

As we know from Kepler's second law, this quantity is constant, and can be obtained by dividing the area of the circle of radius r by the period T of full revolution around C,

$$\frac{dA}{dt} = \frac{\pi r^2}{T}. \tag{12.36}$$

Comparing the two expressions for each of two masses m_1 and m_2 we can express the velocities v_1 and v_2 as functions of respective radii and the common period T:

$$\frac{\pi r_i^2}{T} = \frac{r_i v_i}{2} \rightarrow v_i = \frac{2\pi r_i}{T}, \quad i = 1, 2, \tag{12.37}$$

therefore, substituting v_i into expressions for centripetal force acting on each body, we get

$$F_1 = \frac{m_1 v_1^2}{r_1} = \frac{4\pi^2 m_1 r_1}{T^2}, \quad F_2 = \frac{m_2 v_2^2}{r_2} = \frac{4\pi^2 m_2 r_2}{T^2}. \tag{12.38}$$

From Newton's third law we have $F_1 = F_2$, and we get, simplifying by common factor $4\pi^2/T^2$ simple relationship defining the center of mass:

$$m_1 r_1 = m_2 r_2. \tag{12.39}$$

Let us add up to both sides the same term $m_2 r_1$; we get then

$$m_1 r_1 + m_2 r_1 = m_2 r_1 + m_2 r_2 \rightarrow (m_1 + m_2) r_1 = m_2 (r_1 + r_2),$$

and finally, substituting $r_1 + r_2 = a$, the major axis, we get

$$r_1 = \frac{m_2 a}{(m_1 + m_2)}. \tag{12.40}$$

Now we can recall that the force F_1 is also equal to the gravitational pull of one body on another, according to the inverse square law:

$$F_1 = \frac{Gm_1m_2}{a^2}. \tag{12.41}$$

Comparing with the expression for F_1 in formula (12.38) and expressing r_1 as in (12.40), we arrive at

$$F_1 = \frac{Gm_1m_2}{a^2} = \frac{4\pi^2 m_1 r_1}{T^2} = \frac{4\pi^2 a m_1 m_2}{T^2(m_1 + m_2)}, \tag{12.42}$$

leading to the final formula for the inverse cube of major axis a proportional to the inverse square of the period T:

$$\frac{1}{a^3} = \frac{4\pi^2}{G(m_1 + m_2)} \frac{1}{T^2}. \tag{12.43}$$

The proportionality factor depends on both masses m_1 and m_2, but in the case when one of them is by several orders of magnitude greater, i.e. when $m_1 = M \gg m_2$, the latter can be neglected and the formula takes on the form depending only on the mass of the central body, which can coincide with the center of the system of reference.

$$\frac{1}{a^3} = \frac{4\pi^2}{GM} \frac{1}{T^2}. \tag{12.44}$$

This is the form in which Kepler's third law was postulated, although the physical interpretation of proportionality factor was yet unknown: the law involved only dimensionless relations: for planets with semi-major axes $a_1, a_2, a_3, \ldots$ and respective orbital periods $T_1, T_2, T_3, \ldots$, Kepler observed that

$$\frac{a_1^3}{a_2^3} = \frac{T_1^2}{T_2^2}, \quad \frac{a_2^3}{a_3^3} = \frac{T_2^2}{T_3^2}, \quad \text{and so forth.} \tag{12.45}$$

This happens in our Solar System where the total mass of all planets orbiting around the Sun represents no more than 0.14% of the total, while the mass of the Sun contributes as much as 99.86% of the total, Still during Kepler's life the moons of Jupiter, discovered by Galileo, provided another confirmation of this law, with a different ratio a^3/T^2, which, as we know, is directly related to the mass of the central body.

[* For readers acquainted with parametric representation of the ellipse we give a more rigorous derivation. We shall use polar coordinates (r, φ), with origin of coordinates coinciding with the position of the central mass M (the Sun), and in which the equation of the elliptic orbit is given by

$$r(\phi) = \frac{p}{1 + e \cos \phi},$$ (12.46)

where the constant p, with the dimension of length, is related to the total angular momentum J and the two masses M and m by the formula which we shall not derive here, but which was known to Newton:

$$p = \frac{J^2}{GMm^2}.$$ (12.47)

Let us denote by $v_\perp$ the component of instantaneous velocity perpendicular to the radius vector. The area ΔA of the infinitesimal triangle swept by the radius vector during time Δt is the area of the slim triangle spanned by r and $v_\perp$,

$$\Delta A = \frac{1}{2} r v_\perp \Delta t, \quad \frac{\Delta A}{\Delta t} = \frac{J}{2m} = \text{Constant},$$ (12.48)

where we have set $J = mr^2 \frac{d\varphi}{dt}$ is the *angular momentum* of the planet, a conserved quantity as long as the mass m can be considered as being constant, too.

The constant areal velocity can be thus estimated by dividing the total area of the elliptic orbit by the sidereal period T:

$$\frac{\pi a b}{T} = \frac{J}{2m}, \quad \text{its square being} \quad \frac{\pi^2 a^2 b^2}{T^2} = \frac{J^2}{4m^2}.$$ (12.49)

The relationship between the major and minor semi-axes depends on the eccentricity as follows:

$$b^2 = a^2(1 - e^2).$$ (12.50)

The *aphelion* is reached when $\varphi = 0$ and $\cos 0 = 1$; then we have

$$r = a(1 - e) \quad \text{and} \quad a(1 - e) = \frac{p}{(1 + e)},$$ (12.51)

therefore, as at the same time from the equation of ellipse we have $r = p/(1 + e)$, we get

$$p = a(1 - e^2) = \frac{J^2}{GMm^2} \rightarrow a(1 - e^2) \quad GM = \frac{J^2}{m^2}. \qquad (12.52)$$

Comparing with (12.49) and using (12.50), we can write now:

$$\frac{\pi^2 a^2 a^2 (1 - e^2)}{T^2} = \frac{J^2}{4m^2} = \frac{a(1 - e^2)GM}{4}. \qquad (12.53)$$

Simplifying both sides by common factor $a(1 - e^2)$, one arrives at final result:

$$\frac{4\pi^2}{GM} a^3 = T^2, \qquad (12.54)$$

which is exactly Kepler's third law. Note that the tacit assumption is that the mass of the body following an elliptical trajectory is negligible, and the central mass M coinciding with one of the focal points is motionless. *]

12.8 Precession of Earth's axis and of apsides

The exact solution of the two-body problem by Newton's theory of gravity, with its inverse square law, produces elliptical orbits described by two scalar parameters, e.g. by semi-major axis and excentricity. The motion is planar, and the position of the plane in space depends on the initial position and velocity, which determine also the direction of the semi-major axis in space. This direction, as well as the plane perpendicular to the conserved angular momentum, is called *the line of the apsides*, which in astronomical terms joins the periastron with apastron. The direction of apsids in space remains constant, unless forces other than the gravitational pull influence planet's motion.

However, at least one example of an elliptical orbit whose line of the apsides was not constant was known since always: the Moon follows an elliptical orbit whose apogeum and perigeum change systematically their position as if the elliptical orbit was slowly revolving, too. The complicated motion of our natural satellite has been a major challenge to astronomy since Antiquity, it was treated with variable luck by Hipparchus, Ptolemy, Al-Bitruji, Copernicus, Tycho Brahe and Kepler.

In the Book I of *Principia* Newton brilliantly solves the problem of the variation of the Moon's orbit resulting in the regular displacement of the nodes, obtaining the value of $19°$ per year. He compares the motions of two bodies attracted by the same center, one following an ellipse with its focus in the center of gravitational attraction, another following a similar ellipse which rotates in the same direction with a given angular velocity, which may be called a "revolving orbit". We know that with any central force depending only on the distance r from the center, the trajectories are planar. Therefore both bodies' motions can be parametrized by two polar coordinates, (r_1, θ_1) and (r_2, θ_2) respectively. Suppose now that the second body moves with an angular velocity that is proportional to the angular velocity of the first one,

$$\theta_2 = k\theta_1, \quad \omega_2 = \frac{d\theta_2}{dt} = k\frac{\theta_1}{dt} = k\omega_1. \qquad (12.55)$$

If the coefficient k is greater than 1, then the orbit of the second particle will be precessing in the same direction as particle itself; if k is smaller than 1, the orbit will rotate in the opposite direction. The problem, as Newton posed it, was to find what kind of extra force is needed to produce such an effect on the line of apsides of the second body, while the first one follows an elliptic trajectory with unchanged direction of its semi-major axis. Supposing the masses of two revolving bodies equal with common value m, Newton found the difference between the central gravitational force F_1 acting on the first body and the (also central) force F_2 acting on the second body. In order to produce the effect of rotating orbit, the difference between those two forces was found to be

$$F_2(r) - F_1(r) = \frac{L_1^2}{mr^3}(1 - k^2), \qquad (12.56)$$

where L_1 is the conserved angular momentum of the first planet. Newton lacked the knowledge of the value of L_1, because his estimate

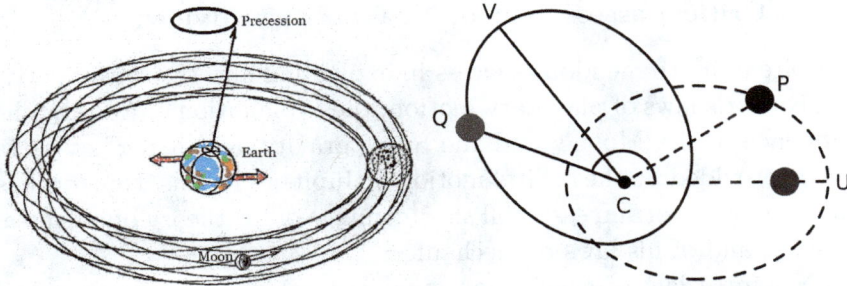

Fig. 12.8 Left: A great number of lunar trajectories fill a torus in space. The Sun's gravity differential action on this torus results in the precession of the lunar orbit; Right: Precession of the apsides, or revolving orbits.

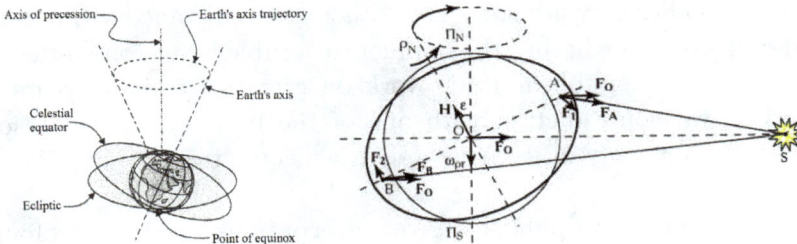

Fig. 12.9 How precession results from differential gravitational pull on a planet that creates the torque.

of the Moon's mass based on the analysis of tides, was erroneous. But he managed to obtain the right order of magnitude. At the end of the Proposition *XLV*, Problem *XXXI* he gives the result of his calculus: *the body, starting from the upper apsis, will arrive at the lower apsis with an angular motion of* 180 *deg.,* 45 *min.* 44 *sec., and this angular motion being repeated, will return to the upper apsis; therefore the upper apsis in each revolution will go forward by* 1 *deg.* 31 *min.* 28 *sec. The apsis of the moon is about twice as swift."*

In Book III, with similar reasoning, Newton gives the explanation of the Earth's axial precession and finds the right value of the precession of equinoxes: $50''$ yearly, corresponding to a full revolution of the Earth's axis after the period of 25 920 years, which is very close to what we know presently, i.e. 25 720 years.

12.9   Critical assessment of Newton's worldview

In spite of its tremendous success in explaining and rigorously deriving Kepler's laws of planetary motion, the explanation of tides by the
influence of the Moon's and the Sun's gravity, and a perfect extra
check provided by the orbital motion of Jupiter's moons, Newton was
fully aware of certain essential shortcomings of his theory of universal
gravity, and of his laws of mechanics.

Newton's law of gravitation supposes not only action at a distance, but what is even less plausible, the action of gravity according
to this law is *instantaneous*. In other words, if somewhere in the
immensity of the Universe a huge mass changes its position, all other
masses immediately feel a variation of its gravitational pull. The idea
that two bodies may influence each other over a distance without any
material agent mediating that influence troubled Newton so deeply,
that he delayed publishing his work on gravity for many years. It
was in total contradiction with one of the fundamental assertions
of physics, not only of its Aristotelian version, stating that "Nature
abhors vacuum".

Even after it was published, Newton continued to have profound
doubts concerning action at a distance, as evidenced by these words
from a letter to Bentley written around 1692: "*It is inconceivable
that inanimate Matter should, without the Mediation of something
else, which is not material, operate upon, and affect other matter
without mutual Contact [...] Gravity must be caused by an Agent
acting constantly according to certain laws*".

This "horror vacui" was the main reason why Descartes introduced the aether and used it to propose a theory of gravity based on
aether vortices. Huygens proved that this theory did not work, and
modified it such that it was harder to disprove. Huygens also believed
light consisted of longitudinal waves in this aether. A theory of gravitational action by an aether of very light, very fast-moving particles
was proposed in 1690 by the Swiss Fatio de Duillier (1664–1753);
he showed that it led to an effective force that went as the inverse
square of the distance because of shadowing effects. His work was
rejected by the French Academy of Sciences, but he gave a lecture to

the Royal Society which was published. The theory was later revived by the Swiss mathematician Le Sage (1724–1803).

Newton proposed that action at a distance was mediated by God, and believed universal attraction (gravity) was a proof of God's existence. He also needed God for the stability of the Solar System, because of the inevitable many-body forces. These ideas became very popular in the 18th century, and were supported and propagated strongly by the French rationalists.

It took almost two centuries until the concept of *field* was introduced by Maxwell and Faraday in order to describe the electromagnetic interactions. Nevertheless they introduced the *aether*, a hypothetical luminoferous substance as medium through which light propagates. How such substance, no matter how thin but permeating all of space, would not stop the motion of planets and stars, remained unexplained.

Another shortcoming of which Newton was equally aware concerned the concept of absolute space and absolute time. That the time flows at the same rate everywhere in the Universe seemed obvious and conformal with common sense, as Newton expressed it in *Principia*:

"I do not define time, space, place, and motion, as they are well known to all. Absolute space by its own nature, without reference to anything external, always remains similar and unmovable."

Newton also needed absolute space, because otherwise God would not have been able to locate the Solar System in the Universe.

However the existence of the unique coordinate system attached to empty space independent of any material bodies and their mutual positions was much less obvious, as follows from another passage:

"It is indeed a matter of great difficulty to discover and effectually to distinguish the true motions of particular bodies from the apparent, because the parts of that immovable space in which these motions are performed do by no means come under the observations of our senses."

Nevertheless, Newton found arguments that seemed to irrefutably prove that although motion with constant speed cannot have absolute meaning, as rightly shown by Galileo, accelerated motion is detectable even without referring to other bodies existing in the Universe. In the first *Scholium* of the first book of *Principia* he discusses the experiment with a bucket filled with water, suspended by a vertical rope, previously twisted. When the suspended bucket is maintained at rest and water in it is at rest, the surface of water remains flat. Now set the bucket free; it will start revolving under the action of untwisting cord.

At first, the bucket is turning with increasing angular velocity, but the water does not turn yet, and its surface remains horizontal. But after a few revolutions the vessel communicates some of its angular momentum to water, and its starts to revolve, too. Immediately one can observe that the rotating water tends to rise at the walls of the bucket and descend at the center, its surface forming a paraboloid of rotation. At this stage water rotates with respect to the outer space, but does not move with respect to the bucket which rotates with the same speed. But the rope continues to twist in the opposite sense, enticing the bucket to follow. Now the bucket starts to turn backwards, but the water inside continues to rotate as before, until the bucket transmits its motion. In the transit, water will recover its flat surface for a while, eventually will rise again towards the walls and descend in the center.

According to Newton, who performed the experiment and made the observations, this issue supports the existence of absolute space with respect to which water rotates. The bucket has no role except for maintaning the liquid inside the recipient. The behavior of rotating water is due to its inertia, which is an intrinsic property of any mass, reacting to acceleration with respect to the absolute space.

Another example of a physical system able to detect absolute acceleration was proposed by Newton in the same *Scholium*. Take two massive balls attached to a short rope at its ends. If the balls are at rest and at a distance smaller than rope's length, the rope does not feel any tension; but let the balls turn around their center of mass, and the rope will be immediately under tension, the stronger the higher the angular velocity of rotation.

This issue was severely criticized by George Berkeley (in his book *De Motu* ("On Motion")). If the water-filled bucket were the unique material body in the Universe, it would be not obvious at all why the liquid should rise towards the walls if there is no material body with respect to which rotation takes place. The same argument was used against the two balls's rotation tending the rope between them: if nothing else existed in the Universe, how to know whether the rope with the balls is rotating or not? According to Berkley's analysis, not only rectilinear uniform motion can not be detected, but also the accelerated motion and its consequence in the form of inertial force would not exist if there were no other massive bodies in the Universe.

So, who is responsible for the inertial forces? — apparently, it cannot be anything else but all masses filling the Universe, the distant stars who play the role of absolute space — answered a century later the Austrian physicist Ernst Mach (1838–1916). As simple common experience shows, two phenomena always coincide: the inertial centrifugal forces and rotation around one's own vertical axis. This is shown in Figure 12.10. A man holding two heavy dumbbells starts to turn rapidly; immediately the dumbbells are pushed away from the axis of rotation, and at the same time the man sees the starry sky

Fig. 12.10   Left: Newton's rotating bucket, Right: Inertial forces caused by acceleration and simultaneous apparent rotation of distant stars.

rotate in opposite direction above his head. Mach's conclusion was that if there were no distant stars, there would be no inertia either.

Another troublesome fact is the strict equality (up to a choice of units) of *gravitational mass* that appears in Newton's law of universal gravity, and the *inertial mass* appearing in Newton's second law of mechanics, $\mathbf{F} = m\,\mathbf{a}$. Why do we need two different masses, if they are always equal? One is enough, and it should be the gravitational mass. If we adopt this point of view, the inertial mass is a manifestation of gravitational forces, albeit different from the fundamental inverse square law.

If Mach's hypothesis is true, the inertial mass of a body should not depend on masses that are close to it, but mostly on masses that are very far away — literally, the most remote stars and galaxies. Supposing a uniform distribution of masses in the Universe with mean density ρ, a spherical layer of radius R and of thickness ΔR contains the amount of mass $\rho \times (4\pi R^2 \Delta R)$. The gravitational influence of each particular mass belonging to this layer is proportional to R^{-2}, therefore the total influence of each such layer is the same, equal to the product $\rho \times (4\pi R^2 \Delta R) \times R^{-2} = 4\pi\rho\Delta R$. But according to Newtonian theory of gravitation, they should cancel if Universe is homogeneous on a big scale and all directions are equivalent.

The situation would be different if a very weak transversal component of gravitation existed, too, dependent on the relative acceleration between masses, and behaving like r^{-1}. Then the influence of successive layers of matter behaves as $\Delta r \times r^2 \times r^{-1} \simeq r\Delta r$, thus making the ultimate faraway layers of apparently fixed galaxies responsible for the force aligned with the relative acceleration of a body, similar to an inertial force opposing the accelerated motion. The theory of General Relativity developed by Einstein in 1915 contains the possibility of transversal components of the gravity field, but it remains unclear how it can produce inertial mass via interaction with faraway matter.

Nobody better than Newton himself expressed the feeling that his theory of gravity does not go deep enough in explaining the mystery of universal attraction. Here is what he wrote in "General Scholium" closing the *Principia*:

"*But hitherto I have not been able to discover the cause of those properties of gravity from phaenomena, and I frame no hypotheses; for whatever is not deduced from phaenomena is to be called a hypothesis; and hypotheses, whether metaphysical or physical, have no place in natural philosophy*".

The statement about framing no hypotheses, sounding in Latin "*Hypotheses non fingo*", has become one of the best known Newton's quotations.[4]

---

[4]This is also one on which he is often attacked.

Chapter 13

The tides

The phenomenon of tides - Joint influence of the Moon and the Sun - Newton's gravity revisited - How the tides can help to determine the solar and lunar mass ratio - Physical consequences of tides - The Earth's rotation slowing down - The Moon slowly drifts away - Destructive tidal forces

13.1 First observation: Variation of the height of tides

The name "tides" denotes the periodic rising and falling of the sea level, controlled by the gravitational influence of the Sun and Moon. The intensity, i.e. the height of tides quite visibly depends on the relative position of the Moon with respect to the Sun. The maximal tides occur during the full Moon, or during the new Moon, i.e. when the Moon, Earth and Sun find themselves on the same straight line, while the height of a tide becomes minimal when the Moon is in its first or last quarter, i.e. when the Moon finds itself on a straight line passing through the Earth's center that is perpendicular to the line that goes from the Earth's center towards the Sun. The first case is called an *opposition* or *conjunction* of the Moon with the Sun, whereas the second case is called a *quadrature*.

Another important observation concerns the frequency of tides. There are tides that happen once a day, twice a day, three times a day, four times a day, and so on. But the twice-a-day tides, which display the biggest energy and height, are the most common ones.

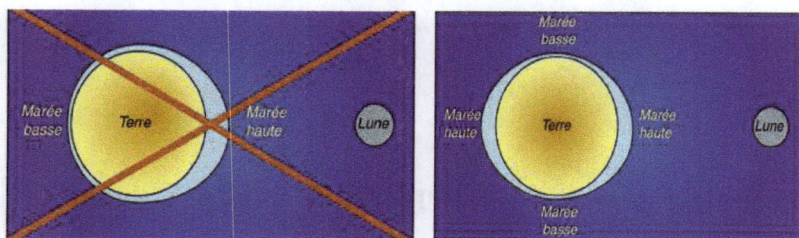

Fig. 13.1   Left: Erroneous view of Moon's influence, with one bulge; Right: Two opposite bulges caused by Moon's gravitational influence due to the differential effect.

They occur *two times* in approximately 24 hours (in fact, 20 hours and 50 minutes on the average): twice a high tide, separated by two low tides in between. This means that on the terrestrial globe there are two zones of high and two zones of low water on the oceans. The Earth rotates with an angular speed of 360° per 24 hours, so that any point on the surface encounters periodically high water zones and low water zones, the average time interval between the low and high tides being equal to 6 hours (more or less). The Moon's influence on the height of tides is clearly more important than that of the Sun. This was observed since Antiquity.

Galileo was convinced that the tides were proving the rotation of the Earth, however he thought that water was high on the side of the Earth turned towards the Moon, and low on the opposite side, thus creating one high and one low tide in 24 hours, contrary to observation. But the tides on the Mediterrenean Sea, especially in the semi-closed Adriatic Sea on which Venice lies, are very weak and superposed with other phenomena causing higher or lower water levels, like winds, sea currents, etc.

In places with important tides the oscillations of the sea level around the average are not only clearly visible, but also measurable with good precision. Gravitational influences of the Sun and Moon are being superposed, sometimes adding to each other, sometimes acting in the opposite sense.

Let us denote the height of tide caused by the Sun by h_S, and the height of the tide caused by the Moon by h_M. Maximal tides occur when both influences add up, so that the resulting high of the tide is $h_M + h_S$. The bulge of the high tide appears then on both sides of the Earth, in the direction along the axis on which Sun and Moon are aligned.

Water is an incompressible liquid, therefore its excess above the average level at some places must be compensated by a similar low level on the perpendicular sides of the globe, with equal size: by $-(h_M + h_S)$ less than the average level. When Sun and Moon are in *quadrature*, i.e. when their directions are perpendicular to each other, the resulting height is the difference of two influences, that of the Moon being preponderant. We have then the smaller high tide equal to $h_M - h_S$ — high lunar tide superposed with low solar tide — and the small low tide equal to $-(h_M - h_S) = h_S - h_M$.

It should be underlined that in order to get the most realistic picture, one should consider the data taken from the observations made at the open ocean, in order to minimize the secondary effects present in many places where the rising sea water during the high tide is rushing through a narrow channel between land masses, creating a natural funnel. The English Channel ("La Manche" in French) separating France from England presents one of the best examples of this funnel effect. The highest tidal effects (the "spring tides" during the equinox) can attain as much as 12 meters in Normandy (Granville).

The average ratio taken from many observations and measurements of tides in the open ocean is well established, and is equal to

$$\frac{h_M + h_S}{h_M - h_S} \simeq 2.5,$$

which leads to the following value of the ratio h_M/h_S:

$$\frac{h_M}{h_S} = \frac{2.5 + 1}{2.5 - 1} = \frac{3.5}{1.5} = 2.33.$$

In the open ocean the *tidal range*, i.e. the difference between the high and low tide is ridiculously small: between 0.6 and 0.7 meters. Comparing the two influences, we get on the average

$$(h_M + h_S) - (h_M - h_S) = 2h_S = 0.65\,\text{m},$$

which leads to the average height of the tide caused by the Sun alone equal to $h_S = 0.33$ meters. Therefore the Moon's contribution to high tide is 2.33 times this value, i.e. $h_M = 0.76$ meters, and the highest tide in the open ocean is just about 1.1 meters high — surprisingly small when compared to high tides observed wherever the funnel effect takes place.

13.2 Explanation of tides by Newton's gravity theory

In what follows, we shall systematically neglect the elliptic character of planet's orbits; both the Earth's and Moon's orbits will be supposed to be strictly circular. This approximation enables us to get results still in excellent agreement with observation.

[* Let us consider the tides induced by the solar gravitational influence first. Let us denote by

- D_S: the distance from the Earth to the Sun, i.e. the radius of Earth's orbit: $150\,000\,000\,\text{km} = 1.5 \times 10^{11}\,\text{m}$;
- R_E: the Earth's radius: $6380\,\text{km} = 6.38 \times 10^6\,\text{m}$.
- ω: circular frequency of the Earth's rotation around the Sun, $\omega = (2\pi)/(365\,\text{days}) = 2 \times 10^{-7}\,\text{sec}^{-1}$.

Let us also denote Newton's gravitational constant by G, by g the gravitational acceleration at the surface of Earth, $g = 9.81\,\text{m}\,\text{sec}^{-2}$; finally, M_S: the mass of the Sun, M_E: the mass of the Earth.

The Earth is so small with respect to the distance to the Sun that it can be treated as a point-like mass; its orbit is supposed to be a circle of radius D_S, and its motion a rotation with a constant angular frequency Ω. We can then assume a perfect equilibrium between the two effects: the gravitational attraction force directed towards the Sun, and the centrifugal force of inertia directed outwards:

$$M_E \omega_S^2 D_S = \frac{G M_E M_S}{D_S^2}. \tag{13.1}$$

According to the *principle of equivalence*, the gravitational mass of the Earth on the right-hand side is equal to its inertial mass on the left; therefore we can simplify the above equation by M_E, and get just the equality of accelerations:

$$\omega_S^2 D_S = \frac{GM_S}{D_S^2}. \qquad (13.2)$$

This is the simplest form of Kepler's third law applied to circular orbits:

$$\omega^2 = \frac{4\pi^2}{T^2} = \frac{GM_S}{D_S^3}, \quad \text{i.e. } T^2 \simeq D^3, \qquad (13.3)$$

the square of the rotation period T is proportional to the cube of the radius of circular orbit.

However, if we take into account the Earth's dimensions which — small as they can be as compared to the distances to the Sun or to the Moon — cannot be totally neglected, the radius R_E being equal to 6380 km, equation (13.1) is satisfied only at the center of our globe, but not on its surface.

As a matter of fact, on the side that is the most remote from the Sun's gravitational influence, i.e. where it is midnight, the distance to the Sun is $D_S + R_E$; as a consequence, the gravitational pull of the Sun is slightly smaller than at the Earth's center, while the inertial centrifugal acceleration due to the annual rotation around the Sun is slightly higher.

$$F_{in} = \omega^2 D_S \rightarrow \omega^2 (D_S + R_E) = \omega^2 D_S \left(1 + \frac{R_E}{D_S}\right). \qquad (13.4)$$

We take this expression with a positive sign, which means that as a vector we project it on the Ox axis pointing from the center of the Sun towards the Earth's center. In the same coordinate frame the gravitational pull (acceleration) of the Sun acting on the night side of the Earth must be taken with a negative sign, because its direction points towards the Sun's center. Its value is slightly lower than the gravitational pull applied to the Earth's center:

$$F_{grav} = -\frac{GM_S}{D_S^2} \rightarrow -\frac{GM_S}{(D_S + R_E)^2} = -\frac{GM_S}{D_S^2} \left(1 + \frac{R_E}{D_S}\right)^{-2}. \qquad (13.5)$$

The ratio $\frac{R_E}{D_S}$ is roughly equal to 4.2×10^{-5}, and we can use the linear approximation of the second expression with almost no error:

$$\frac{GM_S}{(D_S + R_E)^2} = \frac{GM_S}{D_S^2}\left(1 + \frac{R_E}{D_S}\right)^{-2} \simeq \frac{GM_S}{D_S^2}\left(1 - \frac{2R_E}{D_S}\right). \quad (13.6)$$

Comparing the two expressions and taking into account that the main terms $-\frac{GM_S}{D_S^2}$ and $\omega^2 D_S$ balance each other, we see that there is an *excess of acceleration* directed towards the point opposite to the Sun's position, i.e. towards the exterior of the Earth's orbit, on the dark side of the Earth. This excess is given by the expression

$$\frac{3GM_S}{D_S^3}R_E.$$

At the same time on the bright side, at noon, the centrifugal acceleration is slightly lower than the one at the center of the Earth, while the Sun's gravitational attraction is greater because we are closer to the Sun:

$$F_{in} = \Omega^2 D_S \to \Omega^2(D_S - R_E) = \Omega^2 D_S\left(1 - \frac{R_E}{D_S}\right),$$

directed outwards from the Sun. The gravitational pull of the Sun is directed towards it, which is symbolized by the *minus* sign:

$$F_{grav} = -\frac{GM_S}{(D_S - R_E)^2} = -\frac{GM_S}{D_S^2}\left(1 - \frac{R_E}{D_S}\right)^{-2}$$

$$\simeq -\frac{GM_S}{D_S^2}\left(1 + \frac{2R_E}{D_S}\right).$$

producing the excess of acceleration directed *towards the Sun*. Adding up the two forces we get its exact value:

$$F_{in} + F_{grav} = -\frac{3GM_S}{D_S^3}, \quad (13.7)$$

exactly the same as the one found on the dark side of the Earth, but with opposite sign. This explains the symmetry of the two "bumps". *]

We should recall here that these "bumps", extremely small when compared to the Earth's radius, culminate at points of our planet's surface where the Sun is seen at the zenith, and on the opposite side of the Earth at the same time. The ellipsoidal form which the world ocean takes on has its big axis aligned on the direction pointing from the Earth towards the Sun, in the plane of the ecliptic. The ocean water tends to make its surface orthogonal to the acceleration vector which is a sum of the Earth's gravitational acceleration directed towards the Earth's center, and the extra tiny accelerations $\Delta\gamma_S$ resulting from the imbalance between the gravitational pull of the Sun and the centrifugal force caused by the Earth's annual rotation motion. The resulting shape is a three dimensional ellipsoid, with its major axis along the line joining the centers of the Earth and Sun.

We should also remember that this axis lies in the ecliptic plane, while the Earth's own rotation axis is not perpendicular to that plane, but inclined under the angle $23°24'$, so that the height of the tides depends in a complex way on the geographical latitude of the observer. But most of the time, at most latitudes, the two bulges are passed by during 24 hours, creating two high and two low tides per day.

The height of tides caused by the lunar gravitational influence can be computed in a similar manner, but the computation is slightly more complicated, due to the fact that the Earth and Moon rotate around their common center of mass which is located inside our planet, but closer to its surface than to its geometrical center (see Figure 13.2).

The distance between the Earth's geometrical center c and the center of mass G of the Moon-Earth system is equal to 4680 km. This gives us the mass ratio of two celestial bodies: the center of mass of a two-body system is found at the distance x from the center of the more massive body, which in this case is undoubtedly our planet, and at the distance $D - x$ from the center of the Moon, where D is the average distance between the centers of two bodies. The ratio between these two distances is equal to the inverse ratio of the correponding masses, i.e. we must have

$$\frac{D - x}{x} = \frac{M_E}{M_M},$$

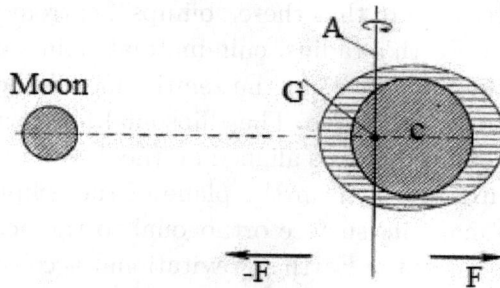

Fig. 13.2   Earth and Moon turning around their center of mass G; c is Earth's geometrical center, A is the common rotation axis; F and $-F$ show the directions of tidal forces.

where for D we take the value $384\,000$ km, the mean Earth-Moon distance, and $x = 4680$ km. Solving the equation we obtain

$$\frac{M_E}{M_M} = \frac{384\,000 - 4680}{4680} = 81.05,$$

which we shall round up to 81. Therefore the mass of the Moon is 81 times less than the mass of the Earth.

     In fact, it is the center of masses of the entire Earth-Moon system that orbits around the Sun following the slightly elliptic orbit (with low eccentricity of about $e = 0.0169$). The Earth's geometrical center is found inside or outside the orbit accordingly to whether the Moon is full (on the far side, over the dark hemisphere), or in the New Moon phase (on the side closer to the Sun, invisible to the observers on Earth at local noon). As a consequence, the Moon's gravitational pull acts in the opposite direction to Sun's gravity, or adds up to it.

     The calculations taking into account the two effects combined are more complicated than the ones we presented above, with an inertial reference frame coinciding with the center of the Sun, and we shall not present them here. Nevertheless, they lead to identical formulae for the extra acceleration, with the mass of the Sun replaced by the mass of the Moon, and the radius of the Earth's orbit replaced by the average radius of the Moon's orbit. The corresponding expressions for two gravitational influences, the solar and the lunar one, are therefore as follows:

$$\Delta\gamma_S = \frac{3M_S G}{D_S^3}R_E, \quad \Delta\gamma_M = \frac{3M_M G}{D_M^3}R_E \tag{13.8}$$

with M_M the mass of the Moon, $D_M = 384\,400\,\text{km}$ the average radius of lunar orbit.

What we discovered here, in fact, are the rules of the elementary differential calculus. If we want to compare the values of a given function $f(x)$ of some real variable x at two points separated by a very small distance, say $\Delta x \ll x$, we should form the difference $f(x+\Delta x) - f(x)$. If the function f is smooth enough in the vicinity of x, then this difference can be very well aproximated by an expression *linear* in the small entity Δx; the corresponding coefficient is called he *derivative* of f at x, and denoted by $\frac{df}{dx}(x)$. Therefore we can write

$$f(x + \Delta x) - f(x) \simeq \frac{df}{dx}(x)\Delta x,$$
$$f(x + \Delta x) - f(x) = \frac{df}{dx}(x)\Delta x + \mathcal{O}((\Delta x)^2)), \tag{13.9}$$

where $\mathcal{O}((\Delta x)^2)$ stays for the terms quadratic in Δx considered as a negligible quantity.

In the computation of the solar tides we had to compare two functions of the distance r from the Sun, the centrifugal force and the gravitational pull. They compensate each other at the Earth's center, where $r = D_S$, but are slightly different at two opposite sides of the Earth's surface, where we have $r = D_S \pm R_E$, with $R_E \ll D_S$. As we have found above, the differences of gravitational pulls of the Sun due to these displacements are, correspondingly,

$$-\frac{M_S G}{(D_S \pm R_E)^2} - \left(-\frac{M_S G}{D_S^2}\right) \simeq \mp\frac{2M_S G}{D_S^3}R_E, \tag{13.10}$$

whereas the difference of centrifugal accelerations is

$$\omega^2(D_S \pm R_E) - \omega^2 D_S = \pm\omega^2 R_E. \tag{13.11}$$

In terms of differential calculus this corresponds to the following simple derivatives:

$$\frac{d}{dx}\left(-\frac{1}{x^2}\right) = \frac{2}{x^3}, \quad \frac{d}{dx}x = 1. \tag{13.12}$$

When multiplied by R_E, which here plays the role of the differential Δx, very small indeed when compared with D_S, the two contribution will have the same sign, which results in the factor 3 in the final expression.

Coming back to the evaluation of the height of the tides, they are measured by the difference between the major and minor semi-axis of the ellipsoid whose form the ocean takes on under the influence of extra accelerations $\Delta\gamma_S$ and $\Delta\gamma_M$ which deform the perfectly spherical shape supposed to be observed under the action of the terrestrial gravity. The difference between the major and minor semi-axis a and b defines the maximal difference of ocean levels during the high and low tides.

13.3   Numerical results and the density ratio

Let us show how these elementary observations concerning the gravitational influence of Moon and Sun on the height of sea tides can be related to the evaluation of relative average densities of these celestial bodies. This estimate does not require any special device or equipment, all the information is available to the naked eye. All that is needed is a solid knowledge of Newton's mechanics and gravitation theory.

As we saw, under the gravitation influence water on the Earth takes on the form that is no more exactly spherical; it can be well approximated by an ellipsoid with semi-major axis a and semi-minor axis b, both being very close to the Earth's radius R_E. Taking into account only the influence of the Sun, the ratio between a and b is given by the ratio of corresponding accelerations on the surface of the ocean:

$$\frac{a}{b} = \frac{R_E + h_S}{R_E - h_S} \simeq \frac{g + \Delta\gamma_S}{g - \Delta\gamma_S} \tag{13.13}$$

where g is the average gravitational acceleration on the surface of the Earth, i.e. $g = 9.81\,\mathrm{msec}^{-2}$.

A similar formula applies to the tides provoked by the Moon, with $\Delta\gamma_M$ replacing $\Delta\gamma_S$. According to the previous estimates the tidal acceleration components are

$$\Delta\gamma_S = \frac{3GM_S}{D_S^3}R_E \quad \text{and} \quad \Delta\gamma_M = \frac{3GM_M}{D_M^3}R_E. \qquad (13.14)$$

The height of the solar tide is given by:

$$h_S = R_E \times \left(\frac{\Delta\gamma_S}{g}\right), \qquad (13.15)$$

and we recall that the lunar tide can be evaluated as $h_M \simeq 2.33\,h_S$. We can also eliminate the mass of the Sun (yet unknown!) by the observable quantity using the relation

$$\frac{GM_S}{D_S^2} = \Omega_S^2 D_S$$

from which we infer that the maximal height of a tide h_{max} is the sum of the two:

$$h_S = \frac{3\Omega_S^2 R_E^2}{g} \simeq 0.49\,\mathrm{m}, \quad h_M = 2.33\,h_S \simeq 1.14\,\mathrm{m}, \qquad (13.16)$$

so that

$$h_{max} = h_S + h_M \simeq 1.63\,\mathrm{m}. \qquad (13.17)$$

Now comes the extra bonus: without knowing the actual values of the solar and lunar masses, due to the ratio of relative magnitudes of tides provoked by both celestial bodies, we can easily find out their relative average densities!

The most important here is to observe that although the gravitational pull provoked by a distant celestial body is proportional to its inverse distance squared, the tidal effect is proportional to the inverse cube of the same distance. This is why although the Moon's gravitational pull is weaker that the Sun's one, the Moon's tidal effect is stronger: the proximity of our satellite, as compared to the enormous distance to the Sun, compensates its incomparably smaller mass. Now, as we have

$$h_S \sim \frac{GM_S}{D_S^3} R_E \quad \text{and} \quad h_L \sim \frac{GM_M}{D_M^3} R_E,$$

we may evaluate the ratio h_L/h_S:

$$\frac{h_M}{h_S} = \frac{M_M D_S^3}{M_S D_M^3}. \tag{13.18}$$

Consider now the average densities of the Moon and Sun:

$$\rho_M = \frac{M_M}{(\frac{4\pi}{3} R_M^3)}, \quad \rho_S = \frac{M_S}{(\frac{4\pi}{3} R_S^3)}, \tag{13.19}$$

therefore

$$\frac{\rho_M}{\rho_S} = \frac{M_M R_S^3}{M_S R_M^3},$$

where R_M is the Moon's radius, and R_S the radius of the Sun. Taking into account that both radii are very small as compared to the distances to the corresponding celestial bodies, $R_M \ll D_M$ and $R_S \ll D_S$.

Due to this circumstance we can replace the angle (in radians) under which both bodies are seen from the Earth by the corresponding sine or tangent functions, given by simple ratios:

$$\alpha_M \simeq \sin \alpha_M \simeq 2\frac{R_M}{D_M}, \quad \text{and} \quad \alpha_S \simeq \sin \alpha_S \simeq 2\frac{R_S}{D_S}. \tag{13.20}$$

At this point we can use the marvellous gift Nature offered us, namely the equal apparent angular size of Sun and Moon:

$$\alpha_M = \alpha_S \simeq 32' = 0.55°. \tag{13.21}$$

Due to this equality, we also have

$$R_M = \frac{\alpha_M}{2} D_M \quad \text{and} \quad R_S = \frac{\alpha_S}{2} D_S \tag{13.22}$$

which enables us to write

$$\frac{\rho_M}{\rho_S} = \frac{M_M R_S^3}{M_S R_M^3} = \frac{M_M D_S^3}{M_S D_M^3} = \frac{h_M}{h_S} \simeq 2.33. \qquad (13.23)$$

Notice that even if the angular sizes of Sun and Moon were not equal, we could find the result by introducing the correcting factor $(\alpha_M^3)/(\alpha_S^3)$, but it would be less spectacular than a simple equation (13.23). Besides, that would require quite precise measurements of both angular sizes, α_M and α_S, whereas the observation of total solar eclipses enabled us to verify that both angular sizes are extremely close, if not exactly equal.

Isaac Newton was the first who understood that the relative masses and densities of Sun and Moon can be estimated using the ratio of the tidal amplitudes provoqued by the two luminaries. The actual values of solar and lunar densities are equal to $1.41\,\mathrm{g/cm}^3$ and $3.3\,\mathrm{g/cm}^3$ respectively; and as we know now, the mass of the Moon is about 81 times smaller than the mass of the Earth, which means that $M_M = 0.0123\,M_E$.

However, in his *Principia*, Newton underestimated this ratio: from his calculations, the mass of the Moon was only 40 times lower than the mass of the Earth. Being aware of the real sizes of both planets, Newton arrived at the conclusion that the Moon's density is about twice higher than the average density of the Earth (the values of those densities being yet unknown in Newton's time). This erroneous estimate was mostly due to the fact that Newton relied upon the data concerning the tides in the estuary of the river Avon, where the funnel effect is quite large.

Let us estimate the ratio of the masses of the Moon and Earth, and of the solar mass as well. From the elementary count of the unknown variables (in this case we have three unknowns, M_E, M_M and M_S) and the number of equations necessary to determine them we see that we need *three* independent equations.

However, if we are interested in their ratios only, *two* independent algebraic equations should suffice: as a matter of fact, if the ratios M_E/M_M and M_S/M_E are known, the third ratio, M_S/M_M is just the product of those two. And we have indeed two such equations at our disposal, thanks to the ratio between the height of the lunar and solar tides on one hand, and the Kepler's third law of orbital motion on the second hand.

Our estimates will be approximate for two different reasons: first, the ratio between the tides, although based on taking the average of many observations, cannot be considered as a well-defined quantity — the error is, at best, no less than 5%.

The second error comes from the simplified version of Kepler's law we shall be using here, namely the assumption that both the Earth's and Moon's orbits are circular, which in the case of the Earth is not very significant, the eccentricity of Earth's orbit being only $e_E = 0.017$, but will cause a non-negligible error in the case of the Moon, whose orbital eccentricity is $e_M = 0.055$, which is more than 5%. Nevertheless we shall see that our estimate will be contained inside the same error bar, within 5% as compared with the present knowledge of the exact values of the masses of the three bodies.

[* Let us recall the notations: M_S the mass of the Sun, M_E the mass of the Earth, M_M the mass of the Moon, G is Newton's gravitational constant. Next, D_S is the radius of Earth's orbit (supposed circular) around the Sun, D_M is the average distance to the Moon from the Earth (i.e. the Moon's orbital radius). Ω_E and Ω_M are the circular frequencies of the Earth and Moon, and T_E and T_M are their periods of orbital rotation, for which we must take the *sidereal* ones: $T_E = 365.25$ days, $T_M = 27.32$ days.

From the Kepler's third law for circular orbits we have, for the Earth's orbit around the Sun and the Moon's orbit around the Earth:

$$\frac{GM_S}{D_S^3} = \Omega_E^2 = \frac{4\pi^2}{T_E^2}, \quad \frac{GM_E}{D_M^3} = \Omega_M^2 = \frac{4\pi^2}{T_M^2}. \tag{13.24}$$

Dividing the first equality by the second one we get

$$\frac{M_S}{M_E} = \frac{D_S^3 T_M^2}{D_M^3 T_E^2}. \tag{13.25}$$

The ratio between the lunar and solar tides, estimated as 2.33, provides another equation involving masses and distances: as found in

(13.14), where we have got

$$\Delta\gamma_S \simeq h_S \simeq \frac{3GM_S}{D_S^3} \quad \text{and} \quad \Delta\gamma_M \simeq h_M \simeq \frac{3GM_M}{D_M^3}. \qquad (13.26)$$

We have therefore, for the ratio between solar and lunar masses,

$$\frac{\Delta\gamma_S}{\Delta\gamma_M} = \frac{3GM_S}{D_S^3}\frac{D_M^3}{3GM_M} = \frac{M_S D_M^3}{M_M D_S^3} = \frac{1}{2.33}. \qquad (13.27)$$

Extracting from the above relations ratios M_S/M_M and M_S/M_E we can write:

$$\frac{M_S}{M_M} = \frac{1}{2.33}\frac{D_S^3}{D_M^3}, \quad \frac{M_S}{M_E} = \frac{D_S^3}{D_M^3}\frac{T_M^2}{T_E^2}. \qquad (13.28)$$

Dividing the first relation by the second one, we get the ratio between Earth's and Moon's masses:

$$\frac{M_E}{M_M} = \left(\frac{1}{2.33}\frac{D_S^3}{D_M^3}\right)\cdot\left(\frac{D_M^3 T_E^2}{D_S^3 T_M^2}\right) = \frac{1}{2.33}\frac{T_E^2}{T_M^2}. \qquad (13.29)$$

This is really nice — we are able to express the ratio of the Earth's and Moon's masses using exclusively the rotation periods, perfectly well known, and the estimate of the ratio of the heights of the solar and lunar tides!

Let us confront this formula with the observational data. We have, in days, $T_E = 365.25$ days and $T_M = 27.32$ days. So we get

$$\frac{T_E^2}{T_M^2} = \frac{(365.25)^2}{(27.32)^2} = 178.73 \quad \text{and} \quad \frac{M_E}{M_M} = \frac{178.73}{2.33} = 76.7, \qquad (13.30)$$

not too bad when compared to the actual value $M_E/M_M = 81.$ *]

13.4 Physical and astronomical consequences of tides

Let us estimate the tidal accelerations occurring on the surface of the Moon under the gravitational influence of Earth.

$$\Delta\gamma_E = \frac{3M_E G}{D^3} R_M \quad \text{versus} \quad \Delta\gamma_M = \frac{3M_M G}{D^3} R_E. \qquad (13.31)$$

The first expression gives the tidal acceleration provoked by the Earth's gravitational field at Moon's surface, the second one corresponds to the tidal acceleration provoked by the Moon on the surface of our planet. The ratio between these two tidal accelerations is therefore

$$\frac{\Delta\gamma_E}{\Delta\gamma_M} = \frac{M_E}{M_M} \cdot \frac{R_M}{R_E}. \tag{13.32}$$

Substituting actual values $M_E/M_M = 81$, $R_M/R_E = (1737/6370) \simeq 0.273$, we get the final result:

$$\frac{\Delta\gamma_E}{\Delta\gamma_M} = 0.273 \times 81 \simeq 22.1, \tag{13.33}$$

which means that if oceans existed on the Moon, the tides provoked by the gravitational pull of the Earth would be 22 times higher than the tides caused by the Moon on Earth. Such tides do not exist on the Moon because of lack of water; nevertheless the mighty gravitational effect induces tidal forces strong enough to stop the rotation of the Moon around its axis with respect to the Earth. This is the reason why we see always the same face of our satellite: its rotation with respect to the distant stars has since long been stabilized and synchronized with its orbital rotation around the Earth.

Similar, and even stronger tidal effects act on the satellites of the big planets, Jupiter and Saturn. All of them turn around their mother planet with one side facing it, exactly as our Moon does with respect to the Earth. The reciprocal action of the satellites on the giant planets is obviously negligible, taking into account the enormous mass discrepancy. But how about the influence of tides on the proper rotation of Earth? In view of our analysis, a similar effect, although very weak, should provoke some slowing down of the Earth's rotation, too.

If the influence of strong tidal forces exercised by the Earth ended up in synchronizing the Moon's spin with its orbital frequency, the Moon's influence and the tides it provokes on Earth should also slow down the Earth's spin. Contemporary atomic clocks are able to measure time with precision up to 10^{-15} second; nevertheless it is not easy to evaluate the deceleration of Earth's diurnal rotation due exclusively to the tidal effects, because global atmospheric and oceanic motions as well as displacements of the mantle under the Earth's crust also provoke detectable variations of the same order.

Nevertheless, astronomers were able to prove that the Earth's spin is slightly diminishing, due to a very precise ancient relation about full solar eclipse that was observed in the city of Babylon on April 15, 136 B.C.E., i.e. more than 2000 years ago. An unknown astronomer noted that the eclipse started shortly after the sunrise, and the full eclipse was observed when the Sun was at 24° above the horizon. With this information it was possible to determine the exact position of the Moon at that time, extrapolating its real motion with respect to the remote stars. It turned out that a full solar eclipse took place at that time, but extrapolating the Earth's rotation as if its angular velocity were constant gave a somewhat paradoxical result: the path of the full shadow Moon casted then could not pass over Babylon, but about 55° westwards, so that the full eclipse should have been observed in South-Western Europe.

The discrepancy between what was observed more than 2000 years ago and the extrapolation made by astronomers led to the conclusion that the Earth's spin is slightly lower now than is was 2000 years ago, which explains the observed time delay. The difference of 55° corresponds to about 3.5 hours (one hour difference corresponds do 15° longitude difference). Now, supposing that the deceleration pace is a negative constant, $d\omega/dt = -C$, then after time T the total difference of rotation angle should be

$$\Delta = -\frac{CT^2}{2}, \qquad (13.34)$$

the same formula that gives the distance covered by a body moving with constant acceleration. Now we are able to determine the constant C:

$$T \simeq 2100 \,\text{years} = 6.62 \times 10^{10} \,\text{sec}, \quad \Delta = -55° \simeq 1 \,\text{radian},$$

$$\text{therefore } C = \frac{2\Delta}{T^2} \simeq -4.55 \times 10^{-22} \,\text{radian} \cdot \text{sec}^{-2},$$

which corresponds to day and night being longer by 1.4 milliseconds after 100 years. Modern atomic clocks confirm the evaluation, and our astronomical time is constantly corrected, although by a very tiny, almost imperceptible amount. The conservation of total angular momentum of the Earth-Moon system results in a slight increment of the Moon's orbital angular momentum, which in turn leads to a change of lunar orbit; its mean distance from the Earth increases by about 3.8 centimeters per year.

13.5   The Roche limit

In some cases tidal forces acting on a satellite or an asteroid which come too close to a very massive planet or star may become strong enough to overcome its own gravitational field and literally tear it into pieces. This is what happened recently with the comet Shoemaker-Levy 9 which crossed Jupiter's orbit very close to the giant planet during the first week of July 1992. It disintegrated, and the fragments of its nucleus fell ino Jupiter's atmosphere with enormous speed, and delivered an amount of energy higher than hundreds of hydrogen bombs. The flashes of light were observed and photographed in real time by the Hubble Space Telescope.

Let us evaluate the distance at which a smaller body approaching a massive planet or star becomes prone to destructive tidal forces. For the sake of simplicity, let us assume that both bodies have spherical shape, and the smaller body is gaseous or liquid (this is what happens when a smaller star comes too close to a much bigger one). Let the approaching satellite's radius be a and its mass m. Let the heavy central star's radius be R and its mass M.

As we know, the tidal force at the surface of the satellite of radius r whose center is at distance D from the center of the massive star, is equal to $\frac{3MG}{D^3} \cdot r$, acting on both sides, towards the attraction center on the side facing it, and outwards on the opposite side. The smaller body itself is kept together by its own gravitational force, whose acceleration on its surface attains the value $\frac{mG}{r^2}$.

The Roche limit (called also the Roche zone) is the distance D_R at which the gravitational attraction towards the central star becomes equal to the directed inwards gravitational forces keeping the small body together: if the satellite crosses the Roche limit continuing to approach the huge central star, the matter from its surface will be literally torn off and start to fall onto the bigger mass. Therefore the equation satisfied by the Roche radius D_R is as follows:

$$\frac{3MG}{D_R^3} \cdot r = \frac{mG}{r^2}. \tag{13.35}$$

Simplifying by G on both sides and dividing by r we get

$$\frac{3M}{D_R^3} = \frac{m}{r^3}. \tag{13.36}$$

The masses M and m can be expressed as products of average densities of each body, ρ_M and ρ_m, by their respective volumes:

$$M = \frac{4\pi R^3 \rho_M}{3}, \quad m = \frac{4\pi r^3 \rho_m}{3}. \tag{13.37}$$

Sunstituting into (13.36) and simplifying by the common factor $\frac{4\pi}{3}$ we get

$$3\rho_M \frac{R^3}{D_R^3} = \rho_m, \quad \text{from which we get} \quad D_R = R\sqrt[3]{3\frac{\rho_M}{\rho_m}}. \tag{13.38}$$

It is remarkable that the final result (13.38) does not depend on the size of the satellite; it depends only on the radius of the heavy star R and the density ratio between the satellite and the star. Obviously enough, the Roche limit is always bigger than the central star's radius R. If the satellite is made of solid matter, the cohesive forces may keep it together somewhat closer than the Roche limit for fluid bodies.

Chapter 14

Huygens and Cassini

Huygens' life - Mathematical achievements - Huygens' principle in optics and wave theory - The isochronous pendulum - Invention of the modern clock - Cassini and the Paris observatory - Astronomical discoveries - Determining distances in the Solar System - Roemer and the speed of light

14.1 Preamble

On October 15, 1997 an extraordinary space research mission named "Cassini-Huygens" was launched from Cape Canaveral, Florida, USA. It was a joint technological and scientific endeavour by the American Space Agency NASA, the European Space Agency ESA and the Italian Space Agency ASI. Its goal was to study Saturn, one of the most fascinating planets of the Solar System. The choice of the name was not accidental: it was a tribute to the scientists who discovered and described Saturn's rings and satellites.

Christiaan Huygens (1629–1695) was the true discoverer of Saturn's rings, which Galileo could not distinguish well enough through his first telescopes, thinking that they were parts of the planet or sister planets. Huygens also discovered Titan, the biggest of Saturn's satellites.

Giovanni Domenico Cassini (1625–1712) noticed that Saturn's rings were not homogeneous, but separated in parts by a dark gap, now called the *Cassini Division*. Continuing his observations, Cassini discovered next the Saturnian satellites Iapetus, Rhea, Tethys, and Dione.

Both men knew each other and met quite often while working in Paris. In 1666 Huygens was invited by the French minister Colbert to come to Paris and to organize the Royal Academy of Sciences. Cassini, who at that time had already become a renowned astronomer, was invited by Colbert three years later, in 1669, to become the head of the newly built astronomical observatory in Paris. Cassini worked there until his death in 1712, contributing greatly to astronomy and geodesy alike. His son and grandson became famous astronomers, too, and worked in the same observatory.

Huygens' contribution to science was much greater than Cassini's, whose field was essentially astronomy and geodesy, although he also made some contributions to mathematics, while Huygens interests and achievements covered mechanics, fluid mechanics, optics, astronomy, chronometry and mathematics. Nevertheless their contributions to astronomy are comparable; this is why they meet again in this chapter.

Fig. 14.1   Left: Christiaan Huygens (1629–1695); Right: Giovanni Domenico Cassini (1625–1712).

14.2   Huygens' life and achievements

The famous Dutch mathematician, physicist and astronomer Christiaan Huygens (1629–1695) is rightly considered as one of the greatest scientists of his time, second only to Newton. Born in the Hague to a rich and distinguished family, he showed exceptional gifts, both mathematical and practical, since his early years. Huygens' father befriended the famous French philosopher René Descartes, who stayed several times in Huygens' household, and was impressed by the young boy's mathematical talents. Later on, Christiaan became a great admirer of Descartes and of his philosophy.

At the age of 16 the young Huygens entered the University of Leyden, to study mathematics and law. Two years later he moved to Breda, where he continued his education, with much attention paid to the new scientific ideas of Descartes. After completing his studies he was assigned a diplomatic mission, but after less than a year he postponed his diplomatic carreer, preferring to devote his time entirely to science. He visited Paris in 1655 and was well received in the scientific and philosophical circles of French Capital. He came back to Paris in 1660 to visit Blaise Pascal with whom he entertained a scientific correspondence since his first visit.

Huygens' scientific acquaintances were incredibly rich. Besides Descartes and Pascal, he met Mersenne and most French mathematicians forming Mersenne's circle. He met Spinoza in The Hague and Amsterdam, and learned from him the technique of grinding glass lenses, which enabled him to produce his own telescope with a magnification of 50. He met Leibniz in Leipzig and befriended him; later on, Leibniz visited Huygens in Paris and learned a lot of mathematics from numerous encounters and discussions. Their correspondence lasted until Huygens' last days. Huygens traveled to London and Cambridge, where he met Newton with whom he had many discussions, not always agreeing with the great Englishman's scientific views.

In 1666, when Huygens' scientific reputation was well established, the First Minister of State Jean-Baptiste Colbert (1619–1683) invited

Huygens to France, asking him to organize the Academy of Sciences similar to the Royal Academy of London to which Huygens already belonged. Accepting the invitation, Huygens asked the members of *Académie Parisienne*, a circle of scientists and philosophers established in 1635 by French polymath Marin Mersenne (1588–1648), to form a new learned society, the *Académie des Sciences*. In fact, Huygens was the first president of this venerable institution which is active until today.[1]

Huygens stayed in France for the next ten years, in spite of the war in which France took part on the side of Holland's enemies. He visited the Hague a few times for health reasons. However, he decided to leave France and return to Holland after King Louis XIV revoked the Edict of Nantes established by his predecessor Henry IV, and guaranteeing equal religious rights to the Protestants. He died in the Hague in 1695.

14.3   Mathematics and physics

Huygens' contribution to mathematics concerned mostly geometry and its applications to mechanics. He was the first to derive the formulae for the centrifugal and centripetal forces using purely geometric constructions. He developed also the theory of evolutes, i.e. curves obtained by developing a string wound on a given curve. Another important Huygens' contribution was a systematic analysis of probabilities applied to the games of chance, to which Huygens devoted an entire book that he wrote encouraged by Pascal to publish his original findings.

In physics, Huygens' contribution to optics and his theory of light is by far the most important one. In opposition to Newton, whose description of light was based on the analysis of rays considered as trajectories of hypothetical particles travelling in empty space, Huygens believed in the wave nature of light, assuming that light

---

[1]It is worthwhile to notice the essential difference between the English and French Academies. The Royal Society of London was an autonomous private organization, sustained exclusively by its members, whereas the Paris Académie Royale was an official body whose members were paid by the government.

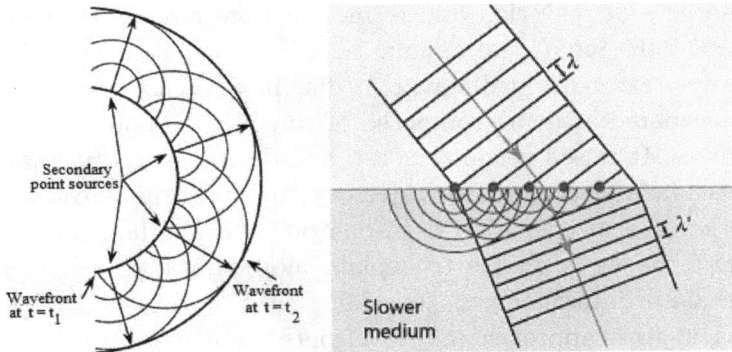

Fig. 14.2   Left: Huygens' principle of wave propagation; Right: Refraction explained using Huygens' principle.

behaves like very short waves propagating in the medium filling the space.

The following principle (bearing his name since then) leads to the construction of wavefronts from a given source or a collection of sources:

Every point on a wave front can be considered as a source of tiny wavelets that spread out in all possible directions at the speed of the wave itself. The new wave front is the envelope of all such wavelets (i.e. the surface tangent to all partial wavefronts).

One can derive from this principle main optical phenomena, like reflection, refraction, diffraction and interference. When a wave hits under certain angle a plane separating two transparent media, e.g. penetrating the surface of water, Huygens' principle explains the Snell-Descartes law of refraction, with inverse velocities replacing optical density parameters n_1, n_2:

$$\frac{1}{v_1}\sin\theta_1 = \frac{1}{v_2}\sin\theta_2 \qquad (14.1)$$

where θ_1 is the angle of incidence formed by the wave vector and the vector normal to the plane of separation of the two media, and θ_2 is the angle of refraction, between the refracted wave vector and the normal to the separation plane.

The laws of reflection and refraction were already explained by Fermat (1607–1665) who announced the principle of least time for light rays travelling in different media. In Fermat's approach light was considered as being composed of tiny particles obeying rules of mechanics. It raised certain uneasy questions, suggesting that light "knows" in advance which trajectory to choose in order to minimize the time of transition from one point to another, distant one. Fermat's theory is unable to explain more subtle phenomena, like diffraction or interference patterns.

In Huygens' approach the wavefronts are constructed from tangents to the most advanced points of partial wavelets. The propagation of light in this way satisfies the principle of causality which was hardly present in the Fermat-Descartes principle of least time. Besides, it explained phenomena which were out of reach for previous models, including differences in refraction of colors.

Huygens made also important contributions to the science of mechanics, correcting Descartes' error in his laws of impact having proved by experiment that the momentum in a fixed direction before the collision of two bodies is equal to the momentum in that direction after the collision. He also proved the conservation of the kinetic energy in elastic collisions.

Although Huygens had a great admiration for Newton whom he met at the Royal Society in England and also privately, at the same time he did not believe the theory of universal gravitation of which he said *it appears to me absurd*. In some sense of course Huygens was right. It is hard to believe that two distant masses attract one another when there is nothing between them, Newton was aware of this problem, too, but it did not prevent him to perform calculations that have an unprecedented predictive power.

Huygens was influenced by Descartes' theory explaining circular motion of planets by aether vortices driving the planets, but later on abandoned it as no less absurd.

14.4 Astronomy

In 1655 Christiaan Huygens constructed a telescope with a flat-convex lens with 5.7 cm diameter and with focal distance of

3.37 meters, and with 50× magnification. He improved the grinding technique and ground the lenses himself. Using this new astronomical tool, he identified the extra bulges of Saturn discovered by Galileo to be a disk surrounding the planet. Being not totally sure whether what he saw was real, imitating Galileo, he published a Latin sentence encoded in the following anagram

"Aaaaaaa, ccccc, d, eeeee, g, h, iiiiiii, llll, mm, nnnnnnnnn, oooo, pp, q, s, ttttt, uuuuu".

The hidden meaning of this collection of letters was revealed by Huygens only three years later, in 1659:

"Annulo cingitur, tenui, plano, nusquam cohaerente, ad eclipticam inclinato",

which means "Surrounded by a ring, thin, flat, nowhere touching, inclined to the ecliptic".

Incidentally, he discovered the largest moon of Saturn, which he named *Saturni Luna* and described it in his tract published a few years later. The moon discovered by Huygens is now known as Titan, and is the second largest (after Jupiter's satellite Ganymede) satellite in the entire Solar System. Saturn's rings were described in a book published in 1659, *Systema Saturnium*.

Huygens' curiosity was raised by the unique shape of the rings, their flatness, and above all, their size relative to Saturn's own

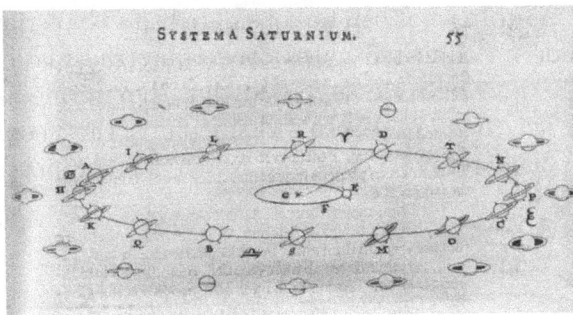

Fig. 14.3 Illustration from Huygens' book *Systema Saturnium*, showing variations in Saturn's rings appearance as seen from the Earth, depending on the relative positions of two planets (credit: sil.su.edu).

dimensions. Perhaps the most important contribution to contemporary astronomy was Huygens' estimate of real distances in the Solar System. Initially, he wanted to determine the real size of Saturn and its rings, the nature of which remained mysterious — are they solid? Do they rotate or stand still? What material are they made of? To have at least a hint to answers to such questions, knowledge of the real sizes was crucial; coincidentally, this meant a reliable estimate of the real distances in the Solar System.

The relative distances were already well known since Copernicus and Kepler, therefore the calibration of distance between any two planets in a well determined configuration would suffice. The closer, the better — this is why Huygens chose Venus as the best object for a measurement of the distance separating us from this planet. During his first sojourn in Paris he observed Venus with his new telescope with magnification 50, and was able to determine its angular size when the crescent's width was about one-third of the disc. Using simple geometry of the Copernican system and knowing that Venus' quasi-circular orbit's radius was 0.72 of Earth's orbital radius, Huygens could express the Earth to Sun distance in terms of Earth's diameters assuming that Venus is of the same size as the Earth. This assumption was later abandoned in favor of another hypothesis, supposing that the size of the Earth is contained between the sizes of Venus and Mars.

Although based on very slippery assumptions, as acknowledged by Huygens himself, his final estimate was surprisingly close to what is known at present. In what can be considered as a Keplerian approach, Huygens made an educated guess concerning the real diameters of Venus, Earth and Mars. He supposed that they formed a geometric progression, i.e. that the Earth's diameter is the geometric mean between the diameters of Venus and Mars. The supposed diameters were as follows:

$$\text{Venus' diameter} = \frac{1}{84} \text{ of Sun's diameter;}$$

$$\text{Earth's diameter} = \frac{1}{111} \text{ of Sun's diameter;}$$

$$\text{Mars' diameter} = \frac{1}{166} \text{ of Sun's diameter.}$$

By a happy coincidence, the supposed ratio between Earth's and Sun's diameters came out extremely close to the actual ratio 1:109.1, as well as the solar parallax 8.6″. The astronomical unit according to Huygens was equal to 24 000 Earth's radii, which corresponds to 152 million kilometers — also very close to the present value 149.6 million kilometers.

Another discovery was made by Huygens with an even better telescope he constructed in 1656 with the help of his brother Constantijn. This new 7-meter long device magnified objects nearly 100 times, and still had a considerable field of view. With this telescope Huygens observed the part of the Orion constellation called "the sword", and discovered the *Orion Nebula*. Here is how he reported on this discovery in 1659:

> *In the sword of Orion are three stars quite close together. In 1656 I chanced to be viewing the middle of one of these with a telescope, instead of a single star twelve showed themselves (a not uncommon occurrence). Three of these almost touched each other, and with four others shone through the nebula, so that the space around them seemed far brighter than the rest of the heavens, which was entirely clear and appeared quite black, the effect being that of an opening in the sky through which a brighter region was visible.*

14.5 Determining longitudes and chronometry

To determine one's latitude it is enough to measure the altitude of the Polaris star, or using appropriate tables, to measure the Sun's height at local noon. Determining the local geographical longitude could be even easier if we had the knowledge of the exact astronomical time in a chosen place on the Earth that could be labeled as "zeroth meridian". Or at least if we could have a clock with initial time calibrated at the departure point, keeping its pace unchanged during the journey, and compare the time of local noon at the arrival point with the time shown by our clock. The difference of times will correspond to the difference of geographic longitudes. In principle this method is even simpler than most latitude measurements, the only missing part was a clock providing universal time, independent of the place on the Earth. Galileo's proposal to use Jupiter's moons as a standard

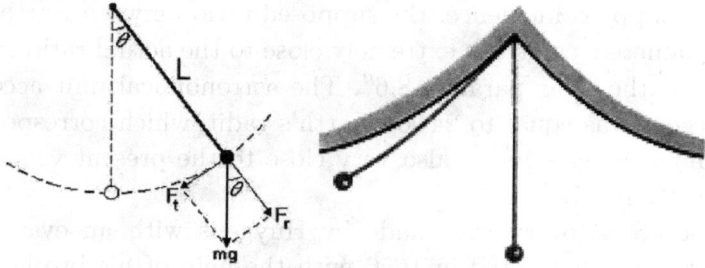

Fig. 14.4 Left: Mathematical pendulum; Right: A pendulum with variable length.

stellar clock was an ingenious idea, but it was not as reliable as he thought (due to the anomalies caused by the finite speed of light, which we shall discuss at the end of this chapter), and above all, the necessity of using powerful telescopes and the impossibility to make observations during daytime or a cloudy night).

Huygens understood that only a good clockwork keeping steady pace will provide the ultimate solution. He was aware of Galileo's conjecture that pendulums have constant period of oscillations, but found that it was true only for very small amplitudes. As shown in Figure 14.4, an ordinary pendulum (on the left) is subject to the gravitational pull **g** directed downwards. Its radial component F_r is canceled by the tension of the string, whereas its tangential component F_t gives rise to tangential acceleration according to Newton's law of dynamics. As it follows from the figure, the absolute value of the force F_t is $mg \sin \theta$, so the equation of motion the pendulum must satisfy is

$$mL\frac{d^2\theta}{dt^2} = -mg \sin \theta, \qquad (14.2)$$

which only for very small amplitudes can be approached by the following *harmonic oscillator* equation

$$\frac{d^2\theta}{dt^2} + \frac{g}{L}\theta = 0.$$

The general solution of this equation is the periodic function requiring two initial values to be fixed, $\theta(0) = \theta_0$, $\frac{d\theta}{dt}(0) = \dot{\theta}_0$.

$$\theta(t) = A\cos(\omega t) + B\sin(\omega t),$$
$$\omega = \sqrt{g/L}, \quad \theta_0 = A, \quad \dot{\theta}(0) = \omega B. \tag{14.3}$$

With growing amplitude the period of the pendulum increases, compromising the measure of time based on the pendulum's supposedly steady pace. On the other hand, in the limit of very small amplitudes, the period of the pendulum is proportional to the square root of its length. Therefore the main problem of pendulum clock is to maintain the period of oscillations independent of their amplitude, which clearly is not true in the case of an ordinary pendulum described by the equation (14.2). If the pendulum's length could be decreased slightly when its departure from vertical position grows, it could compensate the variation of period due to the greater amplitude. Huygens found an ingenious solution: the pendulum should be suspended not by a solid rod, but by a string that should remain tangent to a curved surface, so that when the amplitude grows, the upper part of the string touches the surface, and only the remaining part acts as the shortened pendulum.

The curve described by the extremity of such a pendulum is called the *evolute* of the curve to which the string adheres. The problem posed was to find a symmetric curve with cusp directed upwards, to which the string of the pendulum should remain tangent, ensuring the *isochronous* behavior, i.e. constant period of oscillations totally independent of amplitude. In order to find the unique form of such curve, Huygens replaced the problem of isochronous pendulum by an equivalent problem consisting in finding a concave curve on which a small ball rolling down under the force of gravity will arrive to its lowest point after the same time, independently of its starting point. Such a curve is called a *tautochrone*, and Huygens proved that the solution is given by a *cycloid*.

The evolute of a curve is by definition the locus of all its centers of curvature. In practice this means that at each point of a given curve we can draw not only a straight line tangent to it, but also a circle whose center defines a point of the evolute. Then we pass to the next point and draw a new tangent circle whose center will define the next point of the evolute, and so on. It is clear that the evolute of a circle reduces to one point, which is its own center. The evolute of an ellipse is a symmetric diamond-like *astroid*:

$$\text{the ellipse: } x = a\cos t, \quad y = b\sin t, \quad \frac{x^2}{a^2} + \frac{y^2}{b^2} = 1,$$

$$\text{the evolute: } x = \left(a - \frac{b^2}{a}\right)\cos^3 t, \quad y = \left(b - \frac{a^2}{b}\right)\sin^3 t. \tag{14.4}$$

It can be shown that the unique curve whose evolute is identical with itself is the cycloid: as Figure 14.5 shows, for the cycloid with the parametric representation $(x = r(\theta - \sin\theta), y = r(1 - \cos\theta))$ the evolute will be given by the equation $x = r(\theta + \sin\theta), y = r(\cos\theta - 1)$, which describes a transposed replica of itself.

Despite many efforts and trials, Huygens could not produce a satisfactory cycloidal pendulum clock, and turned his attention to a traditional pendulum whose period of oscillations would be automatically maintained constant by some extra mechanism. By the same token, some amount of extra energy should be added after each oscillation in order to keep the amplitude constant. Primitive versions of similar mechanisms already existed in clocks with verge and foliot. Besides, in 1665 Huygens observed self-synchronization of two pendulums of identical length suspended on a common support, and was fascinated by the discovered phenomenon. As a result, the anchor mechanism was introduced, which became the prototype for

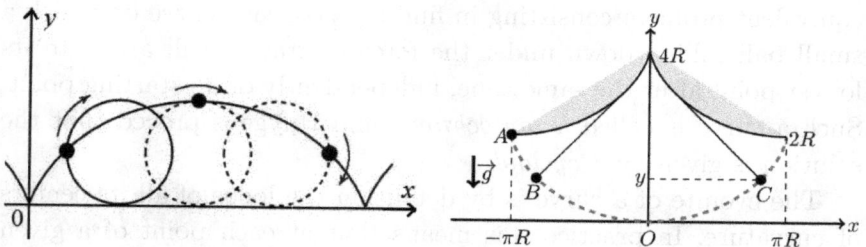

Fig. 14.5   Left: The construction of a cycloid; Right: A cycloidal pendulum.

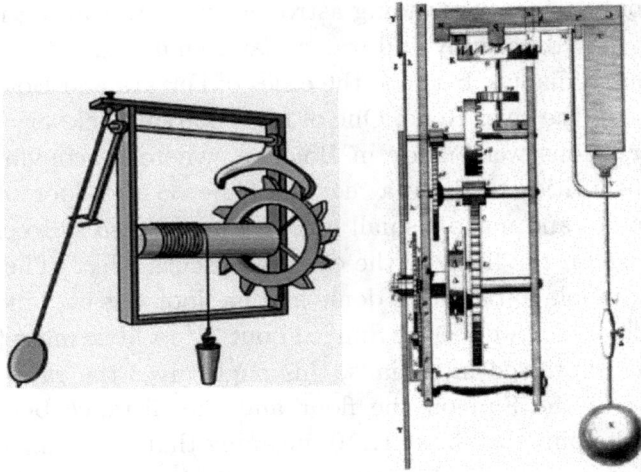

Fig. 14.6 Left: Schematic representation of anchor mechanism, Right: Schematic drawing of Huygens' pendulum clock.

precision clocks called *chronometers*. In 1673, Huygens published his major work on pendulum clocks and timekeeping, *Horologium Oscillatorium sive de motu pendulorum*, i.e. "Theory and Design of the Pendulum Clock", which has become a reference in the field. Incidentally, it contained also many theorems of mechanics, like the conversion of free-fall motion into kinetic energy. Nevertheless, in spite of several experiments with Huygens' pendulum clock at sea, it became clear that this way of time keeping would never work accurately on a heaving ship's deck. A decisive improvement made by Huygens after that was the invention of a watch regulator called a balance spring, which became a standard component for time keeping devices.

14.6 Cassini's life and discoveries

Giovanni Domenico Cassini (1625–1712) was one of the most influential astronomers of his time.

At the age of 20 he was employed as assistant astronomer at Panzano Observatory near Bologna, and after a few years he became professor at the University of Bologna. His reputation grew steadily, and not only as an outstanding astronomer: he was also a renowned specialist in hydrography and water management. In 1657 he helped to arbitrate a dispute between the cities of Ferrara and Bologna over the course of the river Reno. One of his remarkably clever astronomical observations were made in Bologna, where he convinced local church authorities to draw a meridian line on the floor of basilica San Petronio, and make a small hole in the dome so as to create the Sun's image on the floor by the camera obscura effect. The distance from the pinhole in basilica's dome and its floor was 66.8 meters; the angle under which we see the Sun is about 31′ (a little more than half a degree). Expressed in radians, this angle gives the ratio between the image of the Sun on the floor and the distance between the floor and the pinhole: about 1:110, meaning that the diameter of the Sun's image was about 60 cm (in fact, its semi-minor axis, because of the elongation of the circular image due to the latitude angle). The image was large and sharp enough as to enable the observation of annual variations in size, confirming the eccentricity of the Earth's orbit.

But the real breakthrough in his scientific carreer occurred in 1669, when king Louis XIV ordered his minister of state Colbert to invite Cassini to Paris to supervise the construction of a great astronomical observatory and to become its first director.[2]

Supposed to become the most advanced center of astronomical science of that time, the future Paris Observatory was also destined to perform precision mesurements of distances all over the Kingdom of France, to produce reliable maps, and to solve the problem of determining longitudes on land and sea. Cassini's plan was to map the whole of France using triangulations. The chosen meridian was the north-south line on which the chief instrument of the Paris Observatory was aligned. This meridian can be admired today in the great "Cassini Room" of the Observatory.

The triangulation work was continued by Cassini's son and grandson, who took over the directorate of the Observatory after Cassini's

---

[2]As we know from one of the previous chapters, this position was proposed first to Johannes Hevelius from Gdańsk (Danzig), who declined the offer.

death in 1712. They were often referred to as "Cassini II" and "Cassini III", a genuine astronomical dynasty. Although the final version of the *Carte de France* was completed only in 1744, with the meridian running down the center of the country, most of the triangulation results were obtained when Louis XIV was still ruling the country. At the end of the 17th century latitudes were measured with great precision; however, determining the same places' geographical longitude was still very imperfect. Cassini resumed measurements of longitude inspired by Galileo's idea to use Jovian satellites — in particular Io with its 42-hour period — as a celestial clock showing universal time. With longitude determination thus improved, it turned out that France was about 10% "thinner" in the east–west direction that was commonly assumed by previous cartographers. This inspired a famous quip by the King Louis XIV, saying that "his astronomers had cost him more territory than his generals had won" (and they won a lot, especially in the East, including Alsace and Franche-Comté, and in the North, including Dunkirk and Lille).

Performing his duties for which he was hired by Colbert were no impediment for Cassini's astronomical observations leading to many discoveries presented in the next sections. After a long and fruitful life Cassini died in 1712, leaving the direction of the Paris Observatory to his son Jacques Cassini, who continued his father's work, namely in the domain of topography of France.

14.7 Cassini's astronomical discoveries

One of Cassini's first astronomical feats was the description of the comet that appeared in 1652–1653. The account of his observations was published in a booklet dedicated to the Duke of Modena. What is quite amazing is that more than a century after Copernicus' *De Revolutionibus* it appears from the text of his work that he still believed in a geocentric Universe, with comets travelling beyond the orbit of Saturn, nevertheless supposed to originate on the Earth. Later on Cassini opted for the Tychonic geocentric model, and finally let himself be convinced that the heliocentric Copernican model described the reality of solar system.

However, a strange fact concerning Cassini's views is that he did not accept Kepler's first law of elliptic orbits. Instead, he proposed his own version, claiming that the planets move along quartic curves, which nowadays are called the *Cassini ovals*. They are defined as the locus of points whose distances from two focal points, r_1 and r_2, obey the equation $r_1 \cdot r_2 = b^2$ keeping their *product* constant, in contrast with an ellipse which is the locus of points satisfying the relation $r_1 + r_2 = a$. Cassini's oval is given in Cartesian coordinates by the following fourth-order curve:

$$\left((x - a)^2 + y^2\right)\left((x + a)^2 + y^2\right) = b^4 \qquad (14.5)$$

or in polar coordinates (ρ, φ):

$$\rho^4 + a^4 - 2r^2 a^2 \cos(2\varphi) = b^4. \qquad (14.6)$$

This hypothesis was discussed by Cassini in 1680, seven years before the publication of the first edition of Newton's *Principia* containing the constructive proof that universal gravitation with its inverse force law admits planetary motions exclusively along conical sections.

The presence of Christiaan Huygens in Paris was beneficial for the newly founded Observatory, because his help was decisive in the construction of an extraordinary telescope with a 23 cm diameter objective lens and with focal distance of 37 meters. Having at his disposal a new and powerful astronomical tool, Cassini devoted most of his time to observation of the planets. His first target was Saturn and its rings; incidentally, he discovered four new satellites of the planet surrounded by rings (besides Titan discovered earlier by Huygens). Imitating Galileo with his "Medicean Stars", Cassini named Saturn's satellites after the French King Louis XIV "*Sidera Lodoicea*" the ("Louisian stars"). Today they are known as Iapetus (discovered in 1671), Rhea (1672), Tethys (1684) and Dione (1684).

In 1675, Cassini discovered that Saturn's ring was composed of multiple smaller rings with gaps between them; the largest of these gaps now bears the name of the *Cassini Division*. It is a 4800-km-wide region between the two parts of Saturn's rings called consecutively the "A-ring" and the "B-ring".

Fig. 14.7   Saturn and its rings as seen through a good telescope. Cassini's division is clearly visible. (Credit: Webastronomy blog.)

Observing Jupiter, Cassini discovered the giant red spot, and by the same token, he determined the planet's rotation period (about 10 hours). He also noticed the differential rotation of Jupiter's atmosphere. Compiling the results of many years' observations, he established very precise timetables of occultations and transits of Jupiter's moons, the purpose being to use them as universal clock which would be the best way to determine longitudes on Earth, just as Galileo proposed. Cassini realized that this method could not be of any help for sailors in the open sea, especially on cloudy nights, but can be adapted on land, in observatories endowed with good quality telescopes. The systematic delays he noticed on this occasion served a few years later to determine the speed of light by his young assistant Ole Roemer.

Cassini also determined the proper rotation periods of Mars and Venus.

14.8   Determining distances in the Solar System

Strange as it may appear, in spite of the general acceptation of the Copernican heliocentric system and the resulting understanding of the geometry of the solar system, its real size remained unknown, or rather ill-known until the middle of the 17th century. The only realistic estimate known until then was the distance to the Moon, established by Hipparchus as about 60 Earth's radii. Other distances, including all data defining elliptic planetary orbits, were based on ancient estimates more than 15 centuries old; even Copernicus in his *De Revolutionibus* gives the value of the Earth-to-Sun distance only slightly better than Aristarchus' estimate: 15.6 times less than the actual one versus 20 times less by Aristarchus and Ptolemy.

Let us evaluate the maximal distance of a celestial object whose parallax observed from the antipodes (i.e. the maximal distance available on Earth, about 12 000 km) will be 1 angular minute — the limit of resolution accessible to naked-eye astronomers including Tycho Brahe. Let the baseline between two simultaneous observations of a distant planet be d and its distance from the Earth be D; then we should set $d/D = \alpha = 1'$, where 1 arcminute must be expressed in radians. We have:

$$\alpha = \frac{1'}{60 \cdot 57.3} = 2.91 \cdot 10^{-4}\,\text{rad}, \quad D = \frac{d}{\alpha} = \frac{12\,000\,\text{km}}{2.91} \cdot 10^4, \tag{14.7}$$

$$\text{therefore } D = \frac{12\,000\,\text{km}}{2.91} \cdot 10^4 \simeq 41\,000\,000\,\text{km}. \tag{14.8}$$

Given that a realistic baseline would have been rather half of the Earth's diameter, the one arcminute parallax could be observed for planets at a distance of about 20 million km, and as we know at present, no planet of solar system comes even twice that close to the Earth, which explains why no such measurements could be ever performed without telescopes. Even after the first telescopes were constructed by Galileo and Kepler, they still had insufficient magnification and resolving power.

The situation changed when a new generation of telescopes became available, with improved optics, magnifications greater than 50 and higher resolving power. Moreover, it had been discovered how to place a cross-hair at the image plane of the objective lens to facilitate the measurements. And by 1670, micrometer eyepieces with precisely adjustable cross-hairs became available making possible the determination of celestial coordinates and positions with respect to the fixed stars with a precision of a few arcseconds. The greater magnification obtained by the use of objective lenses with very long focal lengths is important not only because it increases the resolving power, but also because larger lenses capture and concentrate more light, making even the very faint stars visible. More stars accessible for observation enable much better precision in determining planet's position, because one can choose a star very close to the observed planet in order to find out its celestial coordinates, and to perform extremely precise measurements of the positions of planets relative to the faint background stars.

Although sometimes Venus comes closer to us than Mars, it happens when it is in the lower conjunction, therefore becoming invisible, and when it reaches maximal brilliance, its distance is about 0.7 A.U.; in contrast, when Mars is in opposition, it becomes very bright and its distance is about 0.38 A.U. Therefore even without being aware of real distances, Mars is the unique planet whose parallax would be hopefully great enough to be observed with new telescopes produced for Cassini and his Observatory with lenses ground with Huygens' help. Although Cassini had little confidence in Huygens' evaluation of distances in the solar system, considering the premises as slippery, he accepted the order of magnitude. With a resolving power of a few arcseconds the challenge of measuring the parallax of Mars during the planet's opposition seemed feasible. Besides, the English astronomer John Flamsteed had predicted that Mars would pass in front of the middle ψ star in the constellation Aquarius on the first of October, 1672, which would give the astronomers an extra opportunity to fix the Mars' parallax with the greatest possible precision.

The baseline chosen was about 6700 km, between Paris and the Cayenne Island, in the French Southern American territory of Guiana. The distance along the great circle between these sites is 7080 km, but what should count here is the segment of straight line joining the two points, which is the genuine baseline to be used in three-dimensional triangulation. An exact knowledge of longitude difference between the two sites was crucial for the success of the forthcoming interpretation of measurements. This is why Cassini sent Jean Richer to Cayenne a year before the opposition in order to determine precisely that island's latitude and longitude. This was necessary in order to make comparison of observations made at different times in Cayenne and in Paris possible. With the exact knowledge of the longitude difference between Paris and Cayenne it was possible to evaluate the universal stellar time of each observation made, and make corrections due to Mars' own motion with respect to the fixed stars, which is accessible due to the precise knowledge of the Martian synodic period of 780 days, i.e. 0.46° per day, or 1.15 arcminute per hour.

After comparison of observations made in Southern America and in Paris, the parallax of Mars was estimated to amount to 22.5 arcseconds. With a baseline of 6700 km, this corresponds to an Earth-Mars distance at opposition of about 57 000 000 km. The opposition of Mars of 1672 was among the closer ones, Mars being at its perihelion, and the relative distance betwen the two planets was about 0.39 A.U., so that the astronomical unit — the average Earth-Sun distance — was calculated to be about 144 000 000 km, a very good result indeed as compared to what we know at present.

At the same time, John Flamsteed, the Royal Astronomer and Director of the freshly constructed Greenwich Observatory, used an alternative method of determining the parallax of Mars when it was at its maximum during opposition. His idea was to use a greater baseline than the distance between Paris and Cayenne chosen by Cassini, and this without leaving England and the town Derby where he was preparing to measure the parallax. The idea was as follows: if two observations are made one in the evening, shortly after sunset, another one the next morning, shortly before sunrise, the rotation of Earth around its axis will move the observer around so that in the morning he will find himself at the antipodes of his evening position — well, the word "antipodes" would be exact for an observer placed somewhere on the equator, while at the latitude 50° the chord between the opposite points of the parallel represents $12\,756 \times \cos 50° = 12\,2756 \times 0.643 = 8200\,\text{km}$ — still greater than 6700 km between Paris and Cayenne. This effect is called *diurnal parallax*. Moreover, the problem of comparing the exact time of observation was much simpler, because Flamsteed could evaluate the local time difference between his evening and morning observations with precision up to a few seconds.

The method chosen by Flamsteed had in principle one serious shortcoming: during the night separating the two observations, both Mars and Earth continue their orbital motion, and the change of apparent position of Mars as seen from Earth can be by far greater than the expected parallax of a fraction of arcminute. However, the effects of relative motion of two planets can be reduced almost to zero if the observations take place during the peak of Mars' retrograde motion, when the Red Planet seems to stand still before changing the direction of its wandering. Such a situation occurred on October 6, 1672, when Flamsteed performed a measurement of the martian *diurnal parallax.* His results were quite close to what Cassini and Richer determined after Richer's return to Paris, which took almost a year since the observations were made. The averaged value of astronomical unit was contained between 135 and 145 million km, about 7% less than the exact magnitude established in 20th century.

14.9 Roemer and the speed of light

When the speed of light is mentioned nowadays, nobody finds the very idea of light propagating with a finite velocity extravagant, and probably the value of this speed, 300 000 km/sec, is widely known, as well as the symbol c denoting it. But the idea of finite velocity was not always obvious. Galileo tried to measure the speed of light using two distant lanterns one of which had to be lit when his assistant saw his lantern ablaze, but the result was that either light was propagating instantly, with infinite velocity, or its velocity was too great to be measured in this primitive way. Descartes who found correct formulas for the reflection and refraction of light, believed that light propagated with infinite speed.

The young Danish astronomer Ole Roemer (1644–1710) arrived in Paris in 1672 to become Cassini's assistant. Working at Paris Observatory founded by Cassini a few years before, Roemer started compiling precision observations of Io, the fastest of Jupiter's satellites (whose orbital period was known since Galileo discovered it, and now with Cassini's telescope was firmly established to be 42 hours 27 minutes and 22 seconds). Comparing this universal clock with the local time measured by observing the passing of the Sun or a given star through the local meridian, the longitude of the place could be easily found, which was theoretically extremely important for navigation.

Although this sophisticated method of determining local longitude turned out to be impractical (mostly because it cannot be performed at daytime, and during the night cloudy weather often makes it impossible), Cassini continued tedious observations of Jupiter's satellites expecting to use them for longitude determination on firm ground and not on the ships in the open sea. Roemer was supposed to analyze Cassini's observations to show that their precision was good enough to determine local longitude by comparing this celestial "clock" with local time.

But Roemer's efforts led to an unexpected result: the first reliable evaluation of the velocity of light.

As seen from the Earth, the satellite Io is eclipsed by Jupiter once every complete orbital revolution. By carefully examining the timing of these eclipses which was noted by Cassini over many years, Roemer noticed something very peculiar. The time interval between successive eclipses became steadily shorter as the Earth in its orbit moved toward Jupiter and became steadily longer as the Earth moved away from Jupiter. These differences accumulated with time. Analyzing Cassini's observations, Roemer found that when the Earth was nearest to Jupiter, the eclipses of Io would occur about eleven minutes earlier than predicted based on the average orbital period over many years. And 6.5 months later, when the Earth was farthest from Jupiter, the eclipses would occur about eleven minutes later than predicted. The maximal difference was therefore 22 minutes.

Assuming that these variations in timing were caused by different retardation between the moments when the sunlight reflected by Jupiter's satellite started its travel towards the Earth and the reception of the same light by the observer, Roemer was able to determine the speed of light dividing the distance difference by the time difference. More precisely, light from the Jovian system has to travel farther to reach the Earth when the two planets are on opposite sides of the Sun than when they are closer together. Roemer estimated that light required twenty-two minutes to cross the diameter of the Earth's orbit.

Luckily enough, the radius of the Earth's orbit was determined independently by Huygens, Flamsteed, Cassini and his collaborator Richer a year before; the average obtained was 146 000 000 km (about 97.6% of the actual value 149 600 000 km). Inserting these data into the equation defining the speed of light, Roemer got the answer:

$$c = \frac{\Delta D}{\Delta t} = \frac{2 \cdot 146 \cdot 10^6 \, \text{km}}{22 \cdot 60 \, \text{sec}} = 221\,200 \, \text{km/sec}. \tag{14.9}$$

Although the speed of light determined by Roemer was still less than its real value by more than 25%, the order of magnitude was correct. It was so incredibly high, that most of Parisian astronomers, including Cassini, did not believe that it could be real. Especially those who were influenced by Descartes' view that light propagates with infinite speed. Roemer did not give up and predicted that the next occultation of Io by Jupiter will occur delayed by about 10 minutes. His prediction turned out to be true, but even then incredulity prevailed among the fellow astronomers, with the noticeable exception of Edmond Halley.

The speed of light remained controversial for more than half century after Roemer's brilliant estimate. The value close to the well known value 300 000 km/sec was obtained by English astronomer James Bradley (1693–1762) who discovered the phenomenon of *stellar aberration*. In 1728, using a new powerful telescope, Bradley tried to determine the annual parallax of the Polaris star, hoping to find out its distance from the Sun. Until then, stellar parallax eluded observation; in any case, it was clear since Tycho Brahe's fruitless attempts, that it must be less than one arcminute. In fact, during one year the Polaris star seemed to travel along an ellipse whose width was about 40 arcseconds. This was observed by the French astronomer Jean Picard (1620–1682) and the English astronomer John Flamsteed (1646–1719), but it was only Bradley who understood the phenomenon after discovering that all stars behaved in the same manner, and that the directions of deviations were the same at

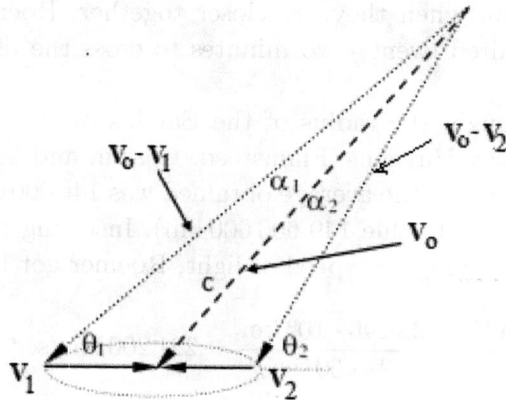

Fig. 14.8   Stellar aberration due to the Earth's orbital velocity. $\mathbf{V}_0$ is the vector parallel to the light ray from a star as observed in the reference frame of the Sun, $\mathbf{V}_1$ is the Earth's velocity in January; $\mathbf{V}_2$ is the Earth's velocity in June.

the same times of the year. The conclusion was that the aberration was due to the combination of the Earth's velocity on its orbit with the velocity of light. The vectorial addition of the two resulted in a vector slightly deviated with respect to the original direction towards a given star. Stellar aberration discovered by Bradley turned out to be one of the most precise methods for determining the mean radius of the Earth's orbit. The maximal departures from stars' mean positions were measured to be of 20.49″ (arcseconds) for stars situated in the direction perpendicular to the ecliptic. This is in fact the ratio of the Earth's speed V in its motion around the Sun, and the speed of light c arriving from a perpendicular direction: $\Delta\alpha = V/c$, expressed in radians. Knowing c we can determine V; if we express the length of tropical year T in seconds, then $V \times T = 2\pi R$ is the circumference of Earth's orbit (supposed circular for simplicity's sake), and the radius R of the Earth's orbit is given by

$$R = (V \cdot T)/(2\pi) = (\Delta\alpha \cdot c \cdot T)/(2\pi). \qquad (14.10)$$

With $\Delta\alpha$ and T well known, what remains to be determined as exactly as possible is the speed of light c.

The final estimate of the speed of light was due to the fine experiments performed by the French physicists Fizeau (1819–1896) in 1849–1850 and Foucault (1819–1868). The main idea was to use the stroboscopic effect, observing the light sent through a rotating gear, or a toothed wheel, to a mirror posted on a hill in the vicinity of Paris, at a distance of 8.613 km, and then reflected back along the same straight line. If the toothed wheel turned fast enough, the time the light needed to go forth and come back could be sufficient for the notch on the wheel to be replaced by the next tooth, and to shut the light signal down. The bright spot will give way to darkness. Knowing the distance $2L$ the light had to cover going to the faraway mirror and back, as well as the angular velocity ω of the wheel at which the occultation of light occurred, and the number of teeth N on wheel's circumference, the speed of light c could be easily found. A schematic representation of Fizeau's experiment is shown in Figure 14.9. The mathematical side of this experiment is extremely simple. Let the angular velocity of the toothed wheel be $\omega = d\theta/dt$; therefore $dt = d\theta/\omega$. In order to obstruct visibility when the reflected ray comes back, the wheel has to turn by the angle corresponding to one slot between its teeth, i.e. we must have $d\theta = 2\pi/N$, and $dt = 2\pi/(N\omega)$.

The distance between the semi-transparent mirror serving as a source of light and the distant reflecting mirror being equal to L, the time between the emission and return of the light ray is $2L/c$.

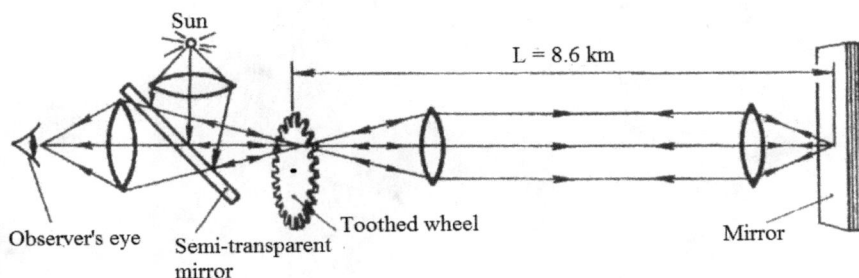

Fig. 14.9 Schematic representation of Fizeau's experiment determining the velocity of light.

Comparing the two expressions of dt we get

$$dt = \frac{2\pi}{N\omega} = \frac{2L}{c} \rightarrow c = \frac{N\omega L}{\pi}. \tag{14.11}$$

In the experiment performed by Fizeau the distance L was 8.663 km, the number of teeth N on wheel's circonference was 720, and the angular velocity of the wheel at which total obstruction of the reflected ray was observed was 25.2 rotations per second. Inserting these values into formula (14.11), Fizeau obtained $c = 313\,000$ km/sec, about 4% higher than what was established later with more accurate experiments. The present value of c is fixed at 299 752 km/sec.

We can now insert this value into the equation (14.10) and obtain the desired value of the average radius of Earth's orbit, i.e. the astronomical unit A.U., with the aberration angle determined by Bradley expressed in radians, the linear velocity of Earth on its orbit V and the speed of light c expressed in km/sec. We have

$$\Delta\alpha = \frac{20.49''}{57.296 \cdot 3600} = 9.934 \cdot 10^{-5}\,\text{rad}. \tag{14.12}$$

Multiplying this number by the speed of light $c = 299\,752$ km/sec and by the annual rotation period $T = 365.2422 \times 24 \times 3600 = 3.1557 \cdot 10^7$ sec, and dividing by 2π, we get the value of A.U. equal to 149 679 000 kilometers.

Chapter 15

Halley and the transit of Venus

Halley's early years - Astronomer, geographer, sailor - The southern sky - Friendship with Newton and Cassini - Halley's comet - Flamsteed's successor as Astronomer Royal - The transit of Mercury - Determining distance to the Sun - Halley's appeal - The post-mortem triumph.

15.1 Edmond Halley

Edmond Halley was born in London in 1656 as a son of a successful soap-maker. Since childhood he showed a great interest in mathematics, geography and astronomy, and was admitted to the Queen's College of the Oxford University at the age of 17. Still as an undergraduate, he published two articles on the sunspots and on the particularities of the solar system. He was introduced to John Flamsteed, who was at that time the Astronomer Royal, heading the newly constructed Observatory in Greenwich near London.

Influenced by Flamsteed's project to compile a catalogue of northern stars, Halley proposed to do the same for the stars of the Southern Hemisphere, which had not been observed at that time yet. Saint Helena island was the southern-most British territory then, so it was the destination chosen for this endeavour.

King Charles II sent a letter to the East India Company desiring that Halley be granted free passage to Saint Helena and, without bothering to take his degree and aged only 20, Halley undertook the journey to Jamestown on the Indiaman Unity.

Fig. 15.1   Left: Edmond Halley, portrait by Richard Phillips around 1720 (National Portrait Gallery, London); Right: The comet bearing Halley's name.

In November 1676 Halley sailed to the Southern Atlantic on a ship belonging to the East India Company. taking along a great specially constructed sextant of five-and-a-half-foot (165 cm) radius fitted with telescopes in place of sights (the "*alidades*"), his own two-foot quadrant, and several telescopes of different focal lengths up to 24 feet (720 cm). By the time he returned home in 1678 he had recorded the celestial positions of 341 stars, which he published in a star catalogue on his return to England, along with a chart of the southern heavens.

On November 7th 1677 he also became the first astronomer ever to observe the complete transit of Mercury across the solar disc, He then conceived the idea how to use the transit of Venus to evaluate the distance to the Sun.

In 1679 the young Edmond Halley (aged 23, but already recognized for his astronomical skills) was commissioned by his scientific masters Robert Hooke and John Flamsteed to pay a visit to Johann Hevelius in order to verify the veracity of his claims concerning the extraordinary precision of astronomical observations. Halley arrived at Hevelius' observatory in Danzig bringing a telescope of the latest construction, and spent a month performing astronomical observations side by side with Hevelius, who preferred observations by naked eye. The results were practically the same, and comforted Hevelius' obstination to neglect telescopic observations.

In 1682 Halley witnessed a great comet that appeared in September that year. Later, with Newton's help, he was able to calculate the parameters of its elliptical orbit, predicting its return in 1758 (which he was sure not be able to witness, being born in 1656).

Halley was so impressed by Newton's mathematical methods and by his theory of universal gravity, that he urged him to make them public. After some hesitation, Newton gathered his mathematical, physical and astronomical writings and handed them to Halley, who asked Cambridge University to publish them.

Unfortunately, at the time it happened, the University had almost no money left, because the year before it funded the publication of a very expensive illustrated "History of the Fishes". That year Halley did not get his wages, instead the University granted him with ten copies of the expensive book. Finally, Halley paid the publication of Newton's *Principia* in 1687 from his own pocket, thus taking part in one of the most important events in the history of science.

Edmond Halley was interested not only in astronomy; he was also a fine engineeer and mathematician. He brought important improvements to diving bells; he also invented a water bed for magnetic compass in order to make it insensitive to ships' swing and wobble at the open sea; In 1693 he published an article on life annuities, a major and seminal contribution to science of demography and finance.

In 1698, Halley was given command of the ship *Paramour*, with mission to carry out investigations in the South Atlantic into the laws governing the variation of the terrestrial magnetic field. He took command of the ship and, in November 1698, sailed on what was the first purely scientific voyage by an English naval vessel. The island Saint Helena was chosen to serve as the first destination and as a basis for further explorations.

Halley thereafter received a temporary commission as Captain in the Royal Navy, recommissioned the Paramour on August 24, 1699 and sailed again in September 1699 to make extensive observations on the conditions of terrestrial magnetism. This task he accomplished in a second Atlantic voyage which lasted until September 6, 1700, and extended from 52 degrees north to 52 degrees south. The results were published in *General Chart of the Variation of the Compass*

(1701). This was the first such chart to be published and the first one on which isogonic, or Halleyan lines, appeared. These are still in use today, in both meteorological (weather) maps and also in the topographic and oceanic cartography.

In 1718 Halley discovered the proper motion of stars, until then believed to be fixed for eternity. In articles published in Philosophical Transactions of 1694 and 1716, he proposed that the length of time taken by a planet to cross the Sun's disk should be observed at a number of suitably selected stations; from the differences in these times, the solar parallax can be inferred. The method has the advantage of not needing elaborate instrumental equipment, but it suffers from the disadvantage that it requires the visibility of both entrance and exit of a planet on the solar disc at the same station. The method was widely used at the transits of Venus of 1761, 1769, 1874 and 1882.

After John Flamsteed passed away in 1720 Halley succeeded him as Astronomer Royal, a position he held until his death in 1742.

15.2   The Halley comet

Halley's name is immortalized by one of his major astronomical discoveries: identifying the spectacular comet of 1682 as the same one that had been observed many times before, in particular in the years 1456, 1531 and 1607. By analyzing their trajectories in the sky and comparing them with the trajectory of the 1682 comet, Halley came to the conclusion that all these appearances represented the same celestial object, orbiting the Sun along an ellipse with a very high eccentricity, extending beyond the orbit of Saturn, but he lacked the adequate mathematical tools to be able identify the elliptical shape of the comets orbit and identify its period using Kepler's third law.

By a strange coincidence, another impressive comet appeared in 1680. Discovered through a telescope by Gottfried Kirch (1639–1710), the Royal Astronomer of the freshly inaugurated Berlin Observatory, it became visible soon after, and could be seen from Europe as one of the brightest comets of the 17th century. Isaac Newton observed and described it later in his *Principia*.

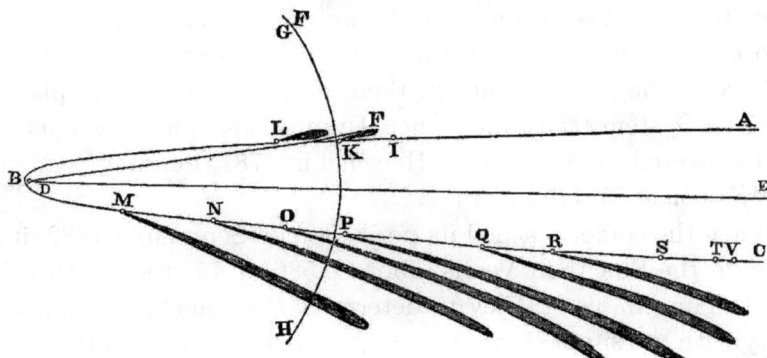

Fig. 15.2 The orbit of the comet that appeared in 1680, as calculated by Newton in his *Principia*.

He was able to determine its orbit's elliptical shape, and gave the right explanation of the phenomenon itself, interpreting the cometary tail as a cloud of gas and dust erupting from comet's head under the influence of the Sun heating the ice present in the comet and provoking the eruption of steam and dust visible in the Sun's rays.

After Halley convinced himself that the 1682 comet was the same object as several previous comets observed earlier, he conjectured that the length of its period must be about 76 years. But to be sure, a mathematical proof was needed. Halley asked Robert Hooke whether the inverse square law of attraction leads to elliptic trajectories, but did not get a convincing proof. Soon after that he paid a visit to Newton, who showed him the mathematical proof and the way to fix orbital parameters from the observation. With this in mind, Halley was able to get both the period and the size of the 1682 comet's orbit.

Applying Kepler's third law, it was easy to calculate the semi-major axis a_H of such an object. Its period $T = 76$ years and a_H are related by proportionality of their square and cube: $T^2 \simeq a^3$, as are the same parameters of other planets orbiting the Sun, The simplest thing is to compare the corresponding quantities for the Earth: $T_E = 1$ year and $a_E = 1$ Earth's orbit radius. One has then $76^2 : T_E^2 = a_H^3 : a_E^3$, therefore $a_H = (76)^{\frac{2}{3}} \simeq 18\, a_E$. On the other hand, the perihelion of Halley's comet was inside the Earth's orbit, about $0.6\,a_E$. This means that the aphelion must be at a distance almost twice the semi-major axis itself, between 35 and $36\,a_E$ — far beyond the orbit of Saturn, then the last and farthest planet of the Solar System (the next planet, Uranus, was not known yet — it was discovered by Sir William Herschel in 1781, hundred years after Halley's comet observations).

When the comet reached its perihelion on September 1682, it was closer to the Sun than Venus, about 0.586 of a_E, the Earth orbit's radius. This enabled Halley to determine the comet's orbital eccentricity with the simple formula $e = (a_H - 0.6)/a_H = 17.4/18 = 0.967$. Until then, such elongated orbits were unknown to astronomers. The enormous aphelion's distance suggested that Solar System extends itself far beyond Saturn.

Halley was also first to discover the peculiar motions of stars, finding that Arcturus and Sirius have moved significantly since Ptolemy's observations, the latter having progressed during 1800 years by 30 arcminutes (about half a degree, the diameter of the Moon) southwards.

15.3 Mercury's transits

In any model of the Solar System, be it Ptolemaic, Tychonic or Copernican, both Mercury and Venus are called the "inner", or "inferior" planets, because their orbits pass between the Earth and the Sun. This statement is true in the Ptolemaic system, in which all celestial bodies turn around the Earth, the Moon's orbit being the closest, the orbits of Venus and Mercury being the next, then the Sun, and finally the outer planets, Mars, Jupiter and Saturn. It remains true in the Tychonic system, with the Earth in the center, the Moon and the Sun revolving around it, the Moon's orbit being closer to the Earth than that of the Sun. The inner planets were supposed to revolve around the Sun, while the orbits of Mars, Jupiter and Saturn were far above the orbit of the Sun and the orbits of inner planets accompanying it. The Copernican system turned out to be by far the closest to reality: all planets, including our Earth, revolve around the motionless central Sun, Mercury the closest, then Venus, the Earth as third with the Moon orbiting around it, then Mars, Jupiter and Saturn; the steady sphere of stars assumed to be very far away.

The inner planets are periodically in *conjunctions*, the superior one when a planet is on the other side of the Sun, practically invisible due to the daylight, and the inferior one, when it passes between our planet and the Sun. They are never in *opposition*, i.e. observed in the place of the sky opposite to the Sun, in the midst of the night. The oppositions occur only for the outer planets, Mars, Jupiter and Saturn (for Uranus and Neptune too, but these two planets were unknown before the end of the 18th century).

The orbits of Mercury and Venus are inclined with respect to the ecliptic plane: among the Solar System planets Mercury's orbit has the largest inclination equal to 7°, while Venus' orbit is inclined by 3°24′ with respect to the ecliptic. The seven angular degrees can cover slightly more than 14 solar discs; and the 3.4° of Venus orbit's inclination can contain seven solar discs. This is why most of the time Mercury and Venus pass over or under the solar disc during their lower conjunction.

The *transit*, i.e. passing of a planet before the solar disc, comparable to a mini-eclipse, can occur only when one of the inner planets passes between us and the Sun in one of the two *nodes*, the points at which the planes of their orbits cross the plane of the ecliptic. The frequency of such an event is not very high — in fact, much lower than the solar eclipse by the Moon, mostly because of much longer periods of revolution, as compared to lunar synodic month.

At first glance one would expect transits of Venus to be more frequent than Mercury's ones, but the reality is just the opposite: the transits of Mercury occur about 13 times per century, whereas the transits of Venus occur about 13 times per millenium, i.e. they are ten times less frequent, in spite of Venus being on the average closer to the ecliptic than Mercury. The most obvious reason is that Mercury's sidereal period is $T_{Merc} = 88$ days, and its synodic period is $T_{Merc}^{syn} = 116$ days, whereas Venus' sidereal period is $T_{Ven} = 225$ days, and its synodic period is $T_{Ven}^{syn} = 587$ days.

Johannes Kepler was the first to foresee the possibility of Mercury's transit in front of the solar disc. The first astronomer who observed this phenomenon when it occurred on November 7, 1631, was Pierre Gassendi (1592–1655). The apparent diameter of Mercury was surprisingly small, which made Gassendi conclude that the Earth was much more distant from the Sun than it was usually believed in his time — in fact, Aristarchus' estimate of the solar size and distance was still considered to be valid.

Nowadays, all transits of Mercury fall within several days of May 8 and November 10. Mercury is near its perihelion in November, and its disk is only 10 arcseconds in diameter. But during May transits Mercury is near aphelion — far from the Sun, therefore closer to Earth, and appears 12 arcseconds across. It is still much smaller than the disc presented by Venus during its rare transits.

15.4 The transits of Venus and solar parallax

There are four major reasons to prefer a transit of Venus over Mercury's transit as a mean to measure true distances in the Solar System. The first is the quasi-circular shape of the orbit of Venus (eccentricity $\simeq 0.0068$, less than 1%), while the orbit of Mercury is an ellipse with an eccentricity equal to 0.206, which makes all trigonometric calculations more complicated.

Secondly, the parallax is inversely proportional to the distance, and the distance to Mercury during transit is almost twice the distance from Earth to Venus in similar circumstance: taking the average distances to the Sun $D_M = 0.4$ for Mercury, $D_V = 0.7$ for Venus and $D_E = 1.0$ for Earth, we get the distance from the Earth in lower conjunction (when the transit occurs) to be $L_E - L_M = 0.6$ for Mercury, and $L_E - L_V = 0.3$ for Venus.

The third reason is that not only Venus is closer to us, but it is also bigger than Mercury, so that it is fairly visible as a round black spot when it crosses the solar disc, while Mercury appears as a tiny dot, hardly noticeable. Finally, the apparent motion of the spot is much slower in the case of Venus, which, as we shall see, is very important for taking the exact measure of the solar parallax.

Halley realized that the total time of a transit provides precious data that can be transformed into angular variables, then into

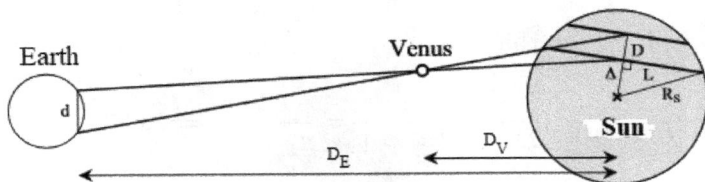

Fig. 15.3 Transit of Venus and the corresponding parallax scheme.

distances using the Thales' theorem if the size of one of the sides of the triangle joining two different observers to Venus is given. This basis should be as large as possible, preferably spanning a line along the North-South direction. Halley's method is illustrated by the scheme in Figure 15.3.

Let us denote the distance from the Earth to the Sun by D_E, and the distance from Venus to the Sun by D_V. Then the distance from Earth to Venus in transit is denoted by $D_E - D_V$.

Consider two observers located at two points of the terrestrial globe at distance d (being as close to the same meridian as possible), observing the transit of Venus roughly at the same time. Due to the tiny difference of angle under which they see Venus (caused by the parallax phenomenon), the paths of Venus they observe on the solar disc will represent two slightly different chords of the disc. Let us denote by D the distance between the two paths; it is strongly exaggerated in Figure 15.3, the real distance being hardly visible if the scale was respected.

By Thales' theorem we have $D : d = D_V : (D_E - D_V)$. Instead of distances which are still unknown, we can use sidereal periods of the Earth and Venus, exploiting Kepler's third law. Let us call these periods T_E and T_V; then we have

$$\frac{D_E^3}{D_V^3} = \frac{T_E^2}{T_V^2}, \quad \text{therefore} \quad \frac{D_E}{D_V} = \left(\frac{T_E}{T_V}\right)^{\frac{2}{3}}. \tag{15.1}$$

Now the dimensionless quantity D/d can be expressed in terms of dimensionless ratio of sidereal periods $\frac{T_E}{T_V} = \frac{365}{225} = 1.622$ and

$$\left(\frac{T_E}{T_V}\right)^{\frac{2}{3}} = 1.381, \text{ so that}$$

$$\frac{D}{d} = \frac{D_V}{D_E - D_V} = \frac{1}{\frac{D_E}{D_V} - 1} = \frac{1}{\frac{T_E}{T_V}^{\frac{2}{3}} - 1} = 2.627. \qquad (15.2)$$

Now we can realize how difficult the measurement of distance between the two paths would be if the estimate was made just optically. Although no consensus was achieved yet concerning the real size of Solar system, it was inferred from measures of parallax of Mars in opposition that the so-called *solar parallax*, i.e. the angle under which the Earth's radius is seen from the Sun, must be contained between 8 and 20 arcseconds, corresponding to distance between 160 million km to 65 million km. Newton hesitated between values of the solar parallax varying from 10 to 20 arcseconds.

Now, in the shortest version (and lesser solar diameter) the Sun would be at least 40 times bigger than Earth. If so, supposing that we succeed to observe Venus' transit from two places on Earth situated on the same meridian, distant by $d = 5000$ km from one another. Then the separation D between the two paths of Venus on the solar disc would be

$$2.627 \times 5000 \, \text{km} = 13\,135 \, \text{km}, \qquad (15.3)$$

close to the Earth's diameter. The diameter of the Sun being at least 40 times the diameter of Earth, the separation D would represent $1/40$ of solar disc's diameter, or in angular units, less than $0.8'$. By the time Halley was elaborating his method, a more reliable estimate of astronomical unit was obtained by French team headed by the French astronomer of Italian origin Jean-Dominique Cassini (1625–1712) and including the Danish astronomer Ole Rømer (1644–1710), Jean Richer (1630–1696) and Jean Picard (1620–1682). The idea, described in the previous chapter, was to get precision measurements of the stellar coordinates of Mars which was in opposition in October 1672. The distance to the Red Planet was then at its minimum, about 0.38 of the distance between the Sun and Earth.

Comparing the observations made at the same time in Paris and in Cayenne (French Guiana in South America) astronomers were able to estimate the parallax of Mars being close to $25''$, i.e. almost a half arcminute, and consequently, the solar parallax close to $9.5''$. This would correspond to the Earth-Sun distance equal to 21 600 Earth's radii, or 138 million kilometers. The radius of the Sun in this case is estimated to be 100 times the Earth radius, the separation D between two paths of Venus as seen from two terrestrial observers distant by 10 000 km less than 1/50th of solar disc, making the measurement quite hopeless. Especially that there was no way to transmit the results of the observations in real time from one point on the Earth to another at distances as great as 5000 km, and photography was not yet invented.

Halley found how to determine the tiny angle D without measuring it directly, but estimating with great precision the times of essential phases of transit shown in Figure 15.4. There are four clearly distinguishable moments of transit: the first outer contact (1), then the moment when Venus penetrates the solar disc entirely (2), then the inner contact with the solar disc (3) and the last outer contact (4). The dark disc of Venus is not to scale in Figure 15.3, in reality its angular diameter at its maximum is about 1 arcminute, and it looks like a small dark spot very slowly crossing the Sun. It is this very slowness of Venus' motion that was brilliantly used by Halley to propose an indirect method of evaluating the tiny angular size of Venus' parallax D.

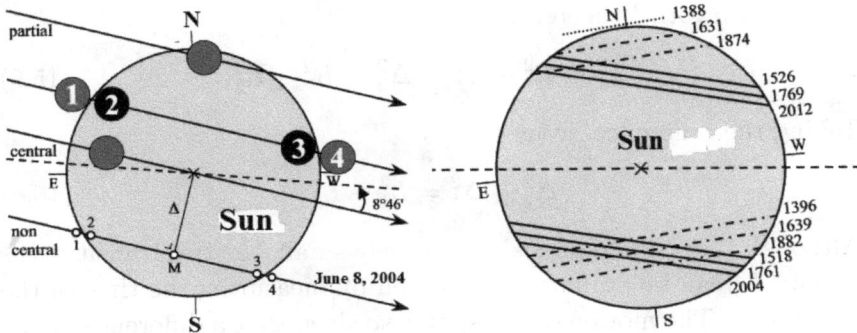

Fig. 15.4   Left: Venus transit scheme; Right: Recent transits of Venus.

In order to determine the visible velocity of Venus' transit across the Sun, one must use this planet's synodic period, which equals 583.92 days. As seen by terrestrial observer, the velocity in angular minutes is

$$v = \frac{\frac{D_V}{D_E - D_V} \times 360 \times 60}{583.92 \times 24} = \frac{2.627 \times 360 \times 60}{583.92 \times 24} \simeq 4'/\text{h}, \qquad (15.4)$$

which is 4 arcseconds per minute. The total duration of transit of Venus in minutes is found by dividing the angular length of its path on the solar disc by this angular velocity. It is maximal when the transit is central, i.e. when Venus has to cover the full diameter of solar disc. Then the duration is $(30')/(4'/\text{hour}) = 7.5$ hours. A central transit is extremely rare; the next transits were to occur in 1761 and 1769, were not central, which made them better adapted for Halley's method.

[* Let us analyse closer the geometry of the transit. Two apparent parallel paths of Venus crossing the solar disc are represented, as seen by two terrestrial observers, supposedly located at two faraway points on the Earth. The angular distance (exaggerated in our figure) between the two paths is denoted by D. Let us denote by Δ_1 and Δ_2 the respective distances of the corresponding paths from the center of the solar disc, so that $D = \Delta_2 - \Delta_1$. Let us denote by $2L_1$ and $2L_2$ the lengths of the respective transit paths, $L_2 < L_1$; let the radius of solar disc be $R \simeq 16'$. Then it is easy to express angular distances Δ_1 and Δ_2 as functions of L_1 or, respectively, L_2, and R, using Pythagoras' theorem:

$$\Delta_1^2 = R^2 - L_1^2, \quad \Delta_2^2 = R^2 - L_2^2. \qquad (15.5)$$

Taking the difference, we get

$$\Delta_2^2 - \Delta_1^2 = L_1^2 - L_2^2. \qquad (15.6)$$

Although these differences are extremely small, the two transit paths L_1 and L_2 can be evaluated fairly well by measuring the time of the full transit. The motion of the spot is so slow, that a difference of one minute between times of total transit corresponds to 4 arcseconds, and with a good chronometer it would be possible to fix the times of first and last contacts within much greater precision, up to a few seconds of time.

Suppose then that L_1 and L_2 have been determined; then we can find $\Delta_1 = \sqrt{R^2 - L_1^2}$ and $\Delta_2 = \sqrt{R^2 - L_2^2}$. Taking the difference, we get the value of D expressed in arcseconds:

$$D = \Delta_1 - \Delta_2.$$

If the distance between the observers on Earth was d, we also have $D = 2.627 \times d$ in kilometers by virtue of (15.2), and finally, dividing D in kilometers by D in angular variables (the arcseconds should be then expressed in radians), obtain the Earth-Sun distance D_E.

Let us follow Halley's computations on a concrete example. A reasonable assumption about d, the distance between the two observers of transit phenomenon, is 5000 km — as projected on plane perpendicular to the line of observation; the distance on Earth's surface might be greater, up to 7000 km. So let $d = 5000$ km. The observer closer to the North Pole will see the lower path L_2, which in this case is longer, while the observer in southern hemisphere will see the upper path L_1, slightly farther from the center of solar disc, and therefore a bit shorter.

All distances on the solar disc will be expressed in angular units, arcminutes and arcseconds. Suppose that the total time of transit along the shorter path L_1 was $t_2 = 6$ h 20 min., while the transit observed from the Northern observatory on Earth took $t_1 = 6$ h 27 min. The corresponding angular lengths of transit paths are:

$$L_1 = 380 \min \times 4''/\min = 25.33', \quad L_2 = 387 \min \times 4''/\min = 25.8'.$$

therefore $L_1 = 12.665'$, $L_2 = 12.90'$. Using these data and by virtue of Pythagoras' theorem applied to the two cases following (15.5) and (15.6) we get, recalling that $R = 16'$:

$$\Delta_1 = \sqrt{R^2 - L_1^2} = 9.465', \quad \Delta_1 = \sqrt{R^2 - L_2^2} = 9.778',$$

$$\Delta_2 - \Delta_1 = 0.313', \tag{15.7}$$

an extremely small angle unaccessible to direct measure by optical means available at Halley's time (and still long after that, too!).

This is the angle under which $D = 13\,135\,\mathrm{km}$ (see Figure 15.3), the distance between the two transition paths on the Sun's surface as seen from the Earth. In order to find the Earth to Sun distance D_E, one should express the corresponding angular size in radians:

$$0.313' \rightarrow \frac{2 \cdot \pi \cdot 0.313}{360 \cdot 60} = 9.105 \cdot 10^{-5}.$$

$$D_E = \frac{D\,\mathrm{km}}{(\Delta_2 - \Delta_1)\mathrm{rad}} = \frac{13\,135\,\mathrm{km}}{9.1 \cdot 10^{-5}} = 144\,340\,660\,\mathrm{km},$$

(15.8)

which is quite realistic according to the present knowledge: the average distance from the Sun is given as 149.6 million km. *]

15.5   Final estimate of the astronomical unit

The history of astronomy is also a history of expanding the limits and the size of our Universe. The distance to the Moon was determined by Aristarchus with reasonable precision; however, his method of evaluating the distance to the Sun was much less reliable: his estimate was only 5% of the real value of the astronomical unit (A.U.) universally accepted since more than two hundred years. The average distance to the Sun remained a constant challenge for astronomers since the first estimates made by the ancient Greeks.

On the occasion of the great Martian opposition of September 1672, several measurements of the parallax of Mars seen from distant places on Earth (French Guyana, Paris, London) were performed by French and English astronomers. The three last items of the Table 15.1 are based on the results of their expeditions. The idea was simple, the realization rather complicated. When Mars is in opposition, especially the great one, its distance from Earth is about 0.38 A.U. Observing Mars from different positions on Earth, as far from each other as possible, one can expect to discover the shift in Mars' positions with respect to the fixed stars due to the parallax effect.

A detailed analysis of the results was presented in the previous chapter; here let us only remind that the parallax angle of 10 arcseconds resulting from the observations of a celestial body from places separated by 6000 km corresponds to a distance of about

Table 15.1   The estimates of the distance to the Sun, from Anaximander till the 17th century.

Astronomer	Year	Parallax	Distance, e.r.	Distance, km
Anaximander	$\simeq$560 B.C.E.	$\simeq$1.06°	$\simeq$54	$3.44 \cdot 10^5$
Eudoxus	$\simeq$360 B.C.E.	$\simeq$25'	$\simeq$150	$1.15 \cdot 10^6$
Aristarchus	$\simeq$250 B.C.E.	9.55'	360	$1.8 \cdot 10^6$
Hipparchus	$\simeq$150 B.C.E.	1.38'	2490	$1.6 \cdot 10^7$
Posidonius	$\simeq$90 B.C.E.	15.76''	$13.1 \cdot 10^3$	$8.35 \cdot 10^7$
Ptolemy	$\simeq$170 C.E.	2.84'	1210	$7.7 \cdot 10^6$
Copernic	1535	2.29'	1500	$9.57 \cdot 10^6$
Kepler	1607	$\leq$1'	$3.45 \cdot 10^3$	$\geq$2.2 $\cdot 10^7$
Huygens	1670	8.6''	$2.3 \cdot 10^4$	$1.48 \cdot 10^8$
Cassini	1672	9.5''	$2 \cdot 10^4$	$1.3 \cdot 10^8$
Flamsteed	1672	10''	$1.9 \cdot 10^4$	$1.24 \cdot 10^8$
Picard	1672	20''	$9.3 \cdot 10^3$	$7.6 \cdot 10^7$

124 million km, which is quite close to the estimates of the A.U. made previously by Huygens on the basis of direct measurements of the Venus' angular size. And Mars being much closer, one could expect the parallax at least twice as big, more than 20 arcseconds, which posed no problem for the telescopes of that time.

Halley was convinced that his method, based on the time of the transit instead of the direct parallax angle measurements, would lead to much better estimates of the astronomical unit. The precise calculations predicting the transits of Venus in the years 1761 and 1769 were presented in his dissertation published by the Royal Society in 1678. In his "advert", published in 1716, he exhorted the astronomers of the next generation to make the necessary observations along the lines he proposed, establishing better estimates of the solar parallax and consequently, the true distances in the Solar System. Being born in 1656 Halley could not expect to live long enough (in 1761 he would have been 105 years old!) to take part in the observations of a transit of Venus.

Halley's appeal was widely followed: many expeditions were organized to observe the transit of June 6, 1761, among others the French astronomers Le Gentil and Pingré who went to an Indian Ocean island, M. Lomonossov in Arkhangelsk, Russia; Cassini in Vienna...

In total, 62 different places were involved, and 120 separate observations made. The points on the globe from which the observations had to be made were chosen in a way to extend the difference in latitudes as much as possible, from South Africa and Madagascar to Siberia and St. Petersburg in Russia and Newfoundland in Canada.

The estimates of solar parallaxes were found to be between 8.5″ and 10.6″ (arcseconds), corresponding to astronomical unit estimates between 125 and 155 million kilometers, which did not represent a real progress with respect to previous estimates based on the measures of Martian parallax. Nevertheless, an important new discovery was made: the black spot travelling across the solar disc was surrounded by a thin aureole, which meant that Venus is endowed with an atmosphere.

The next international cooperation of observations of Venus' transit expected on June 3, 1769, with an even greater number of participants involved, was also better prepared. In total, 177 observations were made, at 76 different points, maximizing latitude differences. The French sent several expeditions, in particular to the Indian Ocean (Le Gentil, who unfortunately could not perform any measurements), Caribbean Islands (Pingré) and Mexico (Father Chappe d'Auroche), performing in parallel observations in Paris. But the most famous adventure was Captain Cook's expedition, commissioned and sponsored by the Royal Academy.

More than six months before the expected transit of Venus, in August 1768, His Majesty's Bark "Endeavour" under Lt. Cook's command weighed anchor to sail to Tahiti, an island in the midst of the Pacific Ocean, discovered only one year before by the French captain Bougainville. There were 94 men on board, including the young naturalist Joseph Banks and the astronomer Charles Green.

The expedition was long and risky, the latitude of Tahiti was well established (17.5° South), but its longitude was much less precisely known. Besides the observation of the transit of Venus, it had also another goal, an experimental test of a remedy against scurvy, the desease that caused serious problems including deaths among sailors after more than 6 weeks at sea.

Fig. 15.5   Left: Captain James Cook; Right: The H.M. bark "Endeavour".

Cook understood that the main reason was the diet deprived of fruits and vegetables. He took on board a great amount of sauerkraut which served as replacement for vegetables and could be conserved for many months without loosing its beneficial qualities (vitamins were unknown at the time, but Cook's intuition was right: the crew remained healthy after 8 months at sea, when the Endeavour finally arrived to the shores of Tahiti on April 1769, with only six men lost).

They were received with hospitality by the population and the King Tàrroa, and disposed of plenty of time to prepare for observations. The weather on June 6 was perfect, the sky almost cloudless, and both Cook and Green were able to measure the time between various stages of transit, and make drawings (Figure 15.6), confirming the presence of an atmosphere on Venus.

Before returning to England, Cook's expedition explored the shores of New Zealand and Australia. On the way back they stayed 10 weeks in Batavia (today's Jakarta), where 15 crew members, including the astronomer Green, died of tropical fever. The Endeavour was back on July 11, 1771.

The results of 1769 campaign were much closer to the final determination of the Sun's parallax and of the astronomical unit that resulted from these data. The average of the parallaxes measured by the Halley's method was now contained in a narrow slot between 8.5″ and 8.9″, which corresponds to an Earth-Sun distance between 155 and 145 million kilometers, or a mean value 150 ± 5 million kilometers.

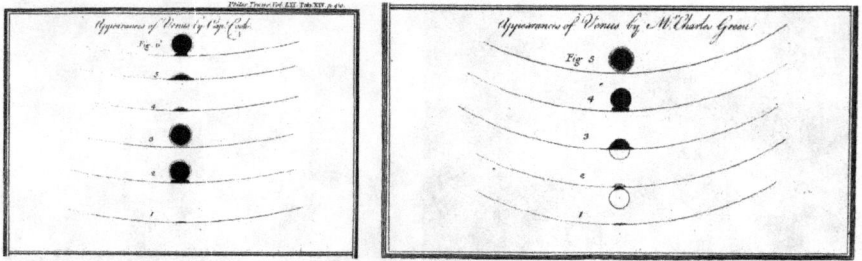

Fig. 15.6 Two sketches of Venus transit observations made on June 3, 1769 by captain James Cook (left) and astronomer Charles Green (right).

Although Halley expected greater precision still, this was probably the limit realizable without use of precision clocks and better telescopes.

Today the officially admitted value of astronomical unit is 149 598 000 km, and the average solar parallax is 8.79″.

Chapter 16

Cavendish and the Earth's mass

The mass puzzle - Henry Cavendish's life - Contribution to chemistry - The description of torsion balance - Cavendish's experiment - Determining the force of gravity - The Earth's average density - The masses of planets, the Sun and the Moon

16.1 Preamble

The experiment intended to determine Earth's average density, and as its corollaries, the Earth's mass and the exact value of Newton's gravitational constant, was performed by Henry Cavendish in Cambridge in 1797–1798. So it took slightly more than 70 years after Newton's death to endow his law of universal gravity with a full predictive power. Just as Kepler's laws for planetary orbits, Newton's law of universal attraction was scale invariant. The heliocentric system with elliptical orbits and planets pursuing their trajectories with periods satisfying Kepler's third law can be realized to any scale without betraying the essential features of the model. Similarly, the gravitational interaction between any two masses can be compared with similar interaction between any other pair, but only the ratio of masses will come out of the comparison, not the masses themselves — until at least one of them is known via some independent measure.

The Kepler's third law is the best way to compare the masses of heavy central bodies endowed with satellites orbiting them. In the Solar System only Mercury and Venus do not possess natural satellites; all other planets do: the Earth one, Mars two, Jupiter

Table 16.1   The estimates of MG for the Sun and the major planets based on their satellites' semi-major axis and periods.

	Satellite	a (in km)	T (in days)	$MG = 4\pi^2 a^3 T^{-2}$
Sun	Earth	$1.49 \cdot 10^8$	365.24	$1.327 \cdot 10^{11}$
Earth	Moon	$3.84 \cdot 10^5$	27.32	$4.0 \cdot 10^5$
Mars	Phobos	$9.38 \cdot 10^3$	2.76	$4.29 \cdot 10^4$
Jupiter	Io	$4.217 \cdot 10^5$	1.77	$1.267 \cdot 10^8$
Saturn	Titan	$1.222 \cdot 10^6$	16.0	$3.82 \cdot 10^7$

and Saturn more than ten satellites each. In order to compare the masses of the planets with satellites and the Sun, we should apply the formula expressing the third law to each case for which the semi-major axis and period of rotation are known.

Table 16.1 provides the comparison between the semi-major axes and periods of several planets and their moons and the similar data for the Earth and Sun. Following Kepler and Newton, the combination $MG = 4\pi^2 a^3 T^{-2}$ is proportional to the mass of the central body. In the last column, this quantity is computed with a given in kilometers, the period T being expressed in seconds.

These data give us the possibility of comparison between masses of central bodies, but they do not give their absolute value in kilograms. However even without knowing the real values of those masses, we can get a lot of interesting information concerning just their ratios. The first and the most striking one being the extraordinary small-ness of planetary masses with respect to solar mass. The mass of Jupiter, by far the heaviest planet of the Solar System, is by *three orders of magnitude* smaller than the mass of the Sun: the exact ratio is $M_J/M_S = 0.955 \cdot 10^{-3}$, which means that Jupiter is about thousand times less massive than the Sun. Moreover, adding the Saturn's mass, which is the second giant planet after Jupiter, does not really change the situation: the ratio between the masses of the two planets taken together and the solar mass remains still as small, just $1.25 \cdot 10^{-3}$. Adding the masses of smaller planets of terrestrial type (the four planets closest to the Sun) will change only the second digit of this ratio.

We can also find out that Mars is not only smaller than the Earth, but its mass is even smaller than it could have been expected from the comparison of their radii: in fact, Mars is almost ten times lighter than our planet.

But the ratios between masses are the only information we can get form astronomical observations as long as we are lacking a "measuring rod" which could be the mass of any planet — but how to "weigh" a planet? This was done by Henry Cavendish who found the way how to determine Earth's average density, and therefore its mass, too.

16.2 Henry Cavendish

Henry Cavendish, one of the most brilliant minds of the 18th century, was born in 1731. He lost his mother at the age of two, and was raised with his baby brother by his father Charles Cavendish, who was interested in politics and in science, and was member of the Royal Society of London. After Henry completed his studies in Cambridge, his father introduced him to the Royal Society, to which Henry was elected in 1760, and where he served in many committees until his death in 1810. He was member of the committee for the assessment of Royal Greenwich Observatory, and of the committee supervising publications in "Philosophical Transactions", the journal of Royal Society, and yet another for the transit of Venus in 1769.

His scientific interest covered physics and chemistry and electricity, to which he made major contributions. His major contribution to chemistry was the discovery of hydrogen (which he called "inflammable air") and of carbon dioxide (which he called "fixed air"). Hydrogen was obtained by the action of acids on metals, and carbon dioxide was the result of action of acids on alkalis (soda). He was also able to prove that air is made of two component, the "dephlogistocated air" which we now call oxygen, and "phlogisticated air" (nitrogen). The names were related to the theory of heat based on the presence of active agant called the "phlogiston", responsible for combustion.

Cavendish was famous for precision of his measurements. After separating oxygen and nitrogen from the atmospheric air, he found a gaseous residue which he estimated to occupy about 1/120th part of nitrogen's volume. He was not sure what it was, but later on another English physicist, Lord Rayleigh (1842–1919), identified this residue as the chemically inert gas argon. However the best known achievement of Henry Cavendish was the experimental measurement of the tiny force of gravity between masses in a laboratory, which resulted in an astonishingly precise estimate of the Earth's average density and mass, and as a corollary, in fixing the value of Newton's gravitational constant G.

16.3   The torsion balance experiment

The experimental proof of existence of the gravitational attraction between masses comparable with everyday life objects down on the Earth and not only between celestial bodies required an extremely sensitive device. In principle it could be just a small mass suspended on a thin light thread which at rest is strictly vertical due to the action of Earth gravity. Then it would be sufficient to bring a huge sphere of radius a and mass M as close as possible, and observe the small departure from verticality due to the tiny horizontal pull created by the gravitational force coming from the mass M. Let the distance between the suspended mass m and the center of the heavy sphere be d (which must be bigger than big sphere's radius a) (see Figure 16.1). The small mass m is subject to two gravitational forces: the terrestrial gravity $\mathbf{g}$ pointing downwards, and the horizontal pull $\mathbf{f}$ directed towards the center of mass M. The resulting vector sum $\mathbf{g} + \mathbf{f}$ is not vertical anymore, due to the tiny horizontal component $\mathbf{f}$, so the new equilibrium position of the pendulum would display a departure from verticality, by a tiny angle $\alpha = \frac{f}{g}$. (for very small angles the tangent can be replaced by the angle expressed in radians).

  [* Let us evaluate whether such an effect can have any chance to be observed and measured. According to Newton's theory, gravitational forces g and f are given by the following formulae:

$$g = \frac{M_E G}{R_E^2}, \quad f = \frac{M G}{d^2}, \tag{16.1}$$

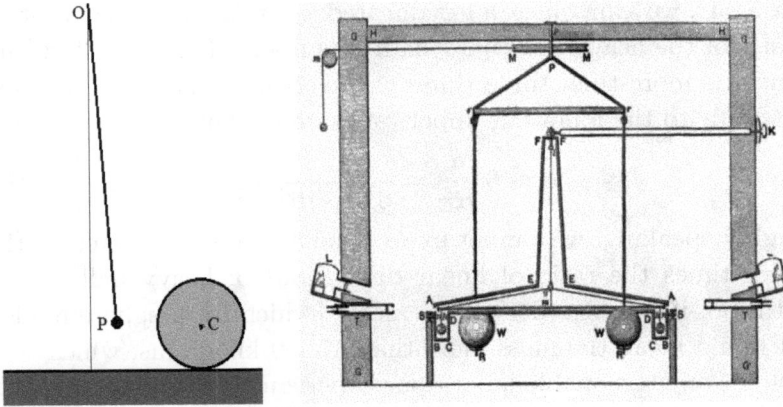

Fig. 16.1 Left: Imaginary experiment with simple pendulum; Right: Torsion balance constructed by Henry Cavendish.

with M_E the mass of the Earth, M the mass of the heavy sphere, $R_E = 6.37 \cdot 10^8$ meters Earth's radius, and d the distance between the centers P and C of masses m and M, respectively. The angle by which the pendulum is deflected is given by:

$$\alpha = \frac{f}{g} = \frac{M}{M_E} \frac{R_E^2}{d^2}. \qquad (16.2)$$

We can eliminate the unknown mass of the Earth M_E and replace it by (also unknown yet) average density ρ_E. Let the density of spherical mass M be ρ_B. Then both spherical masses, M_E and M can be expressed using average densities,

$$M = \frac{4\pi a^3 \rho_B}{3}, \quad M_E = \frac{4\pi R_E^3 \rho_E}{3}. \qquad (16.3)$$

Replacing in (16.2) masses M and M_E by these expressions, we get

$$\alpha = \frac{f}{g} = \frac{\rho_B a^3}{\rho_E R_E^3} \frac{R_E^2}{d^2} \simeq \frac{\rho_B}{\rho_E} \frac{d}{R_E}. \qquad (16.4)$$

In the final formula we replaced the big ball's radius a by distance between its center and the test mass m, because they are not only of the same order of magnitude, but actually do not differ more than by 20%. Another unknown parameter is the ratio of densities ρ_B/ρ_E;

here again we can make an "educated guess", supposing that the density of the heavy ball ρ_B, even if it is made of lead $(11.34\,\mathrm{g/cm}^3)$ cannot be more than three times the average density of our planet. This leads to the following upper estimate of the angle α:

$$\alpha \simeq 3\,\frac{d}{R_E} = \frac{3d}{6.37 \cdot 10^6\,\mathrm{m}}. \tag{16.5}$$

Roughly speaking, we cannot expect the angle α to be greater than several times the ratio of linear dimensions of heavy ball and the Earth. So, if d is of order of one meter (which for a sphere made of lead would mean the mass more than $47\,000$ kilograms, which is too much for any reasonable laboratory experience), the angle α will not bypass $(1\,\mathrm{m})/(6.378 \cdot 10^6\,\mathrm{m})$, which is less than $1.5 \cdot 10^{-7}$ radians, or 0.03 arcseconds, definitely too small to be observed and measured. *]

The only way to measure gravitational pull that weak was to get rid of Earth's gravitational influence, making the horizontal pull of heavy ball the unique external force to be measured. Cavendish has found a solution that was simple and brilliant at the same time. Instead of suspending one small ball on a vertical wire, he suspended a horizontal bar by its center, and placed two identical balls at its ends, creating a *torsion balance*, as shown in Figure 16.1 on the right. After the suspended rod with masses at its ends arrives at the equilibrium position, two heavy leaden spheres are brought in the vicinity of small masses on both sides. The joint gravitational pull f on both sides provokes a weak momentum of forces tending to turn the balance so as to approach the small masses m towards the heavy balls attracting them gravitationally. The torsion balance used by Cavendish was designed and produced by a fellow scientist and a life-long friend John Michell (1724–1793), whose untimely death stopped the experiment that was resumed by Cavendish five years later.

The torsion wire resists the dynamical moment of forces applied to it, and after several small oscillations a new equilibrium position will be attained. The small angle between the initial and final positions of the balance is proportional to the torque of the wire balancing the moment of two gravitational forces, thus enabling its measurement if the rotational elastic coefficient k of the wire is known. This coefficient should be as small as possible in order to increase the sensitivity

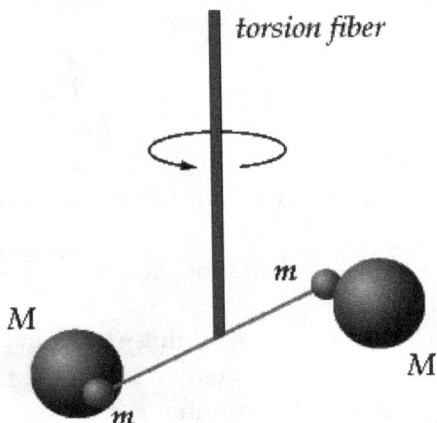

Fig. 16.2   The simplified scheme of Cavendish's experiment with torsion balance.

of the torsion balance. Figure 16.2 describes the experiment in a simplified manner.

Let us now present the detailed description of how Cavendish proceeded, in three stages:

• **1** Calibrating the torsion wire by determining its elastic torsion constant k.

Let the length of the horizontal bar of the balance be L. Let the gravitational pulls of heavy spherical masses applied to the small masses at the ends of the bar be denoted by f. They are perpendicular to the bar and act in opposite directions. The moment of forces created is therefore

$$\mu = 2 \times \frac{L}{2} \times f = Lf. \qquad (16.6)$$

When a moment of forces μ is applied to a torsion wire, the latter responds with a counter-moment, or *torque* τ. For small torsion deformations, measured by an angle θ, the linear dependence is valid: $\tau = -k\theta$. The minus sign tells us that the torque *opposes* the angular deformation of the wire. When the two moments cancel each other, the balance arrives at a new equilibrium state, with angle θ taking the value θ_0:

$$\mu + \tau = 0, \quad Lf - k\theta_0 = 0, \quad f = \frac{k\theta_0}{L}, \tag{16.7}$$

and this will give the exact measure of gravitational pull we want to determine. The force f is, according to the Newton's law of gravity,

$$f = \frac{mMG}{d^2} \quad \text{therefore} \quad k\theta_0 = \frac{mMG}{d^2}L. \tag{16.8}$$

This relation can serve for determining the value of the unknown gravitational constant G, or by comparison with the Earth's gravitational pull at its surface, determining the ratio between the known mass M and the unknown mass of the Earth M_E.

[* What is needed in the first place is the precise value of the elastic torsion coefficient k. To determine it, Cavendish used the equation relating it to the period of small oscillations of the balance. The rotational motion of the balance bar with masses at its ends under the action of a given moment of forces is described by the differential equation which is the angular analog of the fundamental equation of Newtonian mechanics, $F = ma$, where the force F is replaced in this case by the moment of forces δ, the linear acceleration by the angular acceleration, and the mass m by the moment of inertia I:

$$I\frac{d^2\theta}{dt^2} = \delta. \tag{16.9}$$

The moment of inertia of a light bar of length L with two equal masses m at its ends, suspended by the center, is equal to

$$I = 2 \times \left(\frac{L}{2}\right)^2 m = \frac{mL^2}{2},$$

so that our differential equation becomes

$$\frac{mL^2}{2}\frac{d^2\theta}{dt^2} = -k\theta, \quad \text{or} \quad \frac{d^2\theta}{dt^2} + \omega^2\theta = 0, \quad \text{with } \omega^2 = \frac{2k}{mL^2}. \tag{16.10}$$

The directly measurable quantity is of course the period of oscillation $T = 2\pi/\omega$; therefore the most practical formula for determining k is the following:

$$k = \frac{2\pi^2 mL^2}{T^2}. \tag{16.11}$$

The smaller the value of k, the more sensitive the torsion balance, and the more precise the result. *]

• **2** Determining the gravitational attraction of huge massive spheres.

Inserting the expression for k from (16.11) into (16.7), we get the following relation:

$$\frac{mMG}{d^2}L = k \quad \theta_0 = \frac{2\pi^2 mL^2}{T^2}\theta_0. \tag{16.12}$$

Simplifying by m and L and multiplying both sides by d^2 we get the final formula for MG:

$$MG = \frac{2\pi^2 Ld^2}{T^2}\theta_0. \tag{16.13}$$

The length of the bar L and the distance between the centers of the two masses d were known with great precision: in the real experiment performed by Cavendish, L was equal to 186 centimeters (73.3 inches), and d was set at 22.5 centimeters. What remained to be determined were the angular displacement θ_0 from balance's unperturbed equilibrium position and the period T of its natural oscillations around equilibrium. It turned out that both crucial measurements could be done in a single experiment, because when the huge masses were put in the position on both sides of the balance by rotation of special support, the balace reacted very slowly to the tiny moment of forces, and its horizontal bar oscillated around the new equilibrium position several times before coming to rest.

The setup used by Cavendish was so sensitive that in order to eliminate external perturbations coming mostly from slight movements of air, that he placed the apparatus in a sealed room designed so that he could move the weights from outside pulling thin ropes attached to the lower bar bearing the huge balls. Then he observed the balance from outside with a small telescope. Here is what he found by patient and meticulous observations:

The period of balance's oscillations before it came to rest in its new equilibrium position was $T = 14$ minutes and 55 seconds, i.e. $T = 895$ seconds, and the displacement of the arm of the balance was about one inch, which is 2.5 cm.

With these experimental data it became easy to evaluate the unknown quantity MG as follows: first, we have to evaluate the angle θ_0 in radians. The length of balance's arm being $L/2 = 93$ cm, all we have to do is to divide the displacement of its end by $2\pi \times L/2$, which gives

$$\theta_0 = \frac{2.5 \, \text{cm}}{2\pi \, 93 \, \text{cm}} = 4.28 \cdot 10^{-3} \, \text{rad} = 0.245° = 14.7 \, \text{arcminutes}.$$

(16.14)

Using the formula (16.13) we can get the value of MG; inserting the data $L = 0.93$ m, $d = 0225$ m, $T = 895$ seconds, one obtains

$$MG = \frac{2\pi^2 \cdot 0.93 \text{m} \cdot (0.225)^2 \, \text{m}^2}{(895 \, \text{sec})^2} = 1.054 \cdot 10^{-8} \, \frac{\text{m}^3}{\text{sec}^2}.$$

(16.15)

• **3** Determining the Earth's mean density.

The main goal of the experiment was to determine Earth's mean density. This was made possible by calibrating the force of gravity with which a 158-kilogram massive ball made of lead attracted a small spherical mass at a distance $d = 22.5$ cm between their mutual centers. Assuming that our globe is a sphere of radius 6780 km, Earth's mass multiplied by the (yet unknown) gravitational constant G can be evaluated using Newton's law and the known gravitational acceleration at Earth's surface:

$$\frac{M_E G}{R_E^2} = g = 9.81 \frac{\text{m}}{\text{sec}^2} \quad \to \quad M_E G = g \ R_E^2 = 3.99 \cdot 10^{14} \, \frac{\text{m}^3}{\text{sec}^2}.$$

(16.16)

Now we can compare this quantity with similar quantity determined by Cavendish for the massive sphere MG given by (16.15), and by dividing the two eliminate the unknown coefficient G. Here is the dimensionless ratio between the two masses:

$$\frac{M_E G}{MG} = \frac{3.99 \cdot 10^{14}}{1.054 \cdot 10^{-8}} = 3.776 \cdot 10^{22} = \frac{M_E}{M}. \tag{16.17}$$

This is how much the Earth's mass is greater than $M = 158$ kg, the mass of one of the leaden spheres used in the experiment. Of course it is better to replace this mind-boggling number by a more down-to-earth ratio of average densities, which was exactly what Cavendish was looking for. The masses of two spherical objects, i.e. the ball M and the Earth M_E can be expressed with one and the same formula employing their respective densities (in the case of the Earth we can speak of its *average density* only, being ignorant of its internal structure and composition):

$$M = \rho_0 \cdot \frac{4\pi r^3}{3}, \quad M_E = \rho_E \cdot \frac{4\pi R_E^3}{3}, \tag{16.18}$$

with $r = 15.25$ cm the radius of the leaden ball, $M = 158$ kg its mass, $\rho_0 = 10.54\,\mathrm{g/cm^3}$ its density (which was a bit less that density of pure lead, $11.3\,\mathrm{g/cm^3}$). The average radius of Earth was known then to a very good precision due to the latest measurements of French geodesists, so we can put $R_E = 6371$ km. Using these data and the ratio given by (16.17) one can find the only quantity that remains unknown, the average density of the Earth, ρ_E:

$$\frac{M_E}{M} = \frac{\rho_E}{\rho_0}\frac{R_E^3}{r^3} \;\rightarrow\; \rho_E = \rho_0 \cdot \frac{M_E}{M}\frac{r^3}{R_E^3}. \tag{16.19}$$

Inserting the values obtained above, we get readily

$$\frac{r}{R_E} = \frac{15.25 \cdot 10^{-2}\,\mathrm{m}}{6.371 \cdot 10^6\,\mathrm{m}} = 2.394 \cdot 10^{-8}, \quad \left(\frac{r}{R_E}\right)^3$$
$$= 1.371 \cdot 10^{-23}, \tag{16.20}$$

and finally

$$\rho_E = \frac{M_E}{M} \cdot \left(\frac{r}{R_E}\right)^3 \rho_0 = 3.776 \cdot 10^{22} \times 1.371 \cdot 10^{-23}\rho_0$$
$$= 5.177 \cdot 10^{-1}\rho_0 = 0.5177\rho_0 = 5456\,\frac{\mathrm{kg}}{\mathrm{m}^3} = 5.456\,\frac{\mathrm{g}}{\mathrm{cm}^3}. \tag{16.21}$$

In the calculations presented here we have used numerical data expressed in metric system, while Cavendish was using pounds for masses and inches and feet for lengths and distances. This is why the result obtained here is slightly different from the original one. Actually, the value determined by Cavendish in his experiment was $5.49\,\mathrm{g/cm^3}$, closer to what is measured nowadays with more sophisticated devices, and which is $5.51\,\mathrm{g/cm^3}$. The precision with which Cavendish was able to measure Earth's average density is amazing — up to 1% of actual value. His measurement remained unparalleled for almost a century. The average density value found by Cavendish was in itself a big surprise: it was much higher than the average density of the most common minerals and rocks dominant in the composition of continents, like granite $(2.75\,\mathrm{g/cm^3})$, limestone $(2.71\,\mathrm{g/cm^3})$, basalt $(3.0\,\mathrm{g/cm^3})$. On the other hand, the iron, which was then believed to be the most common metal in Earth's crust, has much higher density, $7.87\,\mathrm{g/cm^3}$. The surprisingly high value of Earth's mean density made plausible the hypothesis that Earth's central part contains a huge amount of iron or other heavy metals. The presence of iron is also needed to explain terrestrial magnetic field. Cavendish himself was cautious and preferred not to make hypotheses without possibility of direct measurements.

16.4 Masses and gravitational constant defined

With the average density determined, we can evaluate the mass of our planet, supposing a perfect spherical shape (not true, but the flattening along the poles is very small) with the mean radius $R_E = 6371$ km. The mass of the Earth is equal to $5.968 \cdot 10^{24}$ kilograms. The next step consists in finding out the value of gravitational constant G from the relation $MG = g\,R_E^2$. The value accepted today is $G = 6.674 \cdot 10^{-11}\,\frac{\mathrm{m^3}}{\mathrm{kgsec^2}}$.

At the time when Cavendish completed his experiments determining not only the Earth's mean density, but incidentally its total mass, the real distance to the Sun was firmly established, as well as the overall dimensions of the Solar System. The knowledge of the distances to the planets made possible a very accurate estimate of

Table 16.2 The Solar System objects' radii, masses and average densities.

	Radius, km	mass, kg	ρ, g/cm^3	g, m/sec^2	V_{esc}, m/sec
Mercury	$2.44 \cdot 10^3$	$3.3 \cdot 10^{23}$	5.54	3.73	4.3
Venus	$6.05 \cdot 10^3$	$4.87 \cdot 10^{24}$	5.24	8.87	10.38
Earth	$6.38 \cdot 10^3$	$5.97 \cdot 10^{24}$	5.52	9.81	11.2
Moon	$1.74 \cdot 10^3$	$7.35 \cdot 10^{22}$	3.34	1.62	2.38
Mars	$3.39 \cdot 10^3$	$6.42 \cdot 10^{23}$	3.93	3.76	5.03
Jupiter	$6.99 \cdot 10^4$	$1.9 \cdot 10^{27}$	1.326	24.79	59.5
Saturn	$5.82 \cdot 10^4$	$5.68 \cdot 10^{26}$	0.69	10.4	35.5
Uranus	$2.54 \cdot 10^4$	$8.68 \cdot 10^{25}$	1.27	8.87	21.3
Neptune	$2.46 \cdot 10^4$	$1.02 \cdot 10^{26}$	1.64	11.15	23.5
Sun	$6.96 \cdot 10^5$	$1.99 \cdot 10^{30}$	1.41	274	617.0

their real dimensions, i.e. their radius R and their volume, and as a corollary, their mean density, the gravitational acceleration g on their surface, and the escape velocity $V = \sqrt{2Rg}$.

The comparative data are displayed in Table 16.2. (Note: the masses of Mercury, Venus and Moon, neither of which has satellites, were determined in 20th century using cosmic probes.)

The data contained in the table lead to some surprising conclusions about our planetary system. First of all, they show a clear distinction between the four planets closest to the Sun, including our Earth, and the giant planets, which in Cavendish's time were only two: Jupiter and Saturn. Comparing mean densities we see that Mercury, Venus and Earth are very similar. Their average densities being very close to each other, all three between 5.24 and 5.54, suggests that they must have in common a very dense inner core. In contrast, Mars and Moon have both much lower average density, 3.93 and 3.34 respectively, which suggests that they are deprived of a metallic inner core. All four planets are referred to as "terrestrial type".

The giant planets, whose diameters are of one order of magnitude greater than those of small rocky planets orbiting near the Sun, are much less dense, especially Saturn whose average density is only two-thirds of the density of water. Jupiter is denser, but still surprisingly light: only 1.34 g/cm^3. Uranus and Neptune, discovered only in the 19th century, belong to the same class of "giant gaseous planets".

16.5   Gravity and planets' spherical shape

The sphere was considered by the Greek philosophers as the most perfect shape among all geometrical forms. This belief was reinforced by the shape of the starry sky over our heads, and the apparently spherical form of our two luminaries, the Sun and the Moon. When Galileo Galilei saw Venus, Mars and Jupiter through his telescope, he was not surprised to discover that the major planets of the Solar System are spherical in shape, thus confirming their similarity with the Earth, which according to the heliocentric Copernican model was the third planet orbiting between Venus and Mars.

Newton's theory of gravitation explains very well how huge masses end up acquiering a spherical shape provided they surpass certian critical size, which depends on properties of solid material they are made of. However, to make the comparison between resistance of solid bodies against external forces and their own gravitational pull possible, the knowledge of the exact value of Newton's gravitational constant G is needed. This is what Cavendish's experiment provided; since then, gravitational foces can be evaluated for any massive body, huge and small, and be compared with forces of different nature, mechanical, electric and magnetic as well.

Let us consider a celestial body with mass m and average density ρ. The radius R of a spherically shaped homogeneous solid can be found using the definition of density:

$$\rho = \frac{m}{V} = \frac{m}{\frac{4\pi R^3}{3}} = \frac{3m}{4\pi R^3} \;\rightarrow\; m = \rho \frac{4\pi R^3}{3}, \qquad (16.22)$$

from which we get

$$R = \left[\frac{3m}{4\pi\rho} \right]^{\frac{1}{3}}. \qquad (16.23)$$

In what follows, we shall omit numerical factors if they are not very different from 1; what we are interested in are the orders of magnitude. For example, the third root of $\frac{3}{4\pi}$ is about 0.62, which is sufficiently close to 1 for not impeding on the orders of magnitude we shall be getting from our calculations. Therefore we shall assume that $R \simeq (\frac{m}{\rho})^{\frac{1}{3}}$.

The gravitational acceleration at the surface of a spherical celestial body of mass m can be also expressed in terms of its mass and average density:

$$g \simeq \frac{Gm}{R^2} \simeq Gm^{\frac{1}{3}}\rho^{\frac{2}{3}}. \tag{16.24}$$

The bodies we are speaking about need not to be spherical — it is enough if they form a compact bulk — like a potato or a pear, but not like a cucumber. Then our approximate formula conserves its validity as a tool for obtaining right orders of magnitude.

Our next estimate concerns the maximal pressure inside the body, close to its center of mass. Here again, assuming for the sake of simplicity that ρ is constant, the pressure at the bottom of a homogeneous column of matter of constant density ρ, a given height h and constant cross-section A is defined as its total weight over the area of its base A:

$$P = \frac{\text{weight}}{\text{area}} = \frac{\rho \cdot g \cdot A \cdot h}{A} = \rho \cdot g \cdot h. \tag{16.25}$$

The pressure close to the central parts of a huge celestial body should be thus of the order $P \simeq \rho \cdot gR$, with R denoting its characteristic dimension, which in the case of spherical body would be comparable to its radius. Of course in reality neither the density ρ nor the gravitational acceleration g are constant; they change as one goes deeper towards the center of the bulk. Nevertheless the order of magnitude is right. Replacing g by the expression (16.24), we obtain the following formula for the pressure close to the center:

$$P \simeq \rho gR \simeq Gm^{\frac{2}{3}}\rho^{\frac{4}{3}}. \tag{16.26}$$

In a homogeneous spherical body the pressure due to gravitational forces is also spherically symmetric and directed towards the center of gravity which in this case coincides with the center of mass. The body is uniformly compressed without changing its spherical shape. The degree of deformation depends on the compressibility of the material that particular body is made of. Liquids and solids develop strong resistance to deformation, gases can be compressed much more strongly, until the internal pressure equilibrates the gravitational force.

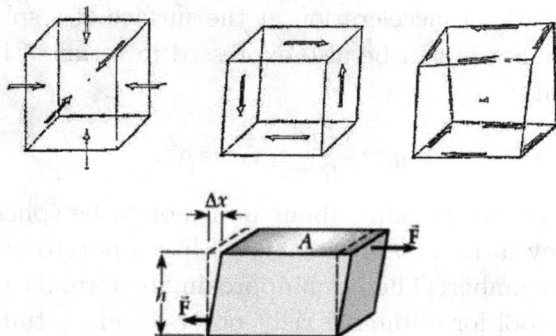

Fig. 16.3   Above: Pure pressure stress, pure shear stress and torsion and shear stress; Below: Deformation of a rectangular volume element under the shear stress action. Relative deformation $\delta x/h$ is proportional to the stress F/A.

But if a huge body departs from the spherical symmetry, gravitational forces developed within it will be asymmetric, too, provoking stresses. These can be of two kinds: pure shear, pure torsion, or the combination of both, as shown in Figure 16.3. The stress has the physical dimension of pressure: force per unit of surface. Shear stress arises when the forces acting on opposite sides of small volume element act in opposite directions. A solid body resists shear stress applied to it, undergoing only small deformation which is proportional to linear departure from natural state divided by the dimension of the side perpendicular to the force, as shown in Figure 16.3 on the right.

We define the *shear strain* σ as the ratio of small shear deformation Δx and the lateral dimension h perpendicular to it. For small deformations the linear dependence between shear stress and shear strain prevails:

$$\sigma = E\frac{\Delta x}{h}, \qquad (16.27)$$

where E is the shear Young modulus measuring the resistance of a given material to deformation imposed on it by external forces. Similar formulae apply for torsion deformations or for bulk deformations of volumes (compression or blow-up). In the case of huge gravitating bodies with irregular shape departing from spherical symmetry the shear deformation and stress are by far the most important ones (after the compression which keeps the bulk together).

What is important to know is that the linear deformation regime has its natural limits for any material. When shear stress σ passes beyond this limit denoted by σ_{cr}, a solid body does not resist anymore and is either crushed into parts, or starts to flow like a piece of play dough. Assuming that shear stress is of the same order as the pressure, replacing P by σ in formula (16.26) we can write

$$\sigma \simeq Gm^{\frac{2}{3}}\rho^{\frac{4}{3}}. \tag{16.28}$$

Inserting into the above expression the critical value σ_{cr} established experimentally, and using the relationship between the mass m, density ρ and characteristic dimension (radius) R we can eliminate either the radius or the mass in order to get the critical values of mass m and radius R for which huge celestial bodies cannot keep their initial form any longer, being crushed or deformed under the influence of their own gravity. Here are the approximate expressions for critical mass and critical radius as functions of density ρ, critical stress σ_{cr} and the Newton's gravitational constant G:

$$m_{cr} \simeq \frac{1}{\rho^2}\left(\frac{\sigma_{cr}}{G}\right)^{\frac{3}{2}}, \quad R_{cr} \simeq \frac{1}{\rho}\left(\frac{\sigma_{cr}}{G}\right)^{\frac{1}{2}}. \tag{16.29}$$

Table 16.3 shows different critical masses and critical radii beyond which huge celestial bodies are bound to acquire spherical shape.

As we can see, all planets and greater asteroids in Solar System have both masses and radii far above the critical ones. But Phobos, the smaller of two Martian satellites, displays the shape that is far from a spherical one: it looks like a giant potato, with

Table 16.3 The critical masses for a planet to become spherical, depending on the composition.

Material	Density ρ, $kg\,m^{-3}$	σ_{cr}, $\frac{N}{m^2}$	m_{cr}, kg	R_{cr}, km
Ice	900	$3 \cdot 10^6$	10^{19}	200
Volcanic	2500	$3 \cdot 10^7$	$3 \cdot 10^{19}$	300
Granite	2700	10^8	$3 \cdot 10^{20}$	500
Iron	7800	10^9	10^{21}	500

Our Celestial Clockwork

dimensions $14 \times 11.5 \times 10$ kilometers. Whatever the rocky material it consists of, its dimensions and mass are below the critical values of Table 16.3. Similar cases are Jupiter's small satellite Amaltea whose length 265 km is almost twice its width, 160 km. This is also below the critical mass and radius. The asteroid belt contains myriads of rocks of under-critical size, without any particular symmetry, confirming the rough estimate given by formula (16.29).

A formula similar to (16.25) can be used to evaluate the maximal height of a mountain on a given planet. Even the granitic rock will start to flow when pressure becomes high enough. A mountain of height h and total mass M having the form of a cone will develop the pressure at its base approximately equal to

$$P = \frac{Mg}{A} \simeq \frac{\rho V g}{A} = \frac{\rho g h}{3}, \quad \text{because} \quad V = \frac{Ah}{3}. \quad (16.30)$$

Here g is the gravitational acceleration at planet's surface. When stress reaches its critical value σ_{cr} in lower parts of a mountain of conical form with a 45° slope, the rocks at the foot of the mountain will start to flow making the mountain go down or desintegrate. Therefore the critical height can be defined as

$$h_{cr} = \frac{3\sigma_{cr}}{\rho g}. \quad (16.31)$$

Let us try the formula on our own planet: on Earth we have $g = 9.81\,\text{m/sec}^2$; for granitic mountains the critical stress value (Table 16.3) is $\sigma_{cr} = 10^8\,\frac{\text{N}}{\text{m}^2}$ and density is $\rho = 2700\,\frac{\text{kg}}{\text{m}^3}$. Inserting these values into formula (16.31) we obtain:

$$h_{cr} = \frac{3 \cdot 10^8\,\text{N/m}^2}{2.7 \cdot 10^3 \text{kg/m}^3 \cdot 9.81\,\text{m/sec}^2} = 1.1 \cdot 10^4\,\text{m} = 11\,\text{km}. \quad (16.32)$$

Applying the same formula to Mars, on whose surface the gravitational acceleration is only $3.7\,\text{m/sec}^2$, we get $h_{cr} \simeq 27\,\text{km}$.

The highest mountain on Earth is Mount Everest, 8887 meters above the sea level. And here are the data gathered by modern observational means: the highest mountain on Venus is the Maxwell volcano, 12 km high; the highest Martian volcanoes are Olympus Mons

(27 km height) and Elysium Mons (14 km). The highest mountain on the Moon is Mons Huygens, 5.5 km — which suggests that our Moon's density and σ_{cr} of rocks it is made of is sensibly lower than those of Earth, Venus and Mars.

16.6 Deformation due to rotation

In the third book of *Principia* Newton addresses the problem of influence of cenrifugal force on the shape of a planet rotating around its axis. The Proposition XVII, Theorem XVI reads: *That the axes of the planets are less than the diameters drawn perpendicularly to the axes*, and the next Proposition XIX, Problem III, reads: *To find the proportion of the axis of a planet to the diameters perpendicular thereto.*

The equilibrium shape of a planet should imitate the surfaces everywhere perpendicular to local acceleration, which is a sum of gravitational pull of the planet directed towards it center, and the centrifugal force perpendicular to the axis of planet's diurnal rotation. The angular velocity of rotation Ω is measured with respect to the fixed stars. The acceleration due to the gravitational pull of the planet g is given by $g = \frac{GM}{R^2}$ where M is its mass and R its radius. It can be considered as the same on the entire surface of the planet, and directed towards its center. The centrifugal force is always perpendicular to the axis of rotation and directed outwards; its absolute value depends on the latitude φ as $F_c = \Omega^2 R \cos \varphi$.

[* Let us evaluate the essential parameters of the ellipsoidal shape of the Earth due to the action of centrifugal force caused by the diurnal rotation. If the Earth were a liquid sphere, its shape would perfectly adapt itself to the field of forces acting on its surface. In the absence of rotation the only force acting on the surface would be Earth's gravitational pull, directed towars the center; the surface everywhere perpendicular to the radius vector at a given point is a perfect sphere. Rotation around the axis imposes the cylindrical symmetry instead of the spherical one; the resulting shape in three dimensions is an ellipsoid whose vertical sections containing the rotation axis are two-dimensional ellipses (see Figure 16.4).

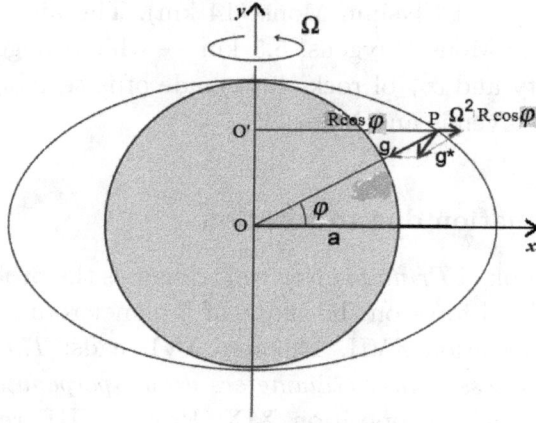

Fig. 16.4   Flattening of a planet due to its rotation. The ellipsoidal shape is strongly exaggerated for the sake of visibility.

In the planar Cartesian coordinates (x, y) an ellipse centered in Earth's center can be parametrized by polar angle φ as follows:

$$x = a \cos \varphi, \quad y = b \sin \varphi, \quad \text{so that} \quad \frac{x^2}{a^2} + \frac{y^2}{b^2} = 1, \qquad (16.33)$$

where a is the semi-major axis and b is the semi-minor axis.

At any point of surface it has to be perpendicular to the force acting on infinitesimal volume elements of the liquid. The vector tangent to the ellipse (16.33) is proportional to $\mathbf{t} \sim [-a \sin \varphi, \ b \cos \varphi]$, therefore the vector normal to the ellipse at the same point is proportional to $\mathbf{n} \sim [b \cos \varphi, \ a \sin \varphi]$. The vectors are not normalized yet.

Two forces act on an infinitesimal volume element on the surface: the corresponding accelerations are the gravitational pull $\mathbf{g}$ directed towards Earh's center of mass, and the centrifugal force $\mathbf{F}$ perpendicular to the axis of rotation $\mathbf{y}$. In Cartesian coordinates (x, y) the first is parallel to the vector $-\mathbf{n}$ and the second aligned along the x-axis. This gives the total force $\mathbf{F}_{tot} = \mathbf{g} + \mathbf{F}$.

We expect that the ellipsoidal form of Earth's surface is not very different from a sphere, which means that the difference between the semi-axes is very small, and let us denote by $\delta = |a - b| \ll a$ that small difference. Writing the unit vector **n** as

$$\mathbf{n} = \frac{1}{\mid \mathbf{n} \mid}[b\cos\varphi,\ a\sin\varphi] = [(a - \delta)\cos\varphi,\ a\sin\varphi],$$

we can safely replace $\mid \mathbf{n} \mid$ by a, although it is slightly different due to the small variation δ. But in linear approximation it will note impede on the numerator of the expression of **n** above.

We can assume that gravitational acceleration at the point of the ellipse parametrized by φ can be written as

$$\mathbf{g} = [-g\cos\varphi,\ -g\sin\varphi], \tag{16.34}$$

being directed towards the center of mass which coincided with the center of Cartesian system (x, y).

The centrifugal acceleration due to rotation around the y-axis is equal to the square of angular velocity multiplied by the distance from the acis of rotation. Here this acceleration is equal to $\Omega^2 R\cos\varphi$ and directed positively along the x-axis. What is left now is to compare the direction of the resulting acceleration and the normal to the ellipse at a given point: the two vectors should be parallel to each other. Therefore, we must have

$$[-g\cos\varphi + \Omega^2 R\cos\varphi,\ -g\sin\varphi]$$
$$\text{parallel to } [-(a - \delta)\cos\varphi,\ -a\sin\varphi]. \tag{16.35}$$

Dividing the first vector by g and the second one by a we obtain

$$\left[-\left(1 - \frac{\Omega^2 R}{g}\right)\cos\varphi,\ -\sin\varphi\right]$$
$$\text{parallel to } \left[-\left(1 - \frac{\delta}{a}\right)\cos\varphi,\ -\sin\varphi\right]. \tag{16.36}$$

The second components of two vectors (16.36) being equal, in order to be parallel the first components must be equal, too. This leads to the simple relation

$$1 - \frac{\Omega^2 R}{g} = 1 - \frac{\delta}{a} \rightarrow \frac{\Omega^2 R}{g} = \frac{\delta}{a} = \frac{\Omega^2 R^3}{MG}. \tag{16.37}$$

In the last form of the ratio δ/a we replaced the acceleration on the surface of Earth by $g = \frac{MG}{R^2}$. This formula can be applied to any planet, provided we know its mass; G is the gravitational constant.

Let us denote by R_{min} the semi-polar axis of rotating planet, and by R_{max} its equatorial radius. We may identify then

$$R_{max} = a, \quad R_{min} = b, \quad a - b = \delta,$$
$$\frac{\delta}{a} = \frac{R_{max} - R_{min}}{R_{max}}. \tag{16.38}$$

This dimensionless quantity characterizing the departure of Earth's form from a perfect sphere should be equal to the ratio of accelerations given in (16.37). Let us check whether this hypothesis is confirmed by observation: Table 16.4 shows the comparison between the two parameters (notice that we do not need to know the masses of other planets neither the Earth mass — only the quantity mG which can be determined by observation of satellites' periods of rotation and applying the Kepler's third law).

In the Book III of *Principia*, Theorems XVIII, XIX and XX, Newton showed how one can determine the flattening of our globe due to its axial rotation, finding the result close to $1/230$ and $R_{max} - R_{min} = 8.5$ miles, or the difference between equatorial and polar diameters equal to 17 miles, or 27.35 kilometers. *]

The very precise modern measures reveal an interesting discrepancy between the actual shape of our planet and its angular velocity. The exact value of deformation for an ideal fluid sphere rotating with agular velocity of 1 rotation in 24 hours gives $1/299.67$, whereas the observed value is $1/298.26$, which is slightly more. Such deformation

Table 16.4   The radii and periods of rotation of major planets of the Solar System.

Planet	R, meters	Period, sec	$\frac{\Omega^2 R^3}{MG}$	$\frac{R_{max}-R_{min}}{R_{max}}$
Earth	$6.37 \cdot 10^6$	86 164	$\frac{1}{289}$	$\frac{1}{298}$
Mars	$3.39 \cdot 10^6$	88 643	$\frac{1}{220}$	$\frac{1}{190}$
Jupiter	$7.14 \cdot 10^7$	35 430	$\frac{1}{11}$	$\frac{1}{15}$
Saturn	$6.03 \cdot 10^7$	36 840	$\frac{1}{6}$	$\frac{1}{9.5}$

corresponds to a slightly higher angular velocity, which is easy to evaluate using the formula (16.37): it should be higher by the factor 1.002. As it was shown in Chapter 13, the Earth's axial rotation slows down due to the loss of kinetic energy dissipated by the tides. The loss of the angular velocity is extremely small, but it can be detected with modern atomic clocks, which show that a modern day is longer by about 1.7 milliseconds than a century ago. It is easy to find out that the Earth's rotation was 1.002 times faster, and the day was 1.002 shorter about 10 million years ago. The shape of our planet will slowly adapt itself to the slower angular velocity, but it takes millions of years — the geological times — to occur.

Epilogue

Our journey through the centuries and astronomical spaces comes to a halt at the dawn of the 19th century. The knowledge of solar system was still far from being complete, the great telescopes and powerful observatories remained yet to be constructed. Announcing the new age rich in astronomical findings, a new object orbiting between Mars and Jupiter was discovered on January 1, 1801, by Italian astronomer Giuseppe Piazzi (1746–1826), exactly in the missing place predicted by the Bode-Titius law. It was given the name *Ceres*, after the Roman goddess of crops. Other similar small objects were discovered later, forming what we call now the *asteroid belt*.

Our narrative followed the development of human understanding of Universe as a whole, with a constant enlargement of its horizon. As of today, the solar system extends up to the *Kuiper's belt*, a faraway flat region beyond Neptune defined by Dutch astronomer Gerard Kuiper (1905–1973), containing supposedly hundreds of thousands of icy bodies larger than 1000 km across, of which Pluto with its satellite Charon, discovered in 1930, was the most eminent member. It was called the ninth planet of Solar System, but downgraded in 2006 after several similar bodies orbiting the Sun at great distances were discovered. Besides these asteroids, the Kuiper's belt contains a trillion of comets, most on circular orbits, and some of them on elongated elliptical orbits with periods of hundred years and more.

The constant enlargement of observable Universe that resulted from the development of astronomy during millenia was accompanied by the parallel increase of estimates of the age of the Earth and the Solar System ([Richet (2007)]). It can be illustrated by the following short list of sizes and distances:

Solar System of 18th century, including Uranus: 37 A.U. across

Kuiper's belt: 30 to 50 A.U. across $\simeq$ 7 light-hours

One light-year = 1 l.y. = 63270 A.U.

The closest star α *Centauri*: 4.37 l.y. away

1 parsec = 1 pc = 3.7 l.y., 1 kpc = 3700 l.y.

Our Galaxy: diameter 10^5 l.y. $\simeq$ 27 kpc

The closest great galaxy M31: $2.2 \cdot 10^5$ l.y. across, distance: 778 kpc

The Virgo cluster: $7.5 \cdot 10^6$ l.y. across, at $6.5 \cdot 10^7$ l.y. distance

Cosmic background radiation horizon: $4.6 \cdot 10^{10}$ l.y.

The size of the Universe conceived by ancient civilizations was quite modest by our standards. The maps of the world found on clay tablets do not go beyond a few thousand kilometers; ancient Greeks' *ekumene*, or inhabited world, was limited by parts of Europe, Western Asia and Northern Africa. At the end of Antiquity, Earth's size was known quite well. It took about two millenia until the size of the solar system was firmly established. The corresponding cognitive leap represents six orders of magnitude on the logarithmic scale: from a few thousands of kilometers on Earth to $5.5 \cdot 10^9$ km including the orbit of Uranus. Great as it was, it became still conceivable to the modern human mind.

What came next is beyond the scope of this book; let us only say that the cognitive leap imposed by the mind-boggling sizes and distances of Universe discovered by means of modern spatial telescopes is much greater, even on the logarithmic scale according to the Weber-Fechner law. Indeed, our Milky Way is 10^8 times bigger than our solar system, and the cosmic background radiation which represents the ultimate horizon accessible to our observation, is roughly 10^8 times greater than our Galaxy — the leap of 16 orders of magnitude in one hundred years as compared to the leap of 7 orders of magnitude which occurred during the period from the Antiquity to the end of the 18th century. No wonder that one feels a kind of vertigo being confronted with such a tremendous cosmic scale.

How deeply this ever growing scale of the Universe influences our perception can be illustrated by the following two comments. The first is taken from the "General Scholium" which closes Newton's *Principia*: [Newton (1687)]

> "*This most beautiful system of the sun, planets and comets could only proceed from the counsel and dominion of an intelligent and powerful Being.*"

The second one, made about 300 years after Newton, comes from Steven Weinberg's (b. 1933) book *The First Three Minutes*: [Weinberg (1977)]

> "*The more the universe seems comprehensible, the more it also seems pointless.*"

However one should not let oneself be discouraged by such pessimistic statements too much. After all, if the Universe, our existence and all our actions were really pointless, this would *ipso facto* apply to Weinberg's remark as well.

Yes, one can but feel similar disappointment after realizing that not only we and our beautiful planet, but even the majestuous Solar System, represent nothing more than a tiny peripheral speck in our Galaxy hosting billions of suns, which itself is a tiny lump of matter lost among billions of galaxies forming large clusters filling space with gigantic filament-like structures. However, the fact that in spite of being such a negligible part of it we are able to grasp all this immensity, think of it and feel awe and bewilderment, is no less mysterious and fascinating than the Universe itself.

Although we still don't understand the goal, it does not mean that there is no purpose at all. It might be that we don't even know how to formulate such questions properly?

Seemingly impersonal and neutral laws governing physical reality do not imply that we cannot find meaning and sense of our own existence. Compared with the age of the Universe, the entire human history is like a short eyeblink. Other intelligent beings may exist somewhere, and may be superior to us, as suggested in Fred Hoyle's science-fiction novel *The Black Cloud* [Hoyle (1957)]. We are still at the beginning of a road full of discoveries.

Perhaps the search for a purpose is the ultimate goal.[1] The Universe could have existed without us, but we could not be there without the Universe. So, if we find the meaning of our own existence, the entire Universe will acquire meaning and purpose in our eyes, too.

---

[1]See e.g. Arhur C. Clarke's story *The Nine Billions Names of God* [Clarke (1967)].

Bibliography

Ariew, Roger, *Descartes as Critic of Galileo's Scientific Methodology* Synthese **67**, pp. 77–90, by D. Reidel Publishing Company (1986).

Arnold, Vladimir I., *Huygens and Barrow, Newton and Hooke*, Birkhäuser (1990).

Asimov, Isaac, *Nightfall* Amazing Stories, N.Y., (1941).

Bergerac, Cyrano de, *Histoire comique par Monsieur de Cyrano Bergerac contenant les Estats & Empires de la Lune*, Charles de Sercy editor, Paris (1657).

Berkley, George, *"De Motu" and "The Analyst"*, Modern Edition, Springer (1992).

Blair, Anne, *Tycho Brahe's critique of Copernicus and the Copernican system.* Journal of the History of Ideas **51**(3): 355–377 (1990).

Brzostkiewicz, Stanisław R., *Nicolas Copernicus — The Birth of Modern Astronomy* (in Polish), Urania, XLIII (9), pp. 229–238 (1972).

Burton, David, *The History of Mathematics: An Introduction*, 7th edition, McGraw Hill (2011).

Chavez, P.S., Remote sensing of Environment, **24**, pp. 459–479 (1988).

Clarke, Arthur C., *The Nine Billion Names of God*, Harcourt (1967).

Cook, Alan, *Edmund Halley: Charting the Heavens and the Seas*, Clarendon Press, Oxford (1998).

Copernicus, Nicolae, *De Revolutionibus Orbium Celestium*, Johannes Petreius ed., Nuremberg (1543); English translation by Edward Rosen: Polish Scientific Editors (*PWN*), Warsaw, (1978).

Couper, Heather and Henbest, Nigel, *New Worlds*, Addison-Wesley (1986).

De Saedeleer, Bernard, *Le transit de Vénus du 8 juin 2004.* Galactée, 41 (2004).

Descartes, René, *Discours de la Méthode*, Ian Maire ed., Leyden (1637).

Diringer, David, *The Alphabet: A key to human history*, Hutchinson (1968).

467

Dutton, Blake D., *Physics and Metaphysics in Descartes and Galileo*, Journal of the History of Philosophy, Vol. **37** (1), pp. 49–71 (1999).

Flammarion, Camille, *Astronomie populaire*, Hachette, Paris (1880).

Flammarion, Camille, *L'atmosphère: météorologie populaire*, Hachette, Paris (1888).

Galilei, Galileo, *Dialogo* di Galileo Galilei, Landini, Firenze (1632) Linceo, Firenze, MDCXXXII.

Galilei, Galileo, *Discorsi e Dimostrazioni Matematiche* (Dialogues Concerning Two New Sciences), Elsevier, Leyden (1638).

Gindikin, Semyon G., *Tales of Physicists and Mathematicians*, Benjamin (1985).

Goodstein, David L. and Goodstein, Judith R., *Feynman's Lost Lecture: The Motion of Planets around the Sun*, Vintage Publishing (1997).

Halley, Edmond, *A New Method of determining the Parallax of the Sun, or his Distance from the Earth*, Phil. Trans. Vol. XXIX, Sec. R. S. No. 348, p.454. Translated from the Latin (1716).

Hartner, Willy, *Copernicus, the Man, the Work, and His History*, Proceedings of the American Philosophical Society, Vol. 117, No. 6, pp. 413–422 (1973).

Haskiel, Daniel, *A Chronological View of the World*, J.H. Colton publisher, New York (1848).

Hofstadter, Douglas R, *Gödel, Escher, Bach: An Eternal Golden Braid*, First Edition: Basic Books, USA (1979).

Hoyle, Fred, *The Black Cloud*, Penguin, London (1960).

Hughes, David W., *Six Stages in the History of the Astronomical Unit*, Journal of Astronomical History and heritage, **4** (1), pp. 15–28 (2001).

Hughes, David W., *Measuring the Moon's Mass*, Observatory, Vol. 122, No. 1167 (2002).

Huygens, Christiaan, *Systema Saturnium*, Adriani Vlacq, The Hague (1659).

Huygens, Christiaan, *Treatise on Light*, Macmillan, London (1912).

Jeans, James, *The Growth of Physical Science*, Cambridge University Press, Second Edition (1952).

Kant, Immanuel, *Critique of Pure Reason*, Cambridge University Press (1998); see also Michael Friedman, *Kant and the Exact Sciencers*, Harvard University Press (1992).

Kepler, Johannes, *Harmonices Mundi* (1596).

Kepler, Johannes, *Astronomia Nova* (1609).

Kepler, Johannes, *Tabulae Rudolphinae* (1610).

Kepler, Johannes, *De nive sexangula*, Teubneri, Lipsiae MDCXI (1611).

Kepler, Johannes, *Somnium* (posthumous, edited by Kepler's son Ludwig Kepler), Frankfurt (1634).

Kerner, R, van Holten, J.-W., Colistete Jr., R *Relativistic Epicycles*, Classical and Quantum Gravity, **18**, pp. 4725–4742 (2001).

Koyré, Alexandre, *From the Closed World to the Infinite Unierse* Library of Alexandria ed., (1957).

Koyré, Alexandre, *The Astronomical Revolution*, Hermann, Paris and Methuen, London (1973).

Lewis, Bernard, *What Went Wrong? — The Clash Between Islam and Modernity in the Middle East*, Oxford University Press (2002).

Mac Cormack, Lesley B., *Flat Earth or Round Sphere: Misconceptions of the Shape of the Earth and the Fifteenth-Century Transformation of the World Ecumene*, Volume: 1 issue: 4, pp. 363–385 (1994).

Mie, G., *Ann. Phys. (Leipzig,* **25** pp. 377–452 (1908).

Miner, Ellis D., Wessen, Randii R. and Cuzzi, Jeffrey N., *Planetary Ring Systems*, Springer, (2007).

Mourelatos, Alexander P.D., *The Pre-Socratics*, Princeton Legacy Library, Princeton, N.J. (1974).

Neugebauer, Otto, The History of Ancient Astronomy Problems and Methods, *Journal of Near Eastern Studies*, Vol. 4, No. 1, pp. 1–38 (1945) Published by: The University of Chicago Press.

Neugebauer, Otto, *A History of Ancient Mathematical Astronomy*, Springer (1975).

Newton, Isaac, *Philosophia Naturalis Principia Mathematica*, Pepys, London, (1687), new English edition: *Snowball Publishing*, 466p. (2010).

Newton, Isaac, *Opticks*, Smith and Walford ed., London (1704).

North, John, *Cosmos: The Illustrated History of Astronomy and Cosmology*, The University of Chicago Press (2008).

Nunes, Pedro, *Obras, Volume II: De Crepusculis*, Foundation Calouste Gulbenkian, Lisboa (2003).

Perelman, Yakov I., *Astronomy for Fun*, Prodinova ed., (2012) Russian original: Moscow, (1949).

Rademacher, Hans and Toeplitz, Otto, *Von Zahlen und Figuren*, (1930), second edition: Springer (2001); English version *The Enjoyment of Mathematics*, Princeton University Press (1957).

Richet, Pascal, *The Natural history of Time*, The University of Chicago Press, (2007).

Richet, Pascal, *The Creation of the World and the Birth of Chronology*, Comptes Rendus Geoscience, Volume 349, Issue 5, pp. 226–232 (2017).

Ridpath, Ian, *Star Tales* Lutterworth Press, Cambridge (1968).

Rodgers, Eric M., *Physics for Inquiring Mind*, Princeton University Press, Princeton (2011).

Steinhaus, Hugo, *Mathematical Snapshots*, 3rd edition, Oxford University Press (1969).

Todorov, Ivan T., *Galileo (1564–1642) and Kepler (1571–1630): The Modern Scientist and the Mystic*, Bulgarian Journal of Physics, **44**, p. 205–220. (2017).

Weinberg, Steven, *The First Three Minutes: A Modern View of the Origin of the Universe*, Basic Books, (1977).

Wells, Herbert George, *The First Men in the Moon*, George Newnes editor, UK (1901).

Index